RF AND MICROWAVE ENGINEERING

RF AND MICROWAVE ENGINEERING
FUNDAMENTALS OF WIRELESS COMMUNICATIONS

Frank Gustrau

Dortmund University of Applied Sciences and Arts, Germany

A John Wiley & Sons, Ltd., Publication

First published under the title *Hochfrequenztechnik* by Carl Hanser Verlag
© Carl Hanser Verlag GmbH & Co. KG, Munich/FRG, 2011
All rights reserved.
Authorized translation from the original German language published by Carl Hanser Verlag GmbH & Co. KG, Munich.FRG.

This edition first published 2012
© 2012 John Wiley & Sons Ltd, Chichester, UK

Registered office
John Wiley & Sons Ltd, The Atrium, Southern Gate, Chichester, West Sussex, PO19 8SQ, United Kingdom

For details of our global editorial offices, for customer services and for information about how to apply for permission to reuse the copyright material in this book please see our website at www.wiley.com.

The right of the author to be identified as the author of this work has been asserted in accordance with the Copyright, Designs and Patents Act 1988.

Library of Congress Cataloging-in-Publication Data

Gustrau, Frank.
  [Hochfrequenztechnik. English]
  RF and microwave engineering : fundamentals of wireless communications / Frank Gustrau.
    p. cm.
  Includes bibliographical references and index.
  ISBN 978-1-119-95171-1 (pbk.)
  1. Radio circuits. 2. Microwave circuits. 3. Wireless communication systems–Equipment and supplies. I. Title.
  TK6560.G8613 2012
  621.382–dc23
                                                                          2012007565
A catalogue record for this book is available from the British Library.

Paper ISBN: 9781119951711

Typeset in 10/12pt Times by Laserwords Private Limited, Chennai, India

For Sabine, Lisa & Benni

Contents

Preface

This textbook aims to provide students with a fundamental and practical understanding of the basic principles of radio frequency and microwave engineering as well as with physical aspects of wireless communications.

In recent years, wireless technology has become increasingly common, especially in the fields of communication (e.g. data networks, mobile telephony), identification (RFID), navigation (GPS) and detection (radar). Ever since, radio applications have been using comparatively high carrier frequencies, which enable better use of the electromagnetic spectrum and allow the design of much more efficient antennas. Based on low-cost manufacturing processes and modern computer aided design tools, new areas of application will enable the use of higher bandwidths in the future.

If we look at circuit technology today, we can see that high-speed digital circuits with their high data rates reach the radio frequency range. Consequently, digital circuit designers face new design challenges: transmission lines need a more refined treatment, parasitic coupling between adjacent components becomes more apparent, resonant structures show unintentional electromagnetic radiation and distributed structures may offer advantages over classical lumped elements. Digital technology will therefore move closer to RF concepts like transmission line theory and electromagnetic field-based design approaches.

Today we can see the use of various radio applications and high-data-rate communication systems in many technical products, for example, those from the automotive sector, which once was solely associated with mechanical engineering. Therefore, the basic principles of radio frequency technology today are no longer just another side discipline, but provide the foundations to various fields of engineering such as electrical engineering, information and communications technology as well as adjoining mechatronics and automotive engineering.

The field of radio frequency and microwave covers a wide range of topics. This full range is, of course, beyond the scope of this textbook that focuses on the fundamentals of the subject. A distinctive feature of high frequency technology compared to classical electrical engineering is the fact that dimensions of structures are no longer small compared to the wavelength. The resulting wave propagation processes then lead to typical high frequency phenomena: reflection, resonance and radiation. Hence, the centre point of attention of this book is wave propagation, its representation, its effects and its utilization in passive circuits and antenna structures.

What I have excluded from this book are active electronic components – like transistors – and the whole spectrum of high frequency electronics, such as the design of amplifiers, mixers and oscillators. In order to deal with this in detail, the basics of electronic circuit design theory and semiconductor physics would be required. Those topics are beyond the scope of this book.

If we look at conceptualizing RF components and antennas today, we can clearly see that software tools for Electronic Design Automation (EDA) have become an essential part of the whole process. Therefore, various design examples have been incorporated with the use of both circuit simulators and electromagnetic (EM) simulation software. The following programs have been applied:

- ADS (Advanced Design System) from Agilent Technologies;
- Empire from IMST GmbH;
- EMPro from Agilent Technologies.

As the market of such software products is ever changing, the readers are highly recommended to start their own research and find the product that best fits their needs.

At the end of each chapter, problems are given in order to deepen the reader's understanding of the chapter material and practice the new competences. Solutions to the problems are being published and updated by the author on the following Internet address:

http://www.fh-dortmund.de/gustrau_rf_textbook

Finally, and with great pleasure, I would like to say thank you to my colleagues and students who have made helpful suggestions to this book by proofreading passages or initiating invaluable discussions during the course of my lectures. Last but not the least I express gratitude to my family for continuously supporting me all the way from the beginning to the completion of this book.

Frank Gustrau
Dortmund, Germany

List of Abbreviations

3GPP	Third Generation Partnership Project
Al_2O_3	Alumina
Balun	Balanced-Unbalanced
CAD	Computer Aided Design
DC	Direct Current
DFT	Discrete Fourier Transform
DUT	Device Under Test
EM	ElectroMagnetic
EMC	ElectroMagnetic Compatibility
ESR	Equivalent Series Resistance
FDTD	Finite-Difference Time-Domain
FEM	Finite Element Method
FR4	Glass reinforced epoxy laminate
GaAs	Gallium arsenide
GPS	Global Positioning System
GSM	Global System for Mobile Communication
GTD	Geometrical Theory of Diffraction
GUI	Graphical User Interface
HPBW	Half Power Beam Width
ICNIRP	International Commission on Non-Ionizing Radiation Protection
IFA	Inverted-F Antenna
ISM	Industrial, Scientific, Medical
ITU	International Communications Union
LHCP	Left-Hand Circular Polarization
LHEP	Left-Hand Elliptical Polarization
LNA	Low-Noise Amplifier
LOS	Line of Sight
LTE	Long Term Evolution
LTI	Linear Time-Invariant
MIMO	Multiple-Input Multiple-Output
MMIC	Monolithic Microwave Integrated Circuits
MoM	Method Of Moments
NA	Network Analyser
NLOS	Non Line of Sight

PA	Power Amplifier
PCB	Printed Circuit Board
PEC	Perfect Electric Conductor
PML	Perfectly Matched Layer
PTFE	Polytetraflouroethylene
Radar	Radio Detection and Ranging
RCS	Radar Cross-Section
RF	Radio Frequency
RFID	Radio Frequency Identification
RHCP	Right-Hand Circular Polarization
RHEP	Right-Hand Elliptical Polarization
RMS	Root Mean Square
SAR	Specific Absorption Rate
SMA	SubMiniature Type A
SMD	Surface Mounted Device
TEM	Transversal Electromagnetic
UMTS	Universal Mobile Telecommunication System
UTD	Uniform Theory of Diffraction
UWB	Ultra-WideBand
VNA	Vector Network Analyser
VSWR	Voltage Standing Wave Ratio
WLAN	Wireless Local Area Network

List of Symbols

Latin Letters

A	Area (m^2)
A_{dB}	Attenuation (dB)
$\vec{A}$	Magnetic vector potential (Tm)
A_{eff}	Effective antenna area (m^2)
$\mathbf{A}$	ABCD matrix (matrix elements have different units)
$\vec{B}$	Magnetic flux density (magnetic induction) (T; Tesla)
B	Bandwidth (Hz; Hertz)
BW	Bandwidth (angular frequency) (1/s)
c	Velocity of a wave (m/s)
C	Capacitance (F; Farad)
$C(\varphi, \vartheta)$	Radiation pattern function (dimensionless)
C'	Capacitance per unit length (F/m)
D	Directivity (dimensionless)
$\vec{D}$	Electric flux density (C/m^2)
$\vec{E}$	Electric field strength (V/m)
f	Frequency (Hz)
f_c	Cut-off frequency (Hz)
$\vec{F}$	Force (N; Newton)
$\vec{F}_C$	Coulomb Force (N)
$\vec{F}_L$	Lorentz Force (N)
G	Conductance ($1/\Omega = $ S)
G	Gain (dimensionless)
G	Green's function (1/m)
G'	Conductance per unit length (S/m)
$\vec{H}$	Magnetic field strength (A/m)
$\mathbf{H}$	Hybrid matrix (matrix elements have different units)
I	Current (A; Ampere)
$\mathbf{I}$	Identity matrix (dimensionless)

j	Imaginary unit (dimensionless)
$\vec{J}$	Electric current density (A/m^2)
$\vec{J}_S$	Surface current density (A/m)
k	Coupling coefficient (dimensionless)
k	Wavenumber (1/m)
k_c	Cut-off wavenumber (1/m)
$\vec{k}$	Wave vector (1/m)
ℓ, L	Length (m)
L	Inductance (H; Henry)
L	Pathloss (dimensionless)
L'	Inductance per unit length (H/m)
p	Power density (W/m^3)
P	Power (W; Watt)
$P_{antenna}$	Accepted power (W)
P_{inc}	Incoming power (W)
P_{rad}	Radiated power (W)
Q	Charge (C; Coulomb)
Q	Quality factor (dimensionless)
r	Radial coordinate (m)
R	Resistance (Ω)
R_{DC}	Resistance for steady currents (Ω)
R_{ESR}	Equivalent series resistance (Ω)
R_{RF}	Resistance for radio frequencies (Ω)
R_{rad}	Radiation resistance (Ω)
R'	Resistance per unit length (Ω/m)
s_{kl}	Scattering parameter (dimensionless)
$\mathbf{S}$	Scattering matrix (dimensionless)
$\vec{S}$	Poynting vector (W/m^2)
$\vec{S}_{av}$	Average value of Poynting vector (W/m^2)
t	Time (s; second)
T	Period (s)
$\tan \delta$	Loss tangent (dimensionless)
U	Voltage (V; Volt)
$\vec{v}$	Velocity (m/s)
v_{gr}	Group velocity (m/s)
v_{ph}	Phase velocity (m/s)
V	Volume (m^3)
w_e	Electric energy density (J/m^3)
W_e	Electric energy (J; Joule)
w_m	Magnetic energy density (J/m^3)
W_m	Magnetic energy (J)
x, y, z	Cartesian coordinates (m)
Y	Admittance (S; Siemens)
$\mathbf{Y}$	Admittance matrix (S)

Z_A	Load impedance (Ω)
Z_F	Characteristic wave impedance (Ω)
Z_{F0}	Characteristic impedance of free space (Ω)
Z_{in}	Input impedance (Ω)
Z_0	Characteristic line impedance (Ω)
	Port reference impedance (Ω)
$Z_{0,cm}$	Common mode line impedance (Ω)
$Z_{0,diff}$	Differential mode line impedance (Ω)
Z_{0e}	Even mode line impedance (Ω)
Z_{0o}	Odd mode line impedance (Ω)
$\mathbf{Z}$	Impedance matrix (Ω)

Greek Letters

α	Attenuation coefficient (1/m)
β	Phase constant (1/m)
δ	Skin depth (m)
Δ	Laplace operator ($1/m^2$)
$\varepsilon = \varepsilon_0 \varepsilon_r$	Permittivity (As/(Vm))
ε_r	Relative permittivity (dimensionless)
$\varepsilon_{r,eff}$	Effective relative permittivity (dimensionless)
η	Radiation efficiency (dimensionless)
η_{total}	Total radiation efficiency (dimensionless)
γ	Propagation constant (1/m)
λ	Wavelength (m)
λ_W	Wavelength inside waveguide (m)
$\mu = \mu_0 \mu_r$	Permeability (Vs/(Am))
μ_r	Relative permeability (dimensionless)
∇	Nabla operator (1/m)
φ	Phase angle (rad)
φ	Azimuth angle (rad)
ϕ	Scalar electric potential (V)
φ_0	Initial phase (rad)
Ψ_e	Electric flux (C)
Ψ_m	Magnetic flux (Wb, Weber) (Vs)
ρ	Volume charge density (C/m^3)
ρ_S	Surface charge density (C/m^2)
σ	Conductivity (S/m; Siemens/m)
σ	Radar cross-section (m^2)
ϑ	Elevation angle (rad)
ϑ_{iB}	Brewster angle (rad)
ϑ_{ic}	Critical angle (rad)
ω	Angular frequency (1/s)

Physical Constants

μ_0	$4\pi \cdot 10^{-7}$ Vs/(Am)	Permeability of free space
ε_0	$8.854 \cdot 10^{-12}$ As/(Vm)	Permittivity of free space
c_0	$2.99792458 \cdot 10^8$ m/s	Speed of light in vacuum
e	$1.602 \cdot 10^{-19}$ C	Elementary charge
Z_{F0}	$120\pi\ \Omega \approx 377\ \Omega$	Characteristic impedance of free space

1

Introduction

This chapter provides a short overview on widely used microwave and RF applications and the denomination of frequency bands. We will start out with an illustrative case on wave propagation which will introduce fundamental aspects of high frequency technology. Then we will give an overview of the content of the following chapters to facilitate easy orientation and quick navigation to selected issues.

1.1 Radiofrequency and Microwave Applications

Today, at home or on the move, every one of us uses devices that employ wireless technology to an increasing extent. Figure 1.1 shows a selection of wireless communication, navigation, identification and detection applications.

In the future we will see a growing progression of the trend of applying components and systems of high frequency technology to new areas of application. The development and maintenance of such systems requires an extensive knowledge of the high frequency behaviour of basic elements (e.g. resistors, capacitors, inductors, transmission lines, transistors), components (e.g. antennas), circuits (e.g. filters, amplifiers, mixers) including physical issues such as electromagnetic wave propagation.

High frequency technology has always been of major importance in the field of radio applications, recently though RF design methods have started to develop as a crucial factor with rapid digital circuits. Due to the increasing processing speed of digital circuits, high frequency signals occur which, in turn, create demand for RF design methods.

In addition, the high frequency technology's proximity to electromagnetic field theory overlaps with aspects of electromagnetic compatibility (EMC). Setups for conducted and radiated measurements, which are used in this context, are based on principles of high frequency technology. If devices do not comply with EMC limits in general a careful analysis of the circumstances will be required to achieve improvements. Often, high frequency issues play a major role here.

Table 1.1 shows a number of standard RF and microwave applications and their associated frequency bands [1–3]. The applications include terrestrial voice and data communication, that is cellular networks and wireless communication networks, as well as terrestrial

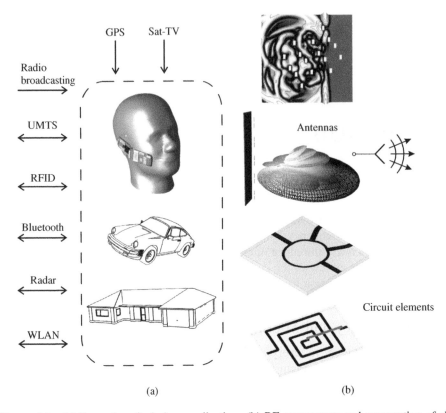

Figure 1.1   (a) Examples of wireless applications (b) RF components and propagation of electromagnetic waves.

and satellite based broadcasting systems. Wireless identification systems (RFID) within ISM bands enjoy increasing popularity among cargo traffic and logistics businesses. As for the field of navigation, GPS should be highlighted, which is already installed in numerous vehicles and mobile devices. Also in the automotive sector, radar systems are used to monitor the surrounding aresa or serve as sensors for driver assistance systems.

1.2   Frequency Bands

For better orientation, the electromagnetic spectrum is divided into a number of frequency bands. Various naming conventions have been established in different parts of the world, which often are used in parallel. Table 1.2 shows a customary classification of the frequency range from 3 Hz to 300 GHz into eight frequency decades according to the recommendation of the *International Telecommunications Union* (ITU) [4].

Figure 1.2a shows a commonly used designation of different frequency bands according to IEEE-standards [5]. The unsystematic use of characters and band ranges, which has developed over the years, can be regarded as a clear disadvantage. A more recent naming convention according to NATO is shown by Figure 1.2b [6, 7]. Here, the mapping of

Table 1.1 Wireless applications and frequency ranges

Cellular mobile telephony		
GSM 900	Global System for Mobile Communication	$880\ldots960\,$MHz
GSM 1800	Global System for Mobile Communication	$1.71\ldots1.88\,$GHz
UMTS	Universal Mobile Telecommunications System	$1.92\ldots2.17\,$GHz
Tetra	Trunked radio	$440\ldots470\,$MHz
Wireless networks		
WLAN	Wireless local area network	$2.45\,$GHz, $5\,$GHz
Bluetooth	Short range radio	$2.45\,$GHz
Navigation		
GPS	Global Positioning System	$1.2\,$GHz, $1.575\,$GHz
Identification		
RFID	Radio-Frequency Identification	$13.56\,$MHz, $868\,$MHz, $2.45\,$GHz, $5\,$GHz
Radio broadcasting		
FM	Analog broadcast transmitter network	$87.5\ldots108\,$MHz
DAB	Digital Audio Broadcasting	$223\ldots230\,$MHz
DVB-T	Digital Video Broadcasting - Terrestrial	$470\ldots790\,$MHz
DVB-S	Digital Video Broadcasting - Satellite	$10.7\ldots12.75\,$GHz
Radar applications		
SRR	Automotive short range radar	$24\,$GHz
ACC	Adaptive cruise control radar	$77\,$GHz

Table 1.2 Frequency denomination according to ITU

Frequency range	Denomination
$3\ldots30\,$kHz	VLF - Very Low Frequency
$30\ldots300\,$kHz	LF - Low Frequency
$300\,$kHz $\ldots3\,$MHz	MF - Medium Frequency
$3\ldots30\,$MHz	HF - High Frequency
$30\ldots300\,$MHz	VHF - Very High Frequency
$300\,$MHz $\ldots3\,$GHz	UHF - Ultra High Frequency
$3\ldots30\,$GHz	SHF - Super High Frequency
$30\ldots300\,$GHz	EHF - Extremely High Frequency

characters to frequency bands is much more systematic. However, the band names are not common in practical application yet.

A number of legal foundations and regulative measures ensure fault-free operation of radio applications. Frequency, as a scarce resource, is being divided and carefully administered [8, 9]. Determined frequency bands are allocated to *industrial, scientific and medical* (ISM) applications. These frequency bands are known as ISM bands and are shown in Table 1.3. As an example, the frequency range at $2.45\,$GHz is for the operation of microwave ovens and WLAN systems. A further frequency band reserved for wireless non-public short-range data transmission (in Europe) uses the 863 to $870\,$MHz frequency band [10], for example for RFID applications.

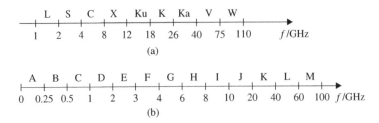

Figure 1.2 Denomination of frequency bands according to different standards. (a) Denomination of frequency bands according to IEEE Std. 521–2002 (b) Denomination of frequency bands according to NATO.

Table 1.3 ISM frequency bands

13.553 . . . 13.567 MHz	26.957 . . . 27.283 MHz
40.66 . . . 40.70 MHz	433.05 . . . 434.79 MHz
2.4 . . . 2.5 GHz	5.725 . . . 5.875 GHz
24 . . . 24.25 GHz	61 . . . 61.5 GHz
122 . . . 123 GHz	244 . . . 246 GHz

1.3   Physical Phenomena in the High Frequency Domain

We will now take a deeper look at RF engineering through two examples that introduce wave propagation on transmission lines and electromagnetic radiation from antennas.

1.3.1   Electrically Short Transmission Line

As a *first example* we consider a simple circuit (Figure 1.3a) with a sinusoidal (monofrequent) voltage source (internal resistance R_I), which is connected to a load resistor $R_A = R_I$ by an *electrically short* transmission line. *Electrically short* means that the transmission length ℓ of the line is much shorter than the wavelength λ, that is $\ell \ll \lambda$. In vacuum–or approximately air–electromagnetic waves propagate with the speed of light c_0.

$$c_0 = 299\,792\,458\,\frac{\text{m}}{\text{s}} \approx 3 \cdot 10^8\,\frac{\text{m}}{\text{s}} \qquad \text{(Speed of light in vacuum)} \qquad (1.1)$$

Therefore, the free space wavelength λ_0 for a frequency f yields:

$$\lambda_0 = \frac{c_0}{f} \gg \ell \qquad (1.2)$$

In media other than vacuum the speed of light c is lower and given by

$$c = \frac{c_0}{\sqrt{\varepsilon_r \mu_r}} \qquad \text{(Speed of light in media)} \qquad (1.3)$$

where ε_r is the relative permittivity and μ_r is the relative permeability of the medium. Typical values for a practical coaxial line would be $\varepsilon_r = 2$ and $\mu_r = 1$, resulting in a speed of light of $c \approx 2.12 \cdot 10^8$ m/s on that line. Given–as an example–a frequency of

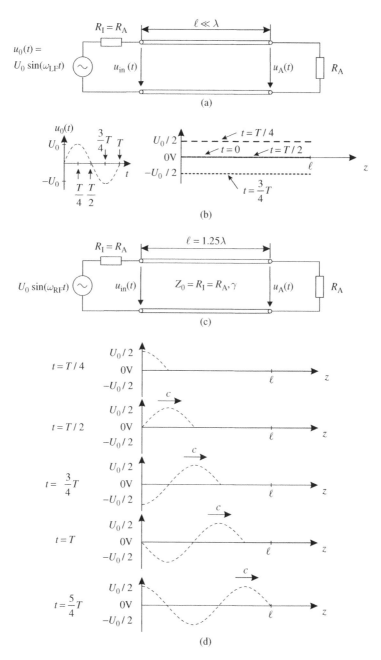

Figure 1.3 Network with voltage source, transmission line and load resistor. Transmission line is *electrically short* in (a), (b) and *electrically long* in (c), (d).

$f = 1\,\text{MHz}$ we get a wavelength of $\lambda_0 = 300\,\text{m}$ in free space and $\lambda = 212\,\text{m}$ on the previously discussed line. A transmission line of $\ell = 1\,\text{m}$ would then be classified as electrically short ($\ell \ll \lambda$). For simplicity[1], we assume further on that the load resistance R_A equals the internal resistance R_I of the source.

Alternatively, *electrically short* can be expressed by the propagation time τ a signal needs to pass through the entire transmission line. Assuming that electromagnetic processes spread with the speed of light c, the transmission of a signal from the start through to the end of a line requires a time span τ

$$\tau = \frac{\text{distance}}{\text{velocity}} = \frac{\ell}{c} \ll T = \frac{1}{f} \qquad \leftrightarrow \qquad \lambda = \frac{c}{f} \gg \ell \qquad (1.4)$$

If the time span τ needed for a signal to travel through the whole line is substantially smaller than the cycle time T of its sinusoidal signal, it seems as if the signal change appears *simultaneously* along the whole line. Signal delay is thus surely negligible.

> A transmission line is defined as being *electrically short*, if its length ℓ is substantially shorter than the wavelength λ of the signal's operating frequency ($\ell \ll \lambda$) or–in other words–if the duration of a signal travelling from the start to the end of a line τ (delay time) is substantially shorter than its cycle time T ($\tau \ll T$).

Let us have a look at Figure 1.3b where the current changes *slowly* in a sinus-like pattern. The term *slowly* refers to the period T that we assume to be much greater than the propagation time τ along the line. The sine wave starts at $t = 0$ with a value of zero and reaches its peak after a quarter of the time period ($t = T/4$). Again after half the time period ($t = T/2$) it passes through zero and reaches a negative peak at $t = 3T/4$. This sequence repeats periodically. Since signal delay τ can be omitted compared to the time period T, the signal along the line appears to be *spatially constant*. According to the voltage divider rule the voltage along the line equals just half of the value of the voltage source $u_0(t)$. The input voltage $u_{\text{in}}(t)$ and the output (load) voltage $u_A(t)$ are–at least approximately–equal.

$$u_{\text{in}}(t) \approx u_A(t) \qquad (1.5)$$

1.3.2 Transmission Line with Length Greater than One-Tenth of Wavelength

In the next step, we significantly increase the frequency f, so that the line is no longer electrically short. We choose the value of the frequency, such that the line length will equal $\ell = 5/4 \cdot \lambda = 1.25\lambda$ (Figure 1.3c). Now signal delay τ compared to period duration T must be taken into consideration. In Figure 1.3d we can see how far the wave has travelled

---

[1] The reason for this determination will become clear when we discuss the fundamentals of transmission line theory in Chapter 3.

at times of $t = T/4$, $t = T/2$ and so forth. The voltage distribution is no longer spatially constant. After $t = 5/4T$ the signal reaches the end of the line.

If the transmission line is not electrically short, the voltage along the line will not show a constant course any longer. On the contrary, a sinusoidal course illustrates the wave-nature of this electromagnetic phenomenon.

Also we can see that the electric voltage $u_A(t)$ at the line termination is no longer equal to that at the line input voltage $u_{in}(t)$. A *phase difference* exists between those two points.

In order to fully characterize the transmission line effects, a transmission line must be described by two *additional parameters* along with its length: (a) the *characteristic impedance* Z_0 and (b) the *propagation constant* γ. Both must be taken into account when designing RF circuits.

In our example we used a characteristic line impedance Z_0 equal to the load and source resistance ($Z_0 = R_A = R_I$). This is the most simple case and is often applied when using transmission lines. However, if the characteristic line impedance Z_0 and terminating resistor R_A are not equal to each other, the wave will be reflected at the end of the line. Relationships resulting from these effects will be looked at in Chapter 3 which deals in detail with *transmission line theory*.

1.3.3 Radiation and Antennas

Now let us take a look at a *second example*. Here we have a geometrically simple structure (Figure 1.4a), which consists of a rectangular metallic patch with side length ℓ arranged above a continuous metallic ground plane. Insulation material (dielectric material) is located between both metallic surfaces. Two terminals are connected to feed the structure.

The geometric structure resembles that of a parallel plate capacitor, which has a homogeneous electrical field set up between the metal surfaces. Therefore, we see *capacitive behaviour* ((Figure 1.4b) Admittance $Y = j\omega C$) at *low frequency* values (geometrical dimensions are significantly below wavelength ($\ell \ll \lambda$)). By further increasing the frequency, we can observe *resonant behaviour* due to the unavoidable inductance of feed lines.

At high frequency levels a completely new phenomenon can be observed: with the structure's side length approaching half of a wavelength ($\ell \approx \lambda/2$), electromagnetic energy will be radiated into space. Now the structure can be used as an antenna (*patch antenna*).

This example clearly illustrates that even a geometrically simple structure can display complex behaviour at high frequency levels. This behaviour cannot yet be described by common circuit theory and requires electromagnetic field theory.

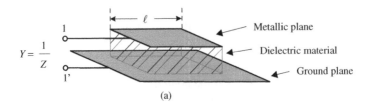

$$Y = \frac{1}{Z}$$

(a)

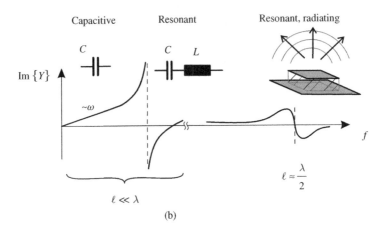

(b)

Figure 1.4 Electrical characteristic of a geometrical simple structure: (a) geometry and (b) imaginary part of admittance.

1.4 Outline of the Following Chapters

The last two examples have given us some insight into the fact that problems involving RF cannot simply be treated with conventional methods, but need a toolset adjusted to the characteristics of RF technology. Chapters 2 to 8 therefore give an in-depth insight into how best to solve RF-problems and show the methods we commonly apply.

First, the principles of electromagnetic field theory and wave propagation are reviewed in Chapter 2, in order to understand the mechanisms of passive high-frequency circuits and antennas. The mathematical formulas used in this chapter mainly serve the purpose of illustrating mathematical derivations and are not intended for further calculations. Nowadays, in work practice, modern RF circuit and field simulation software packages provide approximate solutions based on the above mentioned theories. Nonetheless, an engineer needs to understand these mathematical foundations in order to evaluate such given solutions of different commercial software products with respect to their plausibility and accuracy.

Transmission lines are a major and important component in RF circuits. The simple structure of a transmission line may be used in a variety of very different applications. Chapter 3 will therefore deal with the detailed relationships of voltage and current waves on transmission lines. Calculations in this context can be easily followed and form a safe foundation for treating the ever-occurring issue of transmission lines. This chapter gives a short introduction to the Smith chart as a common instrument of presentation and design in RF technology.

Chapter 4 continues with the characterization of transmission line structures and provides a deeper insight into technically important line types such as coaxial lines, planar transmission lines and hollow waveguide structures. It also gives an overview on common mode and differential mode signals on three conductor lines, since they play a major role in the circuit design of filters and couplers.

Chapter 5 introduces scattering parameters, with which the behaviour of RF circuits is characterized. Scattering parameters link wave quantities at different ports of RF circuits. A great advantage of scattering parameters compared to impedance and admittance parameters (which are preferably used when frequency levels are low) is that they can be directly measured by vector network analysers even at high frequencies.

The basic principles provided in Chapters 2 to 5 now pave the way for the description of more complex practical passive RF circuitry, which is the focus of Chapter 6. It is shown that a thoughtful interconnection of transmission lines may create matching circuits, filters, power splitters or couplers. The aim here is to get to know different important design methods and to apply them to examples and case studies, rather than focusing on their mathematical derivation. The examples given are processed with RF circuit and field simulators and thus show how to employ these tools. Also, as a side issue (as it is not the main purpose of this book), a short preview on electronic circuits and their basic characteristics is given.

In radio communication, an antenna provides the link between aerial radio waves in free space and signals on transmission lines. Therefore, Chapter 7 deals with technically important parameters, which serve to describe the radiation behaviour of antennas. To deepen knowledge, we will mathematically deduce the functioning of a basic antenna element (Hertzian dipole). Further on, we take a look at important practical single and array antenna structures and test various design rules on some illustrative examples.

Finally, when it comes to evaluating wireless systems, it is not enough to only consider the antenna alone. Instead, interference from the surroundings (environment) on the wave propagation between antennas must also be taken into account. Chapter 8 thus introduces basic propagation phenomena and their effects on signal transmission. The book concludes with a short review of empirical and physical path loss models.

References

1. Dobkin DM (2005) RF *Engineering for Wireless Networks*. Newnes.
2. Golio M, Golio J (2008) *The RF and Microwave Handbook*, Second Edition. CRC Press.
3. Molisch AF (2011) *Wireless Communications*. John Wiley & Sons.
4. ITU (2000) *ITU-R Recommendation V.431: Nomenclature of the Frequency and Wavelength Bands Used in Telecommunications*. International Telecommunication Union.
5. IEEE (2002) IEEE Std 521-2002 *Standard Letter Designations for Radar-Frequency Bands*. IEEE.
6. Macnamara T (2010) *Introduction to Antenna Placement and Installation*. John Wiley & Sons.
7. Meinke H, Gundlach FW (1992) *Taschenbuch der Hochfrequenztechnik*. Springer.
8. Bundesnetzagentur (2008) *Frequenznutzungsplan gemaess §54 TKG ueber die Aufteilung des Frequenzbereichs von 9 kHz bis 275 GHz auf die Frequenznutzungen sowie ueber die Festlegungen fuer diese Frequenznutzungen*. Bundesnetzagentur.

9. CEPT/ECC (2009) *The European Table of Frequency Allocations and Utilizations in the frequency range 9 kHz to 3000 GHz*. European Conference of Postal and Telecommunications Administrations, Electronic Communications Committee.

10. Bundesnetzagentur (2005) *Allgemeinzuteilung von Frequenzen fuer nichtoeffentliche Funkanwendungen geringer Reichweite zur Datenuebertragung; Non-specific Short Range Devices (SRD)*. Bundesnetzagentur Vfg 92.

2

Electromagnetic Fields and Waves

In this chapter we will recall the basic electric and magnetic field quantities. We start with the static – non time-varying – case and highlight the relation between electromagnetic field quantities and circuit variables like voltage and current. A full description of time-varying electromagnetic fields is given by Maxwell's equations in combination with additional boundary conditions. Finally we discuss important solutions of Maxwell's equations that are necessary for the understanding of high-frequency behaviour of technical components like transmission lines and antennas.

Our discussion on electromagnetic theory is limited to fundamental aspects. More detailed treatments on this topic can be found in [1–6].

2.1 Electric and Magnetic Fields

In this section we introduce mathematical formulas and physical relations for the static case in order to get a visual understanding of electric and magnetic field quantities.

2.1.1 Electrostatic Fields

2.1.1.1 Field Strength and Voltage

Historically it has been known for a long time that electric charges are the origin of electrical phenomena. Charges produce forces upon each other. We distinguish between positive and negative charges. Charges of opposite signs attract each other whereas charges of same sign push each other away. The absolute value of the *Coulomb force* F_C between two point charges[1] Q_1 and Q_2, separated by a distance r, is given by

$$F_C = \frac{1}{4\pi\varepsilon_0} \cdot \frac{Q_1 Q_2}{r^2} \tag{2.1}$$

where $\varepsilon_0 = 8.854 \cdot 10^{-12}$ As/(Vm) is the permittivity of free space.

---

[1] The concept of *point charges* assumes that the charge is located at a singular point in space.

RF and Microwave Engineering: Fundamentals of Wireless Communications, First Edition. Frank Gustrau.
© 2012 John Wiley & Sons, Ltd. Published 2012 by John Wiley & Sons, Ltd.

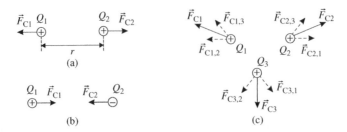

Figure 2.1 Coulomb forces between (a) two positive charges, (b) two charges of opposite polarity and (c) three charges of equal polarity.

The direction of the Coulomb force is defined by a straight line through both charges (see Figure 2.1a and 2.1b). In the case of three or more charges we can add up the vectors of the forces by the principle of superposition (vector sum in Figure 2.1c).

Charges exist in discrete values (integer values of elementary charge $e = 1.602 \cdot 10^{-19}$ C). However, in practical (macroscopic) engineering configurations the number of charges is so huge that we can treat charge as a continuous quantity.

The abovementioned point-charges are assumed to be located at a single point in space. In the case of a continuous distribution of charges in space we introduce a new term called *electric charge density* ϱ. The total charge Q in a certain volume V is then expressed as

$$Q = \iiint_V \varrho \, dv \tag{2.2}$$

Now we will replace the concept of direct forces between distant charges by a new concept of an electric field that emanates from a charge and leads to interaction with other charges. We look at the electric field strength $\vec{E}_1$ of charge Q_1 at the point in space where a charge Q_2 is located. Let us assume a small positive test charge Q_2. The electric field $\vec{E}_1$ is given by the force $\vec{F}_2$ upon the test charge Q_2 divided by the test charge Q_2 itself. Hence, E_1 becomes independent of Q_2.

$$\vec{E}_1 = \frac{\vec{F}_2}{Q_2} \tag{2.3}$$

The vector of the electric field strength $\vec{E}_1$ points in the direction of the force $\vec{F}_2$. Mathematically the electric field is simply the ratio of $\vec{F}_2$ and Q_2. Interestingly we can interpret $\vec{E}_1$ as a *vector field* that exists in the space surrounding charge Q_1, even if no test charge Q_2 is present. We can easily measure the electric field distribution of charge Q_1 by moving a test charge Q_2 around charge Q_1.

The distribution of the electric vector field of a positive charge Q_1 – located at the origin of a spherical coordinate system – is directly derived from Coulomb's law in Equation 2.1 as

$$\vec{E}_1 = \frac{1}{4\pi\varepsilon_0} \cdot \frac{Q_1}{r^2} \vec{e}_r \tag{2.4}$$

A vector field is conveniently visualized by *field lines*. Figure 2.2 shows field line representations of different charge configurations. Field lines represent *direction* and *amplitude*

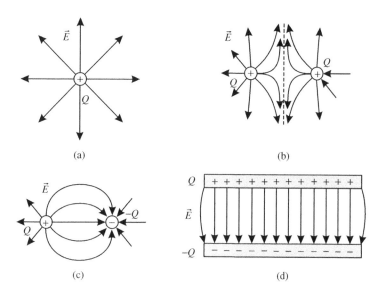

Figure 2.2 Electric field lines for (a) a positive point charge, (b) between two positive charges, (c) between two charges of opposite polarity and (d) in a parallel-plate capacitor (with fringing fields at the edges).

of the vector fields. The direction of the vector field is oriented tangentially to the field lines, and the density of field lines is a measure of the absolute value (denser field lines meaning higher field amplitude).

Looking at the field distributions in Figure 2.2 we notice that the electric field lines always[2] start at positive charges and end at negative ones. If only positive or negative charges are present the field lines extend to infinity. Therefore, we consider a positive charge as a *source* of the electric field and a negative charge as a *sink*.

Field lines of the electrostatic field are *open* field lines, that is they start at positive and end at negative charges. The electrostatic vector field is a *curl-free* or *conservative* field.

Consider a charge Q_2 that moves through the electric field $\vec{E}_1$ of a stationary charge Q_1 from point $\vec{r}_A$ to $\vec{r}_B$. During movement along the path from $\vec{r}_A$ to $\vec{r}_B$ the charge Q_2 experiences a force $\vec{F}_2 = \vec{E}_1 Q_2$. From physics we know that work $W_{\vec{r}_A \vec{r}_B}$ is done.

$$W_{\vec{r}_A \vec{r}_B} = \int_{\vec{r}_A}^{\vec{r}_B} \vec{F}_2 \cdot d\vec{s} = \int_{\vec{r}_A}^{\vec{r}_B} Q_2 \vec{E}_1 \cdot d\vec{s} = Q_2 \underbrace{\int_{\vec{r}_A}^{\vec{r}_B} \vec{E}_1 \cdot d\vec{s}}_{U} = Q_2 U \tag{2.5}$$

---

[2] This is true for the electrostatic case that we are discussing here. We will see later that there can be *closed* electric field lines under time-varying conditions.

The dot product between force $\vec{F}_2$ and differential path vector $d\vec{s}$ indicates that work is maximum for a path tangential to the field lines. A path perpendicular to the force yields zero work. In Equation 2.5 the constant charge Q_2 can be taken outside the integral sign. We interpret the integral over the electric field strength as a new electric quantity called *voltage U*.

$$U = \int_{\vec{r}_A}^{\vec{r}_B} \vec{E} \cdot d\vec{s} \tag{2.6}$$

The electric field is a vector field that is defined for any point in space $\vec{E} = \vec{E}(\vec{r})$. The voltage U is defined between *two* points $\vec{r}_A$ and $\vec{r}_B$ and therefore it is an integral or circuit quantity. The voltage is *no* field quantity.

If we define point $\vec{r}_B$ as a fixed reference point $\vec{r}_0$ we derive a new *field* quantity very similar to the voltage. The electric potential ϕ is given by

$$\phi_{\vec{r}_0}(\vec{r}) = \int_{\vec{r}}^{\vec{r}_0} \vec{E} \cdot d\vec{s} \tag{2.7}$$

The electric potential is a *scalar field*, that is, for any point in space $\phi(\vec{r})$ is given by a single (scalar) value. In the electrostatic case the electric potential ϕ represents the electric voltage between the point $\vec{r}$ and the reference point $\vec{r}_0$.

As discussed earlier the electrostatic field $\vec{E}$ is a conservative field. From mathematics we know that conservative fields can be expressed as a *gradient*[3] of a scalar field. Interestingly the scalar field that leads to the electric field is the negative value of the electric potential

$$\vec{E} = -\text{grad}\,\phi = -\nabla\phi = -\left(\frac{\partial\phi}{\partial x}\vec{e}_x + \frac{\partial\phi}{\partial y}\vec{e}_y + \frac{\partial\phi}{\partial z}\vec{e}_z\right) \tag{2.8}$$

Equation 2.8 shows the definition of the gradient operator in Cartesian coordinates. Definitions of the gradient operator in other coordinate systems are listed in Appendix A.1.

The gradient operator on the scalar electric potential results in an electric vector field. The gradient points in the direction of maximum (positive) change in the scalar potential function. The gradient is normal to an equipotential surface (i.e. surface with constant potential). This feature of the gradient is useful when optimization algorithms are considered: in order to find a local maximum of a function it is most efficient to proceed in the direction of maximum change.

The definition of the potential ϕ in Equation 2.7 is ambiguous. There is an infinite number of potential functions since the position of the reference point is not uniquely defined. However the electric field – defined by a force on a charge – is unique. Potentials with different reference points only differ in an additive constant. In Equation 2.8 we see

---

[3] The gradient operator can be written using the *Nabla* operator ($\text{grad}\,\phi = \nabla\phi$). Both forms are common [3].

that the electric field is calculated from partial derivatives of the scalar potential. Hence, an additional constant has no impact on the resulting value.

2.1.1.2 Polarization and Relative Permittivity

Up to now we have considered charges in free space. In the presence of non-conducting, dielectric materials new expressions can be conveniently defined.

Let us look at a parallel-plate capacitor. The capacitor consists of two ideally conducting plates and a separating dielectric material. In Figure 2.3a the capacitor is filled with vacuum. The charges on both plates are equal in magnitude but have opposite signs: $Q+$ on the upper plate and $Q-$ on the lower plate. If we assume the lateral extensions of the plates are much larger than the spacing between them we can neglect the fringing fields at the edges and regard the electric field as homogeneous. For the air-filled parallel-plate capacitor in Figure 2.3a we can easily calculate the voltage between the plates from Equation 2.6 as $U_0 = E_0 d$, where d is the distance between the plates. If we now put an isolating material (dielectric) between the plates (Figure 2.3b), we observe that the new voltage U_M is reduced $U_M < U_0$. If we remove the dielectric medium the former voltage U_0 is restored. The process is reversible.

In order to understand the macroscopic effect of reduced voltage when a dielectric is placed inside the capacitor, we will now take a look at the microscopic composition of dielectric materials. We first consider a material that consists of polar particles, that is the molecules represent electric dipoles (Figure 2.3d). Without the external field the dipoles are in random order. Adjacent charges cancel each other out. There is no macroscopic net effect.

If we switch on the external field E_0 the dipoles align with the field due to Coulomb forces (Figure 2.3e). In the inner part of the material adjacent charges cancel out. However,

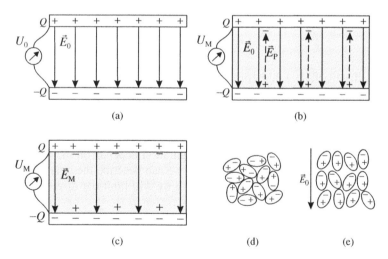

Figure 2.3 Understanding polarization in a dielectric medium: (a) air-filled parallel-plate capacitor with initial electric field E_0, (b) dielectric medium between plates gives rise to an opposing field E_P (c) reduced electric field strength E_M inside the dielectric medium, (d) dipoles in random order without external electric field and (e) dipoles aligned by external field.

at the upper and lower surface of the dielectric material a charge density remains. The surface charge density gives rise to a secondary field E_P that is opposed to the external field. Inside the material we observe a diminished field E_M

$$E_M = E_0 - E_P \tag{2.9}$$

This effect is known as *polarization*.[4] The amplitude of polarization, that is the strength of the opposing field E_P depends on the microscopic structure of the material.

We have shown that materials with molecules that form electric dipoles can be polarized, that is the dipoles align with the field. It can be shown that even materials with non-dipole molecules can be polarized. Let us consider a simple model of a charge-neutral atom with a positive nucleus in the centre and surrounding negative electrons. The centre of the positive and negative charges is now located at the same point in space. If we put such a material into an external electric field E_0 the field exerts opposing forces upon positive and negative charges. This results in a displacement of the charges and induces a dipole moment. Hence we observe a polarizing effect too.

Let us go back now to our initial experiment where a dielectric medium inside the parallel-plate capacitor led to a reduced voltage U_M

$$U_M = \int_{Plate\ 1}^{Plate\ 2} \vec{E}_M \cdot d\vec{s} < U_0 \tag{2.10}$$

In order to get rid of the description of microscopic effects we focus on the net macroscopic effect: the dielectric material reduces the voltage that we measure and this is as a result of the reduced field E_M in the medium. Therefore, we introduce a new term denoted as *relative permittivity*: the relative permittivity ε_r is the ratio between the unperturbed external field E_0 and the reduced field E_M due to polarization

$$\varepsilon_r = \frac{E_0}{E_M} = \frac{U_0}{U_M} \tag{2.11}$$

The relative permittivity is a dimensionless quantity equal to or larger than one. Table 2.1 lists values for common engineering materials. Vacuum has a relative permittivity of one, for air the value is slightly higher ($\varepsilon_r = 1.0006$) [3] but can be considered to be one for nearly all engineering applications.

There are materials with anisotropic properties, that is the relative permittivity that we measure with our parallel-plate capacitor configuration depends on the orientation of the dielectric material in the capacitor. In this case we can describe the polarizing effect by a matrix rather than by a single value. In this book we will limit our discussion to isotropic materials.

Polarization and time-varying fields

We will close our discussion of microscopic phenomena by looking at time-varying fields. Let us connect our parallel-plate capacitor to a harmonic voltage source. The charges now

---

[4] Unfortunately the term *polarization* is also used in the context of plane waves and antennas with a completely different meaning. We will discuss plane waves in Section 2.5.1.

Table 2.1 Relative permittivity ε_r and loss tangent $\tan \delta_\varepsilon$ of common dielectric materials in their typical frequency range

Material	ε_r	$\tan \delta_\varepsilon$	Typical application
Air, vacuum	1	0	Isolator for high precision lines
Polytetraflouroethylene (PTFE)	2.1	0.0002	Dielectric medium for transmission lines
Fibreglass reinforced epoxy laminate (FR4)	3.6–4.5	0.02	Low cost substrate for printed circuits
Alumina (Al_2O_3)	9.8	0.0001	Substrate for microwave circuits
Gallium arsenide (GaAs)	12.5	0.0004	Monolithic Microwave Integrated Circuits (MMIC)

switch polarity periodically. The dipoles of the dielectric material change their orientation with the same frequency as the field. Above a certain frequency the dipoles cannot follow the external field due to their inertia and the effect of polarization is reduced. Therefore, we expect a decrease of the relative permittivity with frequency.

A model for the frequency dependence $\varepsilon_r(f)$ is given by *Debye* formulas where f is the frequency of the external field. A detailed discussion is beyond the scope of this text and can be found elsewhere [13]. Fortunately most RF engineering materials have nearly constant relative permittivity values in their technical frequency range of usage.

Another important effect that occurs when a time-varying field is applied is heating of the material. On a microscopic scale the molecules constantly align with the changing field and their movement produces thermal energy (polarization losses). We will see in Section 2.2.4 that this heat loss can be described macroscopically by a term denoted *loss tangent*. For the purpose of reference we include the loss tangent for important materials in Table 2.1 [8, 9].

2.1.1.3 Electric Flux Density

Using the permittivity and the electric field strength we define the *electric flux density* $\vec{D}$ as a new vector. The letter D comes from the alternate notation of the electric flux density which is *displacement*.

$$\vec{D} = \varepsilon_0 \varepsilon_r \vec{E} \qquad (2.12)$$

For isotropic materials (ε_r is a scalar) the electric flux density $\vec{D}$ and the electric field strength $\vec{E}$ are proportional and point in the same direction. The displacement $\vec{D}$ is an important vector, as we will see when we discuss Gauss' law in Section 2.2. The unit of $\vec{D}$ is C/m^2 (Charge per area). Integrating the electric flux density $\vec{D}$ over an area A gives us the *electric flux* Ψ_e

$$\Psi_e = \iint\limits_A \vec{D} \cdot d\vec{A} \qquad (2.13)$$

2.1.1.4 Electric Energy and Capacitance

Energy is a common and general concept in physics. In an electric field distribution energy can be stored. The electrostatic *energy density* is defined as

$$w_{\mathrm{e}} = \frac{1}{2}\vec{D}\cdot\vec{E} = \frac{1}{2}\varepsilon_0\varepsilon_{\mathrm{r}}\left|\vec{E}\right|^2 \tag{2.14}$$

where $\vec{D}$ is the electric flux density and $\vec{E}$ the electric field strength. The latter term in Equation 2.14 is valid for isotropic materials and most convenient for calculations later in this book. The unit of the energy density is Joule/m^3.

If we integrate the energy density over a volume V we get the electric energy stored in that volume.

$$W_{\mathrm{e}} = \iiint\limits_V w_{\mathrm{e}}\,\mathrm{d}v = \frac{1}{2}\iiint\limits_V \vec{D}\cdot\vec{E}\,\mathrm{d}v = \frac{1}{2}\iiint\limits_V \varepsilon_0\varepsilon_{\mathrm{r}}\left|\vec{E}\right|^2\mathrm{d}v \tag{2.15}$$

Let us assume that the electric field is created by a capacitor with a capacitance C and a voltage U between its electrodes. In this case the energy can be written by using voltage and capacitance as

$$W_{\mathrm{e}} = \iiint\limits_V \varepsilon_0\varepsilon_{\mathrm{r}}\left|\vec{E}\right|^2\mathrm{d}v = \frac{1}{2}CU^2 \tag{2.16}$$

where the volume V includes the complete field between the electrodes.

> Hence, a capacitor is a device which stores electric energy and its capacitance is a measure of how much energy can be stored for a given voltage.

Capacitance in general is defined as the ratio between charge on the electrodes and voltage between them.

$$C = \frac{Q}{U} \quad \text{(Capacitance)} \tag{2.17}$$

For a parallel-plate capacitor with plate area A, plate distance d and dielectric material (relative permittivity ε_{r}) between the electrodes the capacitance is

$$C = \varepsilon_0\varepsilon_{\mathrm{r}}\frac{A}{d} \quad \text{(Capacitance of a parallel-plate capacitor)} \tag{2.18}$$

2.1.2 Steady Electric Current and Magnetic Fields

2.1.2.1 Current Density, Power Density and Resistance

In the previous sections we discussed electrostatic fields, that is stationary fields from non-moving charges. Now we consider charges that move in space. We will see that moving charges give rise to magnetic fields.

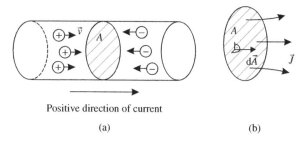

Positive direction of current

(a)   (b)

Figure 2.4   (a) Conduction current I as a result of charges passing through a cross-section area A and (b) current density $\vec{J}$.

Let us first introduce some useful terms for the description of moving charges. In Figure 2.4a we observe charges ΔQ that move through an area A for a given time interval Δt. The *current* I then gives the amount of charge ΔQ per time interval Δt.

$$I = \frac{\Delta Q}{\Delta t} \tag{2.19}$$

Current – like voltage – is an integrated quantity (not a field quantity), because it depends on an area of observation.

The direction of positive charges defines the (positive) direction of the current. In a metallic conductor current flow takes place through (negative) electrons. These electrons flow in the opposite direction of the current. In media where both negative and positive charges move through an area A both charges contribute to the current (Figure 2.4a).

The *current density* $\vec{J}$ describes the movement of charges as a vector field. The direction of the current density vector is given by the velocity $\vec{v}$ of (positive) charges. If we integrate the electric current density over an area A we get the current I through that area.

$$I = \iint\limits_{A} \vec{J} \cdot d\vec{A} \tag{2.20}$$

Current density is a vector field that defines a vector at any point in space $\vec{J} = \vec{J}(\vec{r})$. *Current* is an integral measure that depends on a given area A.

In a conducting material the current density is proportional to the electric field strength inside the medium

$$\vec{J} = \sigma \vec{E} \quad \text{(Ohm's law)} \tag{2.21}$$

where σ is the electric *conductivity*. In an ideal conductor ($\sigma \to \infty$) the electric field strength approaches zero in order to ensure a finite current density. Table 2.2 lists conductivities of good (metallic) conductors. Conductivity values depend on temperature, the values in Table 2.2 are valid at room temperature. If losses are of minor concern good conductors are often approximated by an ideal conductor commonly called PEC for *perfect electric conductor*.

Table 2.2 Conductivity of ideal and technically important conductors at room temperature

Material	$\sigma/10^7$ S/m	Typical application
Ideal conductor	∞	Idealized model for good conductor
Copper (Cu)	5.7	Standard conductor material for transmission lines and circuits
Silver (Ag)	6.1	Circuits
Gold (Au)	4.1	Bondwires in integrated circuits
Aluminum (Al)	3.5	Transmission lines, connectors

Current flow in a conductive material leads to thermal heating due to electric losses. The *power density* p in an electric conductor is given by

$$p = \vec{J} \cdot \vec{E} = \sigma \left| \vec{E} \right|^2 \tag{2.22}$$

Integrating the power density p over a volume V gives us the power P inside that volume

$$P = \iiint\limits_V \vec{J} \cdot \vec{E} \, dv \tag{2.23}$$

From circuit theory we know that a resistor dissipates power.

$$P = UI = \frac{U^2}{R} = I^2 R \tag{2.24}$$

If we divide the power by the square of the current through the volume we get the *ohmic resistance* R

$$R = \frac{P}{I^2} = \frac{1}{I^2} \iiint\limits_V \vec{J} \cdot \vec{E} \, dv \tag{2.25}$$

For a cylindrical conductor with the cross-section area A, the length ℓ and the conductivity σ the resistance[5] is

$$R = \frac{\ell}{\sigma A} \tag{2.26}$$

2.1.2.2 Steady Currents, Magnetic Fields and Vector Potential

In nature there are objects consisting of magnetic materials that attract or repulse other magnetic objects. These forces can be considered as interaction between *magnetic fields* that are created by magnetic materials. The fundamental magnetic field quantity is the magnetic flux density (or induction) $\vec{B}$.

---

[5] This equation is only valid for a current that is constant in time and a current density distribution that is homogeneous over the cross-section area. If we consider time-varying fields the so called skin effect leads to inhomogeneous distribution of current density. We will discuss this in Section 2.4.

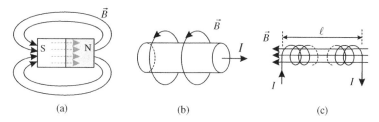

Figure 2.5 Magnetic flux density (a) originating from a permanent magnet, (b) in the vicinity of a straight wire and (c) inside a wire-wound cylindrical coil.

Permanent magnets produce persistent magnetic fields that can be visualized by magnetic field lines of the flux density $\vec{B}$. The region where the magnetic field lines leave the object is called the *north pole* and the region where the field lines enter the object is known as the *south pole* (see Figure 2.5a). If we break a permanent magnet into two pieces, we end up with two magnets each of them having north and south pole. We cannot isolate the north and south poles in the way that we can separate positive and negative electric charges. Magnetic field lines are always *closed* lines that neither start nor end. When considering a permanent magnet the field lines are closed by running back from south to north pole *inside* the magnet.

Field lines of the magnetic flux density $\vec{B}$ neither start nor end, they run in closed loops. The magnetic flux density is a *solenoidal (or divergence-free) vector field*.

We will see shortly what is meant by the term *divergence*. First let us introduce another field quantity the *magnetic field strength* $\vec{H}$. The magnetic field strength and the magnetic flux density are connected by the following relation

$$\vec{B} = \mu_0 \mu_r \vec{H} \tag{2.27}$$

where $\mu_0 = 4\pi \cdot 10^{-7}$ Vs/(Am) is the permeability of free space and μ_r is the *relative permeability*. The relative permeability describes the magnetic property of a material. Its value is nearly $\mu_r \approx 1.0$ for most[6] RF relevant materials like isolators and conductors of transmission lines and antennas.

Permanent magnets are only one way to produce a magnetostatic field. Another source of static magnetic fields are steady (or DC)[7] currents. Currents (or moving charges in general) produce magnetic fields.[8]

A current I that flows through a wire is surrounded by a magnetic field $\vec{B}$ as shown in Figure 2.5b. The magnetic field runs in closed circles. The direction of the current and the direction of the magnetic field lines is defined by the *right-hand rule*: if we grip the wire with the *right hand* so that the *thumb* is oriented in the direction of the positive

---

[6] An exception of this rule are ferromagnetic materials and *ferrites* where the latter are used in circulator networks. Magnetic materials commonly show non-linear and anisotropic behaviour.

[7] *Direct current*.

[8] In a permanent magnet microscopic loops of currents are the sources of the persistent magnetic field.

current flow, the other four fingers of the right hand indicate the direction of the rotating magnetic field vector.

In the electrostatic case we calculated the electric field from a scalar *potential* ϕ by applying the gradient operator on the scalar potential function. In the magnetostatic case we can also introduce a potential: the *magnetic vector potential* $\vec{A}$. The vector potential is an auxiliary quantity that simplifies mathematical calculations. We will use the magnetic vector potential when we determine radiating fields of antennas in Chapter 7. By applying the *curl* operation on the magnetic vector potential $\vec{A}$ we calculate the magnetic flux density $\vec{B}$.

$$\vec{B} = \text{curl } \vec{A} = \nabla \times \vec{A} \tag{2.28}$$

The curl operator can be expressed by the Nabla operator ∇ and the vector cross-product. We will see shortly how to interpret this operator (see Section 2.1.3).

In analogy to the electric flux the *magnetic flux* Ψ_m through an area A is defined by the surface integral of the magnetic flux density.

$$\Psi_m = \iint_A \vec{B} \cdot d\vec{A} \tag{2.29}$$

2.1.2.3 Magnetic Energy and Inductance

We have seen that we can store energy in the electrostatic field. The same is true for the magnetic field. For a linear and isotropic medium the *magnetic energy density* w_m at a point is given by

$$w_m = \frac{1}{2}\vec{B} \cdot \vec{H} = \frac{1}{2}\mu_0\mu_r \left|\vec{H}\right|^2 \tag{2.30}$$

If we integrate the magnetic energy density w_m over a volume V we get the *magnetic energy* which is stored in that volume.

$$W_m = \iiint_V w_m \, dv = \frac{1}{2} \iiint_V \vec{B} \cdot \vec{H} \, dv = \frac{1}{2} \iiint_V \mu_0\mu_r \left|\vec{H}\right|^2 dv \tag{2.31}$$

A device that stores magnetic energy is an inductor or coil. A basic inductor is created from a cylindrical wound wire that carries a steady current (see Figure 2.5c). Inside the cylindrical inductor the magnetic field lines add up constructively.

The amount of magnetic energy that can be stored in an inductor depends on the current and is described by the *inductance L*.

$$W_m = \frac{1}{2} \iiint_V \mu_0\mu_r \left|\vec{H}\right|^2 dv = \frac{1}{2}LI^2 \tag{2.32}$$

For a long cylindrical coil, that is a coil where the length ℓ is much larger than the radius R, the magnetic field inside the coil can be assumed to be homogeneous. Under these conditions the inductance L is given by

$$L = \mu_0 \mu_r n^2 \frac{A}{\ell} \qquad \text{(Inductance of a long cylindrical coil)} \qquad (2.33)$$

where n is the number of turns, A is the cross-section area and μ_r is the relative permeability of the material inside the coil.

2.1.2.4 Lorentz Force

Our discussion in Chapter 2 started with forces (Coulomb force $\vec{F}_C$) between charges. These forces led to the introduction of an electrostatic field. Magnetostatic fields $\vec{B} = \mu \vec{H}$ exert forces on charges too but only if the charges move relative to the field. The *Lorentz force* $\vec{F}_L$ is given by

$$\vec{F}_L = Q\left(\vec{v} \times \vec{B}\right) \qquad (2.34)$$

where $\vec{v}$ is the velocity of the charge Q. The cross-product indicates that the Lorentz force is maximum for a charge moving perpendicular to the field lines. The direction of the force itself is perpendicular to the plane that is defined by $\vec{v}$ and $\vec{B}$.

A charge under simultaneous exposure to an electrostatic field $\vec{E}$ and a magnetostatic field $\vec{B}$ experiences the superposition of Coulomb and Lorentz forces.

$$\vec{F}_{total} = \vec{F}_C + \vec{F}_L = Q\vec{E} + Q\left(\vec{v} \times \vec{B}\right) \qquad (2.35)$$

2.1.3 Differential Vector Operations

In the previous section we encountered a differential operator based on the Nabla operator: the gradient function applied to a scalar field produced a vector field. The gradient operator has a physical meaning since it defines the direction of maximum change of the scalar function. We will now investigate two more differential operators based on the Nabla operator: *divergence* and *curl*. These two operators are applied to vector fields. We will see that they too have a physical meaning.

2.1.3.1 Divergence

The first differential operator we look at is the *divergence*. The divergence operator is applied to a vector field $\vec{V}$ and produces a scalar field S. In Cartesian coordinates we write:

$$S = \text{div}\vec{V} = \nabla \cdot \vec{V} = \frac{\partial V_x}{\partial x} + \frac{\partial V_y}{\partial y} + \frac{\partial V_z}{\partial z} \qquad (2.36)$$

where the Nabla operator ∇ in Cartesian coordinates is defined by the following calculation rule

$$\nabla = \frac{\partial}{\partial x}\vec{e}_x + \frac{\partial}{\partial y}\vec{e}_y + \frac{\partial}{\partial z}\vec{e}_z \qquad (2.37)$$

Formulas for the calculation in cylindrical and spherical coordinates can be found in Appendix A.1.

The divergence has the following physical meaning: at a point $\vec{r}$ where the divergence is non-zero the vector field $\vec{V}$ has a (scalar) source or sink S. If the divergence is zero at any point the vector field has neither source nor sink at that point. A vector field that has *only scalar sources and sinks* is a conservative or irrotational (curl free) vector field.

In the electrostatic case (Section 2.1.1) charges Q are the (scalar) sources and sinks of the electric vector fields $\vec{E}$ and $\vec{D}$. Electric field lines start at positive charges and end at negative ones. In regions where point charges Q or charge density distributions ρ are located the divergence of the electric field has non-zero values. The electrostatic field is a conservative (irrotational) field. The field lines are open.

2.1.3.2 Curl

The second differential operator we will look at is the *curl*. The curl operator is applied to a vector field $\vec{V}$ and produces a vector field $\vec{W}$. In Cartesian coordinates we write:

$$\vec{W} = \text{curl } \vec{V} = \nabla \times \vec{V} \tag{2.38}$$

$$= \vec{e}_x \left(\frac{\partial V_z}{\partial y} - \frac{\partial V_y}{\partial z} \right) - \vec{e}_y \left(\frac{\partial V_z}{\partial x} - \frac{\partial V_x}{\partial z} \right) + \vec{e}_z \left(\frac{\partial V_y}{\partial x} - \frac{\partial V_x}{\partial y} \right)$$

Formulas for the calculation in cylindrical and spherical coordinates can be found in Appendix A.1.

The curl has the following physical meaning: at a point $\vec{r}$ where the curl is non-zero the vector field $\vec{V}$ has a (vector) source $\vec{W}$. If the curl is zero at any point the vector field has no vector source at that point. A vector field that has *only vector sources* is a solenoidal or divergence-free vector field.

In the case of magnetostatic fields (Section 2.1.2) current density distibutions $\vec{J}$ are the (vector) sources of the magnetic vector field $\vec{H}$ and $\vec{B}$. Magnetic field lines circle steady electric currents. In regions where current density distributions $\vec{J}$ are located the curl of the magnetic field has non-zero values. The steady magnetic field is a solenoidal field. The field lines are closed.

2.2 Maxwell's Equations

In the previous sections we reviewed basic electromagnetic field quantities for the static case. Now, Maxwell's equations provide a set of differential equations that describe the time-varying behaviour of electromagnetic phenomena in general. The field quantities in these formulas are real-valued vector functions of space $\vec{r}$ and time t. (If – for instance – we look at the Cartesian coordinates of the electric field vector, the component in x-direction would read $E_x = E_x(x, y, z, t) \in \mathbb{R}$.)

Using Maxwell's equations we can calculate the electric and magnetic fields for a given initial field and a set of boundary conditions (initial boundary-value problem).[9] Maxwell's equations exist in differential and integral form. In the following sections we will investigate both forms and focus primarily on their physical meaning and visual interpretation.

2.2.1 Differential Form in the Time Domain

Maxwell's first equation (Ampere's law) in differential form is given by

$$\boxed{\nabla \times \vec{H} = \vec{J} + \frac{\partial \vec{D}}{\partial t}} \qquad \text{(Ampere's law)} \qquad (2.39)$$

where $\vec{J}$ is the current density in conductive media. It is therefore often called *conduction current density* $\vec{J}_C$. The expression $\partial \vec{D}/\partial t$ has the same unit and is therefore a current density too. This type of current density is not limited to conductive media but can also exist in non-conducting material. The term is therefore called *displacement current density* $\vec{J}_D$.[10] The *true current density* $\vec{J}_{total}$ is given as the sum of conduction and displacement current density.

$$\vec{J}_{total} = \vec{J}_C + \vec{J}_D = \vec{J} + \frac{\partial \vec{D}}{\partial t} = \sigma \vec{E} + \varepsilon_0 \varepsilon_r \frac{\partial \vec{E}}{\partial t} \qquad (2.40)$$

Maxwell's first equation means that the total current density $\vec{J} + \partial \vec{D}/\partial t$ produces a solenoidal magnetic field $\vec{H}$. (The total current density is the vector source of the magnetic field strength.)

Maxwell's second equation (Faraday's law) in differential form is given by

$$\boxed{\nabla \times \vec{E} = -\frac{\partial \vec{B}}{\partial t}} \qquad \text{(Faraday's law)} \qquad (2.41)$$

Maxwell's second equation means that the negative change rate of the magnetic flux $\partial \vec{B}/\partial t$ produces a solenoidal electric field $\vec{E}$. (A changing magnetic field is the vector source of the circulating electric field strength with closed electric field lines.)

---

[9] Only very simple configurations can be solved using pencil and paper. Practical problems are commonly investigated by applying numerical methods. In Section 6.13.2 we will take a look at these EM simulation software tools.

[10] The term was originally introduced by James Clerk Maxwell and is essential for the mathematical treatment of electromagnetic waves as we will see shortly.

We note from our previous discussion of electrostatic fields that the electric field was an irrotational (conservative) field with open field lines that start at positive and end at negative charges. Now in the more general time-varying case we can see something new: electric field lines can be closed. Solenoidal electric fields exist. What is most important to underline here is the fact that in the time-varying case electric and magnetic fields are *interdependent*. Ampere's law says that a time-variant electric field produces a circulating magnetic field and Faraday's law indicates that a time-variant magnetic field in turn produces a circulating electric field. In non-conductive media ($\sigma = 0$) like air Ampere's and Faraday's law are similar in structure and the interdependence of electric and magnetic fields becomes very obvious:

$$\nabla \times \vec{H} = \varepsilon_0 \varepsilon_r \frac{\partial \vec{E}}{\partial t} \quad \text{and} \quad \nabla \times \vec{E} = -\mu_0 \mu_r \frac{\partial \vec{H}}{\partial t} \tag{2.42}$$

Maxwell's third equation (Gauss's law of the electric field) in differential form is given by

$$\boxed{\nabla \cdot \vec{D} = \varrho} \qquad \text{(Gauss's law of the electric field)} \tag{2.43}$$

Maxwell's third equation means that charge density distributions ρ produce irrotational electric fields $\vec{D}$. (Charges are the scalar sources of the conservative (irrotational) electric field with open electric field lines.)

Maxwell's fourth equation (Gauss's law of the magnetic field) in differential form is given by

$$\boxed{\nabla \cdot \vec{B} = 0} \qquad \text{(Gauss's law of the magnetic field)} \tag{2.44}$$

Maxwell's fourth equation means that there are no sources that produce irrotational magnetic fields $\vec{B}$. Therefore, no conservative magnetic fields exist. The magnetic field $\vec{B}$ is always a rotational field with closed field lines.

2.2.2 Differential Form for Harmonic Time Dependence

From basic electrical engineering courses we know that currents and voltages in sinusoidally excited circuits can be represented by using complex-valued amplitudes called *phasors*. In circuit theory these complex amplitudes replace real-valued time functions. In doing so the circuit equations resulting from Kirchhoff's laws turn from differential equations to algebraic equations which are much easier to solve. We remember that the time derivative d/dt becomes a simple factor $j\omega$. For example, if we look at at a capacitor the equations are

$$i(t) = C \frac{du(t)}{dt} \quad \text{(Time functions)} \qquad \rightarrow \qquad I = j\omega CU \quad \text{(Phasors)} \tag{2.45}$$

where j is the imaginary unit, $i(t)$ and $u(t)$ are functions of time and I and U are phasors [10]. From the phasor representation we can go back to the time domain by simply applying two steps:

- First, multiply the phasor with an exponential function that includes the excitation frequency $\omega = 2\pi f$; and
- then take the real part of the product.

$$u(t) = \mathrm{Re}\left\{U \cdot \mathrm{e}^{j\omega t}\right\} \tag{2.46}$$

If we apply the same mathematical procedure to the electric and magnetic field quantities, Maxwell's equations will read

$$\nabla \times \vec{H} = \vec{J} + j\omega \vec{D} \tag{2.47}$$

$$\nabla \times \vec{E} = -j\omega \vec{B} \tag{2.48}$$

$$\nabla \cdot \vec{D} = \varrho \tag{2.49}$$

$$\nabla \cdot \vec{B} = 0 \tag{2.50}$$

where all field quantities $(\vec{B}, \vec{D}, \vec{E}, \vec{H}, \vec{J}, \rho)$ are complex-valued vector functions of the spatial variables x, y, z. That means each component of a vector is complex and depends only on spatial variables. (If – for instance – we look at the Cartesian coordinates of the electric field vector, the component in x-direction would read $E_x = E_x(x, y, z) \in \mathbb{C}$.)

Although the *phasor* representation of fields (components of the field vectors are *complex* and depend on *spatial* variables only) is different from the *time-dependent* field representation (components of the field vectors are *real* and depend on *spatial and temporal* variables) we do *not* introduce a different notation for the two representations. We will see that from the context of the equations it is always absolutely clear which representation has been used. Furthermore, in the scope of this book field variables always represent *peak values*. In the literature RMS (root mean square) values are sometimes used which introduce minor changes in some power related equations.

2.2.3 Integral Form

By using integral theorems from vector algebra we can transform Maxwell's equations from their differential form into the integral form. We will skip the mathematical derivation and focus on the physical interpretation of the results.

Maxwell's first equation (Ampere's law) in integral form is given by

$$\boxed{\oint_{C(A)} \vec{H} \cdot \mathrm{d}\vec{s} = \iint_A \left(\vec{J} + \frac{\partial \vec{D}}{\partial t}\right) \cdot \mathrm{d}\vec{A}} \qquad \text{(Ampere's law)} \tag{2.51}$$

On the right-hand side of Equation 2.51 the true current density is integrated over a surface area A in the direction of the differential surface normal $\mathrm{d}\vec{A}$. This results in the

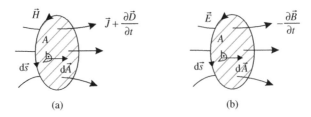

Figure 2.6   Visualization of (a) Ampere's law and (b) Faraday's law.

total current I_{total} that flows through the area A. On the left-hand side the magnetic field strength $\vec{H}$ is integrated along the closed contour line $C(A)$ of surface A in the direction of the differential line element $d\vec{s}$. The direction of the differential surface normal $d\vec{A}$ and the direction of the differential line element $d\vec{s}$ are linked by the right-hand rule: if the thumb points in the direction of the surface normal the other fingers indicate the orientation of the contour path. Figure 2.6a gives us a graphical interpretation of the surface area and contour path.

> The total current through the surface A is equal to the closed line integral of the magnetic field $\vec{H}$ over the contour line of the surface. The total current considers both conduction current density $\vec{J}$ and displacement current density $\partial \vec{D}/\partial t$. Conduction currents and displacement currents are the sources of circulating magnetic fields.

Maxwell's second equation (Faraday's law) in integral form is given by

$$\oint_{C(A)} \vec{E} \cdot d\vec{s} = -\frac{d}{dt} \iint_{A} \vec{B} \cdot d\vec{A} \qquad \text{(Faraday's law)} \qquad (2.52)$$

The integral on the right-hand side is the negative derivative of the magnetic flux $\Psi_m = \iint \vec{B} d\vec{A}$ through area A. On the left-hand side the electric field is integrated along the closed path $C(A)$ around the surface area A. The unit of both integrals is Volt. Figure 2.6b gives us a graphical interpretation of the surface and contour path of the integrals.

> The negative change rate of the magnetic flux Ψ_m through the surface A is equal to the closed line integral of the electric field $\vec{E}$ over the contour line of the surface. A time-variant magnetic flux is the source of circulating electric fields.

Maxwell's third equation (Gauss's law of the electric field) in integral form is given by

$$\oiint_{A(V)} \vec{D} \cdot d\vec{A} = \iiint_{V} \varrho \, dv = Q \qquad \text{(Gauss' law for the electric field)} \qquad (2.53)$$

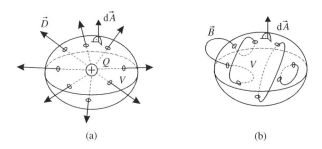

Figure 2.7   Visualization of Gauss' law for the (a) electric and (b) magnetic field.

On the left-hand side of Equation 2.53 the electric flux $\Psi_e = \iint \vec{D} \cdot d\vec{A}$ is evaluated over a closed surface area $A(V)$ of a volume. On the right-hand side we see the integral of the charge density ρ over the volume V which gives us the charge Q in that volume (see Figure 2.7a).

> The electric flux Ψ_e that passes through the closed surface of a volume is equal to the total charge Q in that volume. The charges are the sources of the conservative (irrotational) electric field.

Maxwell's fourth equation (Gauss's law of the magnetic field) in integral form is given by

$$\oiint\limits_{A(V)} \vec{B} \cdot d\vec{A} = 0 \qquad \text{(Gauss' law for the magnetic field)} \qquad (2.54)$$

When we evaluate the magnetic flux $\Psi_m = \iint \vec{B} \cdot d\vec{A}$ through a closed surface area $A(V)$ of a volume the resulting value is zero. Every magnetic field line that moves away from the volume will enter it again, so that there is no net flux through the closed surface (see Figure 2.7b).

> The net magnetic flux through a closed surface is always zero. Magnetic field lines are closed field lines. There is no such thing as a conservative magnetic field.

2.2.4   Constitutive Relations and Material Properties

In our review of static electric and magnetic fields we have seen that the field strengths ($\vec{E}$ and $\vec{H}$) are linked by material properties (relative permittivity ε_r and relative permeability μ_r) to the flux densities ($\vec{D}$ and $\vec{B}$). Furthermore, the electric current density $\vec{J}$ is related to the electric field strength $\vec{E}$ by a material property called conductivity σ. We summarize these equations (known as *constitutive relations*) here for convenience.

$$\vec{B} = \mu_0 \mu_r \vec{H} \qquad (2.55)$$

$$\vec{D} = \varepsilon_0 \varepsilon_r \vec{E} \tag{2.56}$$

$$\vec{J} = \sigma \vec{E} \tag{2.57}$$

Typical values for material properties of technically important isolators and conductors are given in Table 2.1 and 2.2.

Materials can show quite complex behaviour when they are exposed to electromagnetic fields. In order to understand the phenomena involved we will introduce common terms for the classification of materials.

Linear material: A material is *linear* if the material properties (relative permittivity ε_r, relative permeability μ_r and electric conductivity σ) do not depend on the field strengths in the material. The material parameters are not functions of the magnitude of the electric and magnetic field, that is $\varepsilon_r \neq \varepsilon_r(\vec{E}, \vec{H})$, $\mu_r \neq \mu_r(\vec{E}, \vec{H})$ and $\sigma \neq \sigma(\vec{E}, \vec{H})$. If the material is not linear it is called *non-linear*.

Time-invariant material: A material is *time-invariant* if the material parameters do not change with time. The material parameters are not functions of time, that is $\varepsilon_r \neq \varepsilon_r(t)$, $\mu_r \neq \mu_r(t)$ and $\sigma \neq \sigma(t)$. If the material properties depend on time the material is called *time-variant*.

Isotropic material: A material is *isotropic* if the material parameters are independent of the direction in space. The material parameters are then defined by a scalar value. If a material shows different material properties when oriented in different directions it is called *anisotropic*.

Non-dispersive material: A material is *non-dispersive* if the material parameters do not vary with frequency. The material parameters are not functions of frequency, that is $\varepsilon_r \neq \varepsilon_r(f)$, $\mu_r \neq \mu_r(f)$ and $\sigma \neq \sigma(f)$. If the material properties depend on frequency the material is called *dispersive*.

Homogeneous material: A material is homogeneous if the material parameters do not vary in space. The material parameters are not functions of the spatial coordinates, that is $\varepsilon_r \neq \varepsilon_r(\vec{r})$, $\mu_r \neq \mu_r(\vec{r})$ and $\sigma \neq \sigma(\vec{r})$. If the material properties are not constant in space the material is called *inhomogeneous*.

Fortunately, most RF engineering materials are *simple materials* which are linear, time-invariant and isotropic. Furthermore, at least in narrow frequency bands, they can be regarded as non-dispersive[11] and in confined regions they are homogeneous. However

---

[11] Mathematical models for the frequency dependence of material properties are given by *Debye* formulas. A detailed discussion is beyond the scope of this text and can be found elsewhere [7].

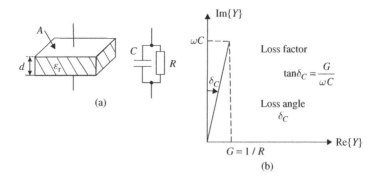

Figure 2.8 Admittance of a lossy capacitor and definition of loss angle and loss factor.

an inhomogeneous distribution of materials in space is a central design element when considering RF components like filters and couplers.

When sinusoidal excitations are applied we commonly use the phasor representation of electromagnetic fields where the fields have complex-valued amplitudes. The relative permittivity and the relative permeability of lossy materials are then described by complex-valued numbers too. Since dielectric losses are more important than magnetic losses, we will look at complex permittivity in detail now.

Let us consider a parallel plate capacitor filled with a lossy dielectric material. In Section 2.1.1 we found a capacitance of $C = \varepsilon_0 \varepsilon_r A/d$ (see Equation 2.18), where A is the area of the plates and d is the separating distance. The relative permittivity ε_r describes how polarizable the dielectric material is. We remember that inside the material electric dipoles align with the external field and reduce the electric field strength inside the dielectric material.

In the case of sinusoidal excitations the orientation of the electric field inside the capacitor switches periodically, forcing the dipoles in the material to constantly realign with the changing field. This movement of particles gives rise to heating and therefore causes *polarization losses*. We want to include these losses into the relative permittivity of the material. In order to do so we take a look of a circuit theory representation of a lossy capacitor.

An equivalent circuit model of a lossy capacitor is given by a parallel arrangement of a lossless capacitor C and a resistor R (see Figure 2.8b). The admittance of this parallel circuit is

$$Y = j\omega C + G \qquad \text{with} \qquad G = \omega C \tan \delta_C \qquad (2.58)$$

where we will write the conductance G of the resistor as a product of ωC and a new factor $\tan \delta_C$ which we call the *loss tangent* or *loss factor* of the capacitor. The loss factor of the capacitor is defined as the ratio between the real and imaginary part of the admittance of the lossy capacitor. The more lossy the capacitor becomes the greater the loss factor. Figure 2.8b shows the geometric interpretation of the loss factor and the *loss angle* of the capacitor in the complex admittance plane.

The losses of the capacitor can be associated with the losses in the dielectric material by including the capacitance of a parallel plate capacitor from Equation 2.18.

With $C = \varepsilon_0 \varepsilon_\mathrm{r} A/d$ we write

$$Y = j\omega C + \omega C \tan \delta_C$$

$$= j\omega\varepsilon_0\varepsilon_\mathrm{r} \frac{A}{d} \left(1 - j \tan \delta_C\right) = j\omega\varepsilon_0 \frac{A}{d} \underbrace{\varepsilon_\mathrm{r} \left(1 - j \tan \delta_\varepsilon\right)}_{\substack{\text{complex relative} \\ \text{permittivity}}} \qquad (2.59)$$

In order to indicate that the losses in the capacitor are caused by the losses in the dielectric material we will replace the loss factor of the capacitor $\tan \delta_C$ with the loss factor of the dielectric material $\tan \delta_\varepsilon$. The complex relative permittivity is then

$$\varepsilon_\mathrm{r,complex} = \varepsilon_\mathrm{r} \left(1 - j \tan \delta_\varepsilon\right) \qquad (2.60)$$

where the loss tangent $\tan \delta_\varepsilon$ of the relative permittivity is a measure of how lossy the material is. The loss tangent of the material is defined as the negative imaginary part of the complex relative permittivity. A loss tangent of zero indicates a lossless material. Typical loss tangents for RF engineering materials are given in Table 2.1.

In Equation 2.60 we used an extra subscripted text 'complex' to distinguish between complex and real relative permittivity. We will omit this long winded notation in the future and use the same symbol ε_r for both complex and real relative permittivity.

2.2.5   Interface Conditions

If the electric and magnetic field lines pass from one medium into another, special interface conditions apply. We will use indexes $i \in \{1, 2\}$ to identify the two different materials. Furthermore, we will use the subscripted letters 't' and 'n' to indicate *tangential* and *normal* components of the fields. Tangential components are *parallel* to the interface and normal components are *perpendicular* to the interface.

Figure 2.9 shows the electric field strength $\vec{E}_1$ in medium 1 with a component E_{1t} that is tangential to the interface and a component E_{1n} that is normal to the interface. We can calculate the electric field in medium 2 by using the following interface conditions between dielectric and magnetic materials.

The tangential component of the electric and magnetic field strength is constant

$$E_{1t} = E_{2t} \qquad (2.61)$$

$$H_{1t} = H_{2t} \qquad (2.62)$$

The normal component of the electric and magnetic flux density is constant

$$D_{1n} = D_{2n} \qquad (2.63)$$

$$B_{1n} = B_{2n} \qquad (2.64)$$

We can now derive formulas for the tangential components of the electric and magnetic flux D_{it} and B_{it} by using the constitutive relations $\vec{D} = \varepsilon_0 \varepsilon_\mathrm{r} \vec{E}$ and $\vec{B} = \mu_0 \mu_\mathrm{r} \vec{H}$ as well as Equations 2.61 and 2.62.

$$\frac{D_{1t}}{\varepsilon_{r1}} = \frac{D_{2t}}{\varepsilon_{r2}} \tag{2.65}$$

$$\frac{B_{1t}}{\mu_{r1}} = \frac{B_{2t}}{\mu_{r2}} \tag{2.66}$$

Furthermore, we will derive formulas for the normal components of the electric and magnetic field strength E_{in} and H_{in} by using the constitutive relations as well as Equations 2.63 and 2.64.

$$\varepsilon_{r1} E_{1n} = \varepsilon_{r2} E_{2n} \tag{2.67}$$

$$\mu_{r1} H_{1n} = \mu_{r2} H_{2n} \tag{2.68}$$

Figure 2.9 shows as an example the behaviour of the electric field strength for an interface between two dielectrics. The tangential component of the electric field strength is constant (Equation 2.61) and the normal component are discontinuous at the interface (Equation 2.67).

In the special case where surface charge densities ρ_S or surface current densities $\vec{J}_S$ are present at the interface Equation 2.62 and Equation 2.63 change to

$$\left| H_{1t} - H_{2t} \right| = \left| \vec{J}_{S\perp} \right| \tag{2.69}$$

$$\left| D_{1n} - D_{2n} \right| = \left| \varrho_S \right| \tag{2.70}$$

Unlike charge densities ρ that are distributed in a three-dimensional region (charges per volume; unit: C/m^3) *surface charge densities* are located in a two-dimensional region (charges per area; unit: C/m^2). A current density $\vec{J}$ is defined as current (that flows in a three-dimensional object) per cross-section area (unit: A/m^2). A *surface current density* $\vec{J}_S$ represents a current that flows *in a two-dimensional plane*, so the cross-section is a

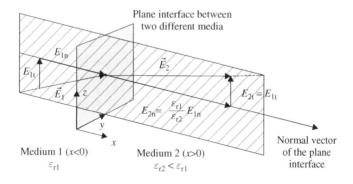

Figure 2.9 Plane material interface between two dielectrics: electric field strength $\vec{E}_1$ in Medium 1 (ε_{r1}) and electric field strength $\vec{E}_2$ in Medium 2 ($\varepsilon_{r2} < \varepsilon_{r1}$).

one-dimensional line (unit: A/m). The subscripted symbol ⊥ indicates that the direction of the current density is perpendicular to the direction of the tangential magnetic field [11].

2.3 Classification of Electromagnetic Problems

Generally, electromagnetic fields can be categorized into three groups: *static fields*, *quasi-static fields* and *coupled electromagnetic fields*. Only coupled electromagnetic fields require the full set of Maxwell's equations to be solved. In the case of static and quasi-static fields simplifications apply.

2.3.1 Static Fields

We can distinguish between electrostatic fields produced by non-moving charges and magnetostatic fields caused by steady currents. In the static case electric and magnetic fields are *decoupled*. Static field problems are therefore the least complex.

In the *electrostatic* case only electric fields exist. The magnetic fields are zero. The electric fields are caused by charges and can be described most efficiently by the scalar electric potential ϕ as seen in Section 2.1.1. Maxwell's coupled partial differential equations reduce to a second order differential equation (Poisson's equation).

$$\Delta\phi = \nabla^2\phi = -\frac{\varrho}{\varepsilon_0} \qquad \text{(Poisson's equation)} \qquad (2.71)$$

Software tools for electrostatic problems usually look for solutions of Poisson's equation and calculate the electric field strength from the scalar electric potential by using the gradient operator (see Equation 2.8).

In the case of *magnetostatic fields* from steady currents we will first calculate the steady current density distribution in conductive media; that is, we solve the electric part of the problem. In the second step we calculate the magnetic fields that are caused by the current density distribution. The magnetic fields themselves do not retract on the primary electric field. If the steady currents flow in a thin wire *Biot–Savart's law* gives us a simple formula to calculate the magnetic field in the vicinity of the current I

$$\vec{B}(\vec{r}) = \frac{\mu_0 I}{4\pi} \int_S \frac{\vec{ds}' \times (\vec{r} - \vec{r}')}{|\vec{r} - \vec{r}'|^3} \qquad \text{(Biot-Savart's law for free space)} \qquad (2.72)$$

where S is the path of the wire that carries the current I, $\vec{ds}'$ is the differential line element along the wire, the vector $\vec{r}'$ points to the location of the line element and the vector $\vec{r}$ is the point in space, where we calculate the field [11].

2.3.2 Quasi-Static Fields

Quasi-static fields are electric or magnetic fields that vary *slowly* with time. Let us consider a parallel-plate capacitor connected to a DC voltage source U_0. In the electrostatic case the electric field E_0 between the plates is given by

$$E_0 = \frac{U_0}{d} = \text{const.} \qquad \text{(Static case)} \qquad (2.73)$$

where d is the distance between the plates.

Now, we replace the DC voltage source by a *slowly varying* sinusoidal voltage source $U(t) = U_0 \cos(\omega t)$. The electric field *at any instance in time* is now given by Equation 2.73 if we use the corresponding voltage at that instant in time. We write

$$E(t) \approx E_0 \cos(\omega t) = \frac{U_0}{d} \cos(\omega t) \qquad \text{(Quasi-static case)} \qquad (2.74)$$

where $\omega = 2\pi f$ is the angular frequency. Obviously in the quasi-static case the original static formula in Equation 2.73 is simply combined with the factor $\cos(\omega t)$ that describes the time dependence. Such an approach produces approximations that are adequate for practical applications as long as the rate of change (or frequency) does not become too high.

In the upper approach two physical aspects have been neglected:

- First, time delay due to finite propagation speed and
- Second, coupling of electric and magnetic fields.

Electromagnetic phenomena propagate with a finite speed. In free space the propagation velocity v_{P} is equal to the speed of light c_0. So any physical dimension leads to a time delay. In the sinusoidal case time delay is associated with phase shift. In order to neglect phase shifts the maximum geometric extension $s_{\max}$ has to be much smaller than the corresponding wavelength $\lambda = c/f$.

$$s_{\max} \ll \lambda = \frac{c}{f} \qquad (2.75)$$

Quasi-static approximations are valid as long as the geometrical dimensions of a device are much smaller than wavelength and mutual coupling of electric and magnetic fields can be neglected. Therefore, technical components must become smaller with increasing frequency (greater wavelength).

We will use the quasi-static approach when calculating the electric and magnetic field in a cross-section of a coaxial transmission line (see Section 4.2.1).

2.3.3 *Coupled Electromagnetic Fields*

In RF and microwave engineering rapidly varying electromagnetic fields show quite complex behaviour because of the *mutual coupling* between the electric and the magnetic field components. Furthermore, the *propagation effects* produce substantial time delays in time-domain signals and phase shifts in harmonic (phasor) amplitudes. Coupling and time delay give rise to two important electromagnetic phenomena that will accompany us through the rest of the book: *skin effect* and *propagation of electromagnetic waves*.

Skin effect: Only in the static case is the current density in a wire homogeneously distributed over the cross-section area. However, if the currents vary with time the *current density* is *maximal at the conductor surface* and it falls towards the centre of the wire. As the current flows mainly at the surface of the conductor this effect is known as the *skin effect*.

Electromagnetic waves: Due to the mutual coupling of electric and magnetic fields electromagnetic waves can propagate through space or along a transmission line. The speed of propagation is finite and equals the speed of light. The maximum speed $c_0 = 3 \cdot 10^8$ m/s occurs in vacuum.

We will investigate these essential phenomena in more detail in the following Sections and derive important parameters.

2.4 Skin Effect

In order to mathematically analyse the skin effect we start with Maxwell's first and second equation in differential form which are repeated here for convenience

$$\nabla \times \vec{H} = \vec{J} + \frac{\partial \vec{D}}{\partial t} \tag{2.76}$$

$$\nabla \times \vec{E} = -\frac{\partial \vec{B}}{\partial t} \tag{2.77}$$

Since skin effect takes place in good conductors we are interested in the electric fields in conductors with high values of the conductivity σ. Let's further assume non-dielectric non-magnetic material ($\varepsilon_r = \mu_r = 1$). In these conductors current flow is mainly due to conduction current density $\vec{J}$. The displacement current density $\partial \vec{D}/\partial t$ is of minor significance. We will consider a conductive halfspace ($z \geq 0$) in order to ease our calculations by having a simple geometry. First we apply the curl operator[12] and the constitutive relation $\vec{B} = \mu_0 \vec{H}$ on Equation 2.77

$$\nabla \times \nabla \times \vec{E} = -\nabla \times \frac{\partial}{\partial t} \left(\mu_0 \vec{H} \right) \tag{2.78}$$

Next, we will insert Equation 2.76 and Ohm's law $\vec{J} = \sigma \vec{E}$ in Equation 2.78. Since in our halfspace the material is homogeneous we can put the material parameters σ and μ_0 in front of the partial derivative.

$$\nabla \times \nabla \times \vec{E} = -\mu_0 \sigma \frac{\partial \vec{E}}{\partial t} \tag{2.79}$$

---

[12] We remember that curl can be written as $\nabla \times$ (see Section 2.1.3).

We replace the double curl operator by a relation from mathematics [3]

$$\nabla \times \nabla \times \vec{V} = \nabla \left(\nabla \cdot \vec{V} \right) - \Delta \vec{V} \tag{2.80}$$

where $\Delta = \nabla^2$ is the *Laplace operator*. Now we will insert this relation into Equation 2.79 and get

$$\underbrace{\nabla \left(\nabla \cdot \vec{E} \right)}_{=0} - \Delta \vec{E} = -\mu_0 \sigma \frac{\partial \vec{E}}{\partial t} \tag{2.81}$$

Our conductive halfspace is electrically neutral so there is no charge density distribution ($\varrho = 0$). According to Gauss's law of the electric field (see Equation 2.43) the divergence of a charge-free region is zero ($\mathrm{div}\, \vec{D} = \varepsilon_0 \mathrm{div} \vec{E} = \varepsilon_0 (\nabla \cdot \vec{E}) = 0$). So we end up with the following partial differential equation

$$\boxed{\Delta \vec{E} = \mu_0 \sigma \frac{\partial \vec{E}}{\partial t}} \qquad \text{(Diffusion equation)} \tag{2.82}$$

Mathematically this partial differential equation is a *diffusion equation*. It describes processes that show diffusion-like behaviour (unlike processes that show wave-like behaviour as we will see later).

In the case of time-harmonic excitation we use phasor representation and replace the partial derivative with respect to time by the factor $j\omega$

$$\Delta \vec{E} = j\omega\mu_0\sigma \vec{E} \tag{2.83}$$

The electric field strength in the conductive halfspace is given by the following solution[13]

$$\vec{E}(z) = E_0 e^{-z/\delta} e^{-jz/\delta} \vec{e}_x = E_0 e^{-(1+j)z/\delta} \vec{e}_x \tag{2.84}$$

where δ is the so called *skin depth*. The skin depth is an important measure when describing the penetration of the electromagnetic field in conductive regions. It is given by

$$\boxed{\delta = \sqrt{\frac{2}{\omega\sigma\mu_0}}} \qquad \text{(Skin depth)} \tag{2.85}$$

Figure 2.10a shows the conduction current density $\vec{J} = \sigma\vec{E}$ in the conductive region ($z \geq 0$). The current density has its maximum value J_0 at the surface of the halfspace and decays exponentially with depth z. At a distance of $z = \delta$ from the surface the current density is reduced to approximately $37\% = 1/e$ of its maximum value. At a distance of $z = 5\delta$ the current density is less than 1% of its maximum value. The skin depth is therefore a measure of how deep the electromagnetic fields penetrate a conductor.

---

[13] We can easily prove it by inserting the solution from Equation 2.84 into Equation 2.83 (see Problem 2.6).

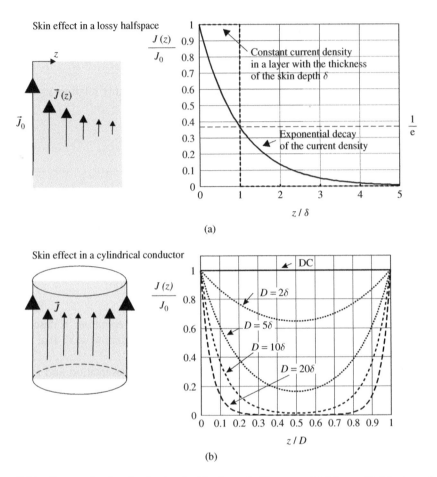

Figure 2.10 Current density distribution in (a) a lossy halfspace and (b) in a cylindrical conductor (Diameter D).

Example 2.1 *Skin Depth of Copper and Gold Material*

At a frequency of $f = 100$ MHz the skin depth of copper ($\sigma = 5.7 \cdot 10^7$ S/m) is $\delta = 6.6$ µm. Figure 2.11 shows a diagram for the skin depth of copper and gold in the frequency range from 1 MHz to 50 GHz. With increasing frequency the skin depth decreases. The current flows nearer and nearer the surface of the conductor.

In Figure 2.10a we see an exponential decay of the current density as the depth z increases. The current that flows through a given cross-section of infinite depth is equal to the current that flows through a surface layer with the thickness equal to the skin depth δ when we assume that the current density in the finite thickness layer is equal to the maximum current density J_0.

The skin depth can be used to calculate the effective resistance of a conductor if the lateral dimensions of a conductor are much larger than the skin depth. Under these

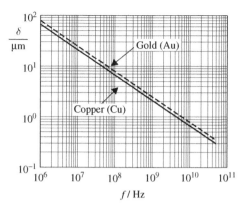

Figure 2.11   Frequency dependence of skin depth δ for gold and copper (see Example 2.1).

circumstances we assume that a homogeneous current flows through a superficial layer with a thickness of the skin depth.

In the case of steady (DC) currents the electric resistance of a cylindrical conductor is given by

$$R_{DC} = \frac{\ell}{\sigma A} = \frac{\ell}{\sigma \pi r^2} \tag{2.86}$$

where $A = \pi r^2$ is the cross-section area and ℓ is the length of the conductor.

In the case of a time-varying (radio-frequency) current with lateral dimensions much greater than the skin depth we assume a constant current flow in a superficial layer with a thickness of δ. Therefore, we can approximate the effective area A_{RF} for current flow by the product of the circumference ($C = 2\pi r$) and skin depth δ (see Figure 2.10b).

$$R_{RF} = \frac{\ell}{\sigma A_{RF}} = \frac{\ell}{\sigma C \delta} = \frac{\ell}{\sigma 2\pi r \delta} \tag{2.87}$$

Finally, Figure 2.10b shows the current density distribution in a cylindrical conductor for different ratios of diameter D and skin depth δ.

2.5   Electromagnetic Waves

2.5.1   Wave Equation and Plane Waves

Now we will take a look at wave propagation in homogeneous, lossless,[14] non-conducting media. Again we start with Maxwell's first and second equation in differential form

$$\nabla \times \vec{H} = \vec{J} + \frac{\partial \vec{D}}{\partial t} \tag{2.88}$$

$$\nabla \times \vec{E} = -\frac{\partial \vec{B}}{\partial t} \tag{2.89}$$

---

[14] In order to keep this review of electromagnetic theory simple we do not consider losses here. However in Section 3.1 we investigate the more general case of waves on lossy transmission lines.

We are interested in non-conducting media so the conductivity is zero ($\sigma = 0$) and there is no conduction current density $\vec{J}$. On the right-hand side of Ampere's law (Equation 2.88) only the displacement current density $\partial \vec{D}/\partial t$ remains. First we apply the curl operator and the constitutive relation $\vec{B} = \mu \vec{H}$ (with $\mu = \mu_0 \mu_r$) on Equation 2.89.

$$\nabla \times \nabla \times \vec{E} = -\nabla \times \frac{\partial}{\partial t}\left(\mu \vec{H}\right) \tag{2.90}$$

Now we insert Equation 2.88 into Equation 2.90 by using the constitutive relation $\vec{D} = \varepsilon \vec{E}$ (with $\varepsilon = \varepsilon_0 \varepsilon_r$). Since we assume the medium to be time-invariant we can write the material parameters outside the derivative.

$$\nabla \times \nabla \times \vec{E} = -\mu \varepsilon \frac{\partial^2 \vec{E}}{\partial t^2} \tag{2.91}$$

As in our discussion of the skin effect (see Equation 2.80) we replace the double curl operator by using the following relation from mathematics

$$\nabla \times \nabla \times \vec{V} = \nabla \left(\nabla \cdot \vec{V}\right) - \Delta \vec{V} \tag{2.92}$$

So we get

$$\underbrace{\nabla \left(\nabla \cdot \vec{E}\right)}_{=0} - \Delta \vec{E} = -\mu \varepsilon \frac{\partial^2 \vec{E}}{\partial t^2} \tag{2.93}$$

Since we consider a charge-free medium the divergence of the electric field is zero ($\mathrm{div}\,\vec{D} = \varepsilon_0 \,\mathrm{div}\,\vec{E} = \varepsilon_0 (\nabla \cdot \vec{E}) = 0$). Eventually, we achieve the following partial differential equation.

$$\boxed{\Delta \vec{E} = \mu \varepsilon \frac{\partial^2 \vec{E}}{\partial t^2}} \qquad \text{(Wave equation in the time domain)} \tag{2.94}$$

This equation is a *wave equation*. It describes phenomena that show wave-like behaviour (in lossless media).

If we consider sinusoidal signals we can turn to phasor representation of the electric field. The second derivative with respect to time is substituted by a factor $(j\omega)^2 = -\omega^2$.

$$\boxed{\Delta \vec{E} = -\omega^2 \mu \varepsilon \vec{E}} \qquad \text{(Wave equation in the frequency domain)} \tag{2.95}$$

A simple but important solution of the wave equation is the *plane wave*. The following phasor $\vec{E}$ describes a plane wave that propagates in positive x-direction and is polarized in z-direction; that is, the electric field vector points to the z-direction. $E_0 = |E_0| e^{j\varphi_0}$ is the complex amplitude of the electric field strength

$$\boxed{\vec{E}(x) = E_0 e^{-jkx} \vec{e}_z} \qquad \text{(Plane wave phasor)} \tag{2.96}$$

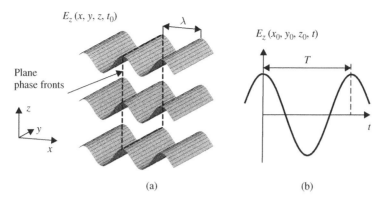

(a)   (b)

Figure 2.12   (a) Instantaneous value of the electric field strength: spatial distribution for a specific instant in time t_0. Points of equal phase (e.g. maxima of field strength) form planes. (b) Time-variant value of the electric field strength at point $\vec{r}_0 = (x_0,\ y_0,\ z_0)$.

where k is the wave number (or real propagation constant) of the wave. If we put the solution – the phasor of the plane wave from Equation 2.96 into Equation 2.95 we calculate k as

$$k = \omega\sqrt{\varepsilon\mu} \tag{2.97}$$

In order to get a more visual understanding of the plane wave we transform the phasor back into the time domain. First, we multiply by $e^{j\omega t}$, then we take the real part of the expression. Our result is

$$\boxed{\vec{E}\,(x,t) = \left|E_0\right|\cos\left(\omega t - kx + \varphi_0\right)\vec{e}_z} \qquad \text{(Plane wave in the time domain)} \quad (2.98)$$

Figure 2.12a shows the electric field strength at a fixed instant in time ($t = t_0$) as a function of spatial coordinates x, y and z. We see a 'snapshot' of a wave travelling in x-direction. The amplitude varies periodically in that direction but is constant in y- and z-direction. The period in space, that is the distance between two maxima, is called *wavelength*. If we consider the local distribution of points where the wave has its maxima we find that those points are located on plane surfaces that move in the direction of travel. We call those faces *wave fronts* or *phase fronts* since they oscillate with the same phase.

Another way to look at a wave is to regard it as a function of time. If we select a fixed spatial location $\vec{r}_0 = (x_0, y_0, z_0)$ (see Figure 2.12b) we observe a sinusoidal time signal $E(x_0, y_0, z_0, t)$ with a *time period* T. From the period T we can determine the frequency f (number oscillations per time interval) and the angular frequency (radians per time interval) as

$$\omega = 2\pi f = \frac{2\pi}{T} \tag{2.99}$$

An important parameter that is associated with the spatial period (wavelength) of a wave is its *wave vector*. The wave vector has the same magnitude as the wave number and points in the direction of travel. In our case it is

$$\vec{k} = k\vec{e}_x = \frac{2\pi}{\lambda}\vec{e}_x \qquad \text{(Wave vector)} \qquad (2.100)$$

An electromagnetic wave travels at a certain speed c. In a time interval T the electromagnetic wave covers a distance of one wavelength λ. Hence, the speed of propagation c – often denoted v_P for *phase velocity* – is

$$c = v_{\text{ph}} = \frac{\lambda}{T} = \frac{\omega}{k} = \lambda f \qquad (2.101)$$

We can determine the absolute value of the propagation speed in a medium from its relative permittivity and permeability and the *speed of light* in vacuum c_0.

$$c = \lambda f = \frac{1}{\sqrt{\varepsilon_0 \varepsilon_r \mu_0 \mu_r}} = \frac{c_0}{\sqrt{\varepsilon_r \mu_r}} \qquad \text{(Phase velocity of a plane wave)} \qquad (2.102)$$

The speed of light in vacuum is given by

$$c_0 = \frac{1}{\sqrt{\varepsilon_0 \mu_0}} = 299\,792\,458\frac{\text{m}}{\text{s}} \approx 3 \cdot 10^8 \frac{\text{m}}{\text{s}} \qquad \text{(Speed of light in vacuum)}$$

$$(2.103)$$

A plane wave is an electromagnetic wave with electric and magnetic field components. We can calculate the magnetic field's strength $\vec{H}$ from our solution of the electric field in Equation 2.96 by using Faraday's law (Equation 2.48). We solve the equations in Problem 2.5. Here we only write down the result that is

$$\vec{H}(x) = -H_0 e^{-jkx}\vec{e}_y \qquad (2.104)$$

Equations 2.96 and 2.104 describe a plane wave that is travelling in positive x-direction. In the case of a plane wave travelling in negative x-direction the phasors of the electric and magnetic field strength would read

$$\vec{E}(x) = E_0 e^{jkx}\vec{e}_z \qquad \text{and} \qquad \vec{H}(x) = H_0 e^{jkx}\vec{e}_y \qquad (2.105)$$

    The ratio between the amplitude of the electric field E_0 and the amplitude of the magnetic field H_0 is defined by the medium in which the wave propagates. It is the *characteristic wave impedance* or *intrinsic impedance* of the medium Z_F. The intrinsic impedance is not a circuit component, it represents a *ratio of field amplitudes*. In order to indicate this, we use the subscripted letter 'F' for 'field'

$$Z_F = \frac{E_0}{H_0} = \sqrt{\frac{\mu_0 \mu_r}{\varepsilon_0 \varepsilon_r}} = Z_{F0}\sqrt{\frac{\mu_r}{\varepsilon_r}} \qquad \text{(Characteristic wave impedance)} \qquad (2.106)$$

where Z_{F0} is the characteristic wave impedance (intrinsic impedance) of vacuum.

$$Z_{F0} = \sqrt{\frac{\mu_0}{\varepsilon_0}} = 120\pi\ \Omega \approx 377\ \Omega \qquad (2.107)$$

Electromagnetic waves that propagate in a certain direction carry energy in that direction. The instantaneous power density per unit area in the time domain is given by the *Poynting vector* $\vec{S}(t)$.

$$\vec{S}(t) = \vec{E}(t) \times \vec{H}(t) \tag{2.108}$$

When using complex notation the complex Poynting vector is

$$\boxed{\vec{S} = \frac{1}{2}\vec{E} \times \vec{H}^*} \quad \text{(Complex Poynting vector)} \tag{2.109}$$

In order to evaluate the real or *effective power* carried by a plane wave we need the Poynting vector averaged over one period in time.

$$\overline{\vec{S}(t)} = \overline{\vec{E}(t) \times \vec{H}(t)} = \frac{1}{2}\text{Re}\left\{\vec{E} \times \vec{H}^*\right\} \tag{2.110}$$

In the case of our plane wave in Equation 2.96 the effective power density is

$$\vec{S}_{av} = \frac{1}{2}\text{Re}\left\{\vec{E} \times \vec{H}^*\right\} = \frac{1}{2}E_0 H_0 \vec{e}_x = \frac{E_0^2}{2Z_{F0}}\vec{e}_x = \frac{1}{2}Z_{F0} H_0^2 \vec{e}_x \tag{2.111}$$

Finally we are going to summarize the most important properties of plane waves:

- The electric and magnetic field strengths are *in phase*; that is, the electric and magnetic field strength have the same phase, in other words they exhibit the same time-dependence $\cos(\omega t - kx + \varphi)$
- The electric field strength $\vec{E}$ and the magnetic field strength $\vec{H}$ are perpendicular to each other ($\vec{E} \perp \vec{H}$). Furthermore, the direction of propagation given by the wave vector $\vec{k}$ is perpendicular to both electric and magnetic field strength ($\vec{k} \perp \vec{E}$ and $\vec{k} \perp \vec{H}$). Hence, a plane wave is a TEM wave (TEM = transverse electromagnetic).
- Surfaces of equal phase (wave fronts) are planes perpendicular to the direction of travel. Within a wave front the magnitude of the field strength is constant.

Figure 2.13 shows – as an example – the electric field oriented in z-direction and the magnetic field oriented in y-direction for a fixed instant in time ($t = t_0$). Wave fronts are yz-planes. The plane wave propagates in positive x-direction.

Homogeneous plane waves that propagate everywhere in space cannot exist due to the infinite power they would carry. However, they can exist in a volume of finite extent and are of great importance when electromagnetic fields are considered far away from their sources. We will refer to these waves as *locally plane waves* (Figure 2.22).

2.5.2 Polarization of Waves

The electric field vector determines the *polarization* of the electromagnetic wave. Considering our plane wave example in Figure 2.13 the electric field oscillates in a vertical plane (xz-plane). Hence, the wave is called vertically polarized. Vertical polarization is a special case of *linear polarization*.

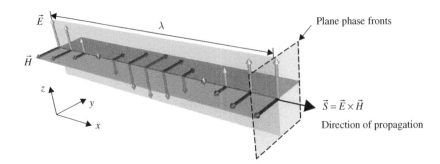

Figure 2.13 Plane wave: spatial distribution of the electric and magnetic field strengths for a given instant in time.

In order to investigate the general case of linearly polarized waves and other types of polarization we superimpose two plane waves $\vec{E}_1$ and $\vec{E}_2$ that propagate in the same direction but have electric fields perpendicular to each other.

$$\vec{E}_1 (x) = E_1 e^{-j(kx+\varphi_1)} \vec{e}_y \qquad (2.112)$$

$$\vec{E}_2 (x) = E_2 e^{-j(kx+\varphi_2)} \vec{e}_z \qquad (2.113)$$

The total electric field strength is simply the sum of both orthogonal components

$$\vec{E} (x) = \vec{E}_1 (x) + \vec{E}_2 (x) \qquad (2.114)$$

We distinguish between three different polarizations for the total field: linear, circular and elliptical polarization.

2.5.2.1 Linear Polarization

In the case of equal phases ($\varphi_1 = \varphi_2$) the total field vector $\vec{E}$ oscillates in a plane defined by the direction of propagation $\vec{e}_x$ and the direction of the total electric field $E_1 \vec{e}_y + E_2 \vec{e}_z$. This scenario is called *linear polarization* (see Figure 2.14a). Depending on the orientation of the plane of oscillation we use the terms *vertical polarization* (the electric field vector oscillates in the z-direction) or *horizontal polarization* (there is no z-component of the electric field strength).

2.5.2.2 Circular Polarization

Circular polarization occurs if the two waves are out of phase by $90°$ ($\varphi_1 = \varphi_2 \pm 90°$) and have equal amplitudes ($E_1 = E_2$). As the wave moves in the direction of propagation the electric field strength rotates on the surface of a cylinder (see Figure 2.14b). At a fixed point in space the electric field vector permanently changes its direction but its length remains constant: the electric field vector rotates in a circle. Looking in the direction of propagation the polarization is called *right-hand circular polarization* (RHCP) if the electric field vector rotates *clockwise*. Otherwise, the polarization is called *left-hand circular polarization* (LHCP) if the electric field vector rotates *counter-clockwise*.

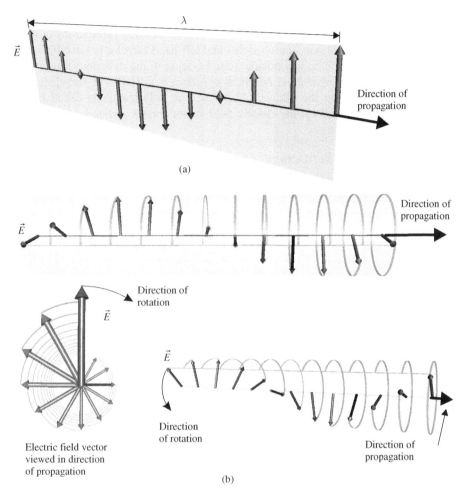

Figure 2.14 Vectors of the electric field strength for an instant in time t_0: (a) linearly polarized wave and (b) right-hand circularly polarized wave.

2.5.2.3 Elliptical Polarization

Elliptical polarization is the more general case when the amplitudes of the orthogonal components are different ($E_1 \neq E_2$) but the phase difference is still 90° ($\varphi_1 = \varphi_2 \pm 90°$). At a fixed point in space the vector of the electric field now describes an ellipse. If we assume – without loss of generality – that $E_1 > E_2$ then E_1 is the major axis of the ellipse and E_2 is the minor axis. An important measure is the *axial ratio AR*

$$AR = \frac{\text{major axis}}{\text{minor axis}} = \frac{E_1}{E_2} \tag{2.115}$$

Furthermore, elliptical polarization occurs if the phase difference deviates from the previously assumed 90°. Now the ellipse is tilted and the major and minor axis of the ellipse are more complicated functions of the orthogonal components E_1 and E_2 and the phase difference $\Delta\varphi = \varphi_1 - \varphi_2$. The interested reader will find detailed expressions in [12].

In accordance with our discussion on right- and left-hand circular polarization we can define right- and left-hand elliptical polarization. Looking in the direction of propagation the polarization is called *right-hand elliptical polarization* (RHEP) if the electric field vector rotates *clockwise* on the ellipse. Otherwise, the polarization is called *left-hand elliptical polarization* (LHEP) if the electric field vector rotates *counter-clockwise*.

2.5.3 Reflection and Refraction

A plane wave that passes through an interface that separates two different media is going to be reflected and transmitted. We summarize the essential relations that describe the wave behaviour at the material boundary. We look at two different cases. First, the less complex case of normal incidence where the direction of propagation is perpendicular to the material interface. Second, the more general case of oblique incidence.

2.5.3.1 Normal Incidence

First let us consider the case of normal incidence on a plane interface. Figure 2.15 shows the orientation of the electric and magnetic components of the incident, reflected and transmitted wave. We assume that the incident plane wave has its source in medium 1 and propagates in positive x-direction. At the interface the incident wave is partially reflected back into medium 1. Therefore, a plane wave in negative x-direction occurs weighted by a *reflection coefficient* r. In medium 1 we see a superposition of plane waves in the opposite directions.

$$\vec{E}_1 = \underbrace{E_0 e^{-jkx}\vec{e}_z}_{\text{forward}} + \underbrace{rE_0 e^{jkx}\vec{e}_z}_{\text{reflected}} \quad \text{and} \quad \vec{H}_1 = \underbrace{-\frac{E_0}{Z_{F1}}e^{-jkx}\vec{e}_y}_{\text{forward}} + \underbrace{r\frac{E_0}{Z_{F1}}e^{jkx}\vec{e}_y}_{\text{reflected}} \qquad (2.116)$$

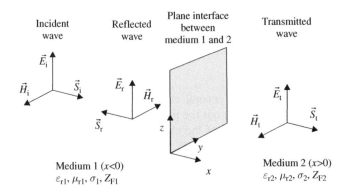

Figure 2.15 Normal incidence of a plane wave on a material interface. Definition of directions of field strengths and Poynting vectors.

A portion of the incident wave enters medium 2 and moves on in a positive x-direction. We see a transmitted wave weighted by a *transmission coefficient t*.

$$\underbrace{\vec{E}_2 = t\,E_0 \mathrm{e}^{-jkx}\vec{e}_z}_{\text{transmitted}} \quad \text{and} \quad \underbrace{\vec{H}_2 = -t\frac{E_0}{Z_{\mathrm{F}2}}\mathrm{e}^{-jkx}\vec{e}_y}_{\text{transmitted}} \tag{2.117}$$

The reflection coefficient determines the ratio of the electric field strength of the reflected (E_r) and incident (E_i) wave and can be calculated from the characteristic wave impedances of the two media as

$$\boxed{r = \frac{E_r}{E_i} = \frac{Z_{\mathrm{F}2} - Z_{\mathrm{F}1}}{Z_{\mathrm{F}2} + Z_{\mathrm{F}1}}} \quad \text{(Reflection coefficient)} \tag{2.118}$$

Furthermore, the transmission coefficient determines the ratio of the electric field strength of the transmitted (E_t) and incident (E_i) wave and can be calculated from the characteristic wave impedances of the two media as

$$\boxed{t = \frac{E_t}{E_i} = \frac{2Z_{\mathrm{F}2}}{Z_{\mathrm{F}2} + Z_{\mathrm{F}1}} = 1 + r} \quad \text{(Transmission coefficient)} \tag{2.119}$$

In the case of lossless materials the characteristic wave impedances of the media ($i \in \{1, 2\}$) are real numbers and are already known from Equation 2.106 given by

$$\boxed{Z_{\mathrm{F}i} = \sqrt{\frac{\mu_0 \mu_{ri}}{\varepsilon_0 \varepsilon_{ri}}} = Z_{\mathrm{F}0}\sqrt{\frac{\mu_{ri}}{\varepsilon_{ri}}} \in \mathbb{R}} \quad \text{(Characteristic wave impedance)} \tag{2.120}$$

In lossy (conductive) media the characteristic wave impedance is complex and depends on frequency even if the material itself is not dispersive.

$$Z_{\mathrm{F}i} = \sqrt{\frac{j\omega\mu_0\mu_{ri}}{\sigma_i + j\omega\varepsilon_0\varepsilon_{ri}}} \quad \text{(Characteristic wave impedance for lossy media)} \tag{2.121}$$

Example 2.2 *Plane Dielectric Interface: Reflection and Transmission Coefficients*

Figure 2.16a shows the reflection and transmission coefficient for normal incidence of a plane wave on a dielectric interface. Medium 1 has a relative permittivity of $\varepsilon_{r1} = 1$. The relative permittivity of Medium 2 ε_{r2} varies in the range of $1 \leq \varepsilon_{r2} \leq 40$. We observe two different scenarios:

1. The plane wave starts in medium 1. In Figure 2.16a the corresponding reflection and transmission coefficients are denoted as $r(1 \to 2)$ and $t(1 \to 2)$, respectively. For a value of $\varepsilon_{r2} = 1$ the reflection coefficient starts at $r = 0$ and the transmission coefficient at $t = 1$. As the relative permittivity ε_{r2} increases, the values for r and t decrease. The reflection coefficient r shows negative values, but the absolute value increases.

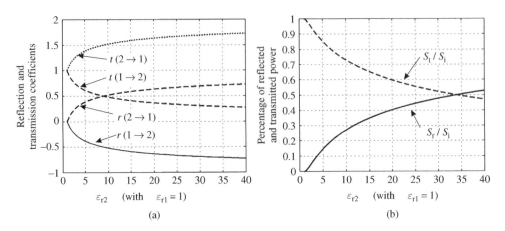

Figure 2.16 Normal incidence: (a) Reflection and transmission coefficients and (b) normalized value of reflected and transmitted Poynting vectors.

2. *The plane waves starts in medium 2. In Figure 2.16a the corresponding reflection and transmission coefficients are denoted as $r(2 \rightarrow 1)$ and $t(2 \rightarrow 1)$, respectively. For a value of $\varepsilon_{r2} = 1$ the reflection coefficient starts at $r = 0$ and the transmission coefficient at $t = 1$. As the relative permittivity ε_{r2} increases the values for r and t increase. At first sight it may seem odd that the transmission coefficient t shows values greater than one. Indeed the electric field strength of the transmitted wave is greater than the electric field strength of the incident wave. This effect does not contradict the principle of energy conversation, since the transmitted power density S_t is smaller than the incident power density S_f.*

Figure 2.16b shows the normalized values of the reflected and transmitted Poynting vectors S_r/S_i and S_t/S_i, respectively. The diagrams are valid for both directions of propagation $(1 \rightarrow 2)$ and $(2 \rightarrow 1)$.

Example 2.3 *Normal Incidence on a Perfect Electric Wall*
    Let us consider a plane wave impinging on an ideal conductive halfspace $(\sigma_2 \rightarrow \infty)$. From Equation 2.121 we calculate a characteristic wave impedance of $Z_{F2} = 0$. Hence, the reflection coefficient is $r = -1$ and the transmission coefficient yields $t = 0$. There is no transmission into the conductive halfspace, the wave is fully reflected back into the first medium.

2.5.3.2 Oblique Incidence

We will start by defining a *plane of incidence* as a plane that contains the wave vector $\vec{k}$ and the surface normal of the material interface (see Figure 2.17a).

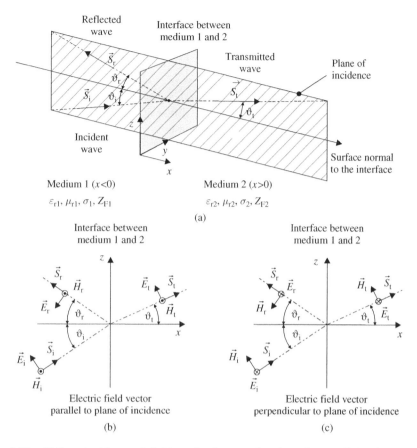

Figure 2.17 Oblique incidence: definition of reference directions for parallel and perpendicular polarization with respect to the plane of incidence.

Furthermore, we decompose the incident wave into two components with orthogonal polarization:

- a *parallel polarized wave* where the electric field vector is parallel ($\parallel$) to the plane of incidence (see Figure 2.17b), and
- a *perpendicular polarized wave* where the electric field vector is perpendicular ($\perp$) to the plane of incidence (see Figure 2.17c).

The reference directions for both polarizations are given in Figure 2.17. Both polarizations are treated separately and added up afterwards (superposition).

The angles of reflection and transmission are identical for both polarizations. The angle of reflection ϑ_r equals the angle of incidence ϑ_i.

$$\boxed{\vartheta_r = \vartheta_i} \tag{2.122}$$

The angle of transmission ϑ_t is given by *Snell's law* as

$$\boxed{\frac{\sin\left(\vartheta_i\right)}{\sin\left(\vartheta_t\right)} = \frac{n_2}{n_1} = \frac{\sqrt{\mu_{r2}\varepsilon_{r2}}}{\sqrt{\mu_{r1}\varepsilon_{r1}}}} \quad \text{(Snell's law)} \qquad (2.123)$$

where $n_1 = \sqrt{\varepsilon_{r1}\mu_{r1}}$ and $n_2 = \sqrt{\varepsilon_{r2}\mu_{r2}}$ are the indices of refraction[15] for lossless media. The reflection and transmission coefficients are different for parallel and perpendicular polarization. Let us look at parallel polarization first.

$$r_\parallel = \frac{Z_{F2}\cos\left(\vartheta_t\right) - Z_{F1}\cos\left(\vartheta_i\right)}{Z_{F2}\cos\left(\vartheta_t\right) + Z_{F1}\cos\left(\vartheta_i\right)} \qquad (2.124)$$

$$t_\parallel = \frac{Z_{F2}}{Z_{F1}}\left(1 - r_\parallel\right) = \frac{2Z_{F2}\cos\left(\vartheta_i\right)}{Z_{F2}\cos\left(\vartheta_t\right) + Z_{F1}\cos\left(\vartheta_i\right)} \qquad (2.125)$$

In the case of perpendicular polarization the reflection and transmission coefficient are given by

$$r_\perp = \frac{Z_{F2}\cos\left(\vartheta_i\right) - Z_{F1}\cos\left(\vartheta_t\right)}{Z_{F2}\cos\left(\vartheta_i\right) + Z_{F1}\cos\left(\vartheta_t\right)} \qquad (2.126)$$

$$t_\perp = 1 + r_\perp = \frac{2Z_{F2}\cos\left(\vartheta_i\right)}{Z_{F2}\cos\left(\vartheta_i\right) + Z_{F1}\cos\left(\vartheta_t\right)} \qquad (2.127)$$

Figure 2.18 shows the absolute value of the electric field strength of a wave crest that hits a dielectric interface ($\varepsilon_{r2} > \varepsilon_{r1}$) with an angle of incidence of $\vartheta_i = 45°$. The plane wave has perpendicular polarization. The black and white arrows indicate the direction of propagation of the incident, reflected and transmitted wave components. We can see the different amplitudes and the changing direction of the transmitted wave due to reduced propagation velocity in medium 2.

From the expressions above we can derive two technically important special cases: total transmission (Brewster angle) and total reflection.

Brewster Angle

We consider two dielectric media, the second medium having higher relative permittivity ($\varepsilon_{r2} > \varepsilon_{r1}$). Furthermore, we assume parallel polarization. For a certain angle of incidence ϑ_{iB} the reflection coefficient $r_\parallel$ becomes zero and the wave fully enters the medium 2 (total transmission).

$$\vartheta_{iB} = \arctan\left(\sqrt{\frac{\varepsilon_{r2}}{\varepsilon_{r1}}}\right) \quad \text{for} \quad \varepsilon_{r1} < \varepsilon_{r2}; \; \mu_{r1} = \mu_{r2} = 1 \quad \text{(Brewster angle)} \quad (2.128)$$

This angle is known as *Brewster angle* and is useful in generating linear polarized optical waves. (If a plane wave with arbitrary polarization impinges at an angle of incidence that equals the Brewster angle, only the parallel polarized wave component is reflected. Hence, after reflection the wave is linearly polarized.)

---

[15] Refractive indices are commonly used in optics. A related term is *optical density*. A medium is called (optically) *denser* than another medium if it exhibits a higher refractive index and therefore has a slower speed of light. The least optically dense medium is vacuum.

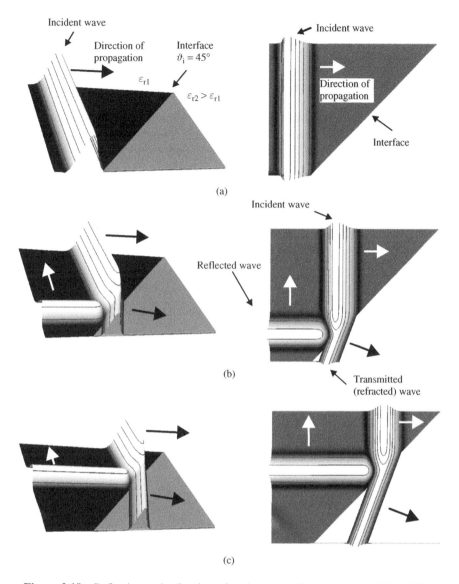

Figure 2.18   Reflection and refraction of a plane wave for $\varepsilon_{r1} < \varepsilon_{r2}$ and $\vartheta_i = 45°$.

Critical Angle

Now, we consider a dielectric interface with $\varepsilon_{r2} < \varepsilon_{r1}$. According to Snell's law (Equation 2.123) the angle of transmission is always greater than the angle of incidence. For a certain angle ϑ_{ic} (called the *critical angle*) the angle of transmission becomes 90°.

$$\vartheta_{ic} = \arcsin\left(\sqrt{\frac{\varepsilon_{r2}}{\varepsilon_{r1}}}\right) \quad \text{for} \quad \varepsilon_{r1} > \varepsilon_{r2}; \ \mu_{r1} = \mu_{r2} = 1 \quad \text{(Critical angle)} \quad (2.129)$$

If the angle of incidence is equal to or greater than the critical angle the electromagnetic wave is fully reflected back into the first medium, regardless of the polarization of the wave.

Although there is full reflection (and therefore no transmission) the second medium is not free of electromagnetic fields. In medium 2 a wave propagates along the interface without attenuation. This wave shows an exponential decay perpendicular to the direction of travel and is called an *evanescent wave*. In order to allow this evanescent wave to travel along the interface, medium 2 has to have a specified minimum thickness. However, this thickness is rather small due to the rapid exponential decay [5].

Example 2.4 *Oblique Incidence*
Figure 2.19 shows the angles of reflection and transmission for two cases

- *wave propagation from air to dielectric ($\varepsilon_{r1} = 1$ and $\varepsilon_{r2} = 6$), and*
- *wave propagation from dielectric to air ($\varepsilon_{r1} = 6$ and $\varepsilon_{r2} = 1$).*

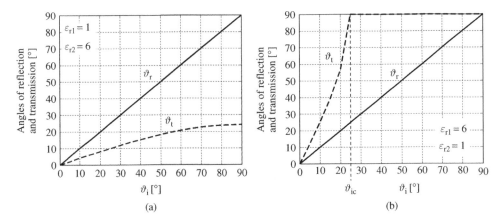

Figure 2.19 Reflection and transmission angles for (a) $\varepsilon_{r1} = 1$ and $\varepsilon_{r2} = 6$ and (b) $\varepsilon_{r1} = 6$ and $\varepsilon_{r2} = 1$.

The angle of reflection equals the angle of incidence at all times (Figure 2.19). A wave passing from the less to the more dense medium is refracted towards the normal, the refraction angle is smaller than the angle of incidence. On the contrary, a wave that travels from the denser to the less dense medium is bent away from the surface normal, the refraction angle exceeds the angle of incidence. For angles equal to the critical angle ϑ_{ic} the angle of transmission becomes 90°. Therefore, if the angle of incidence exceeds the critical angle total reflection occurs.

In Figure 2.20 we see the reflection and transmission coefficient of a wave passing from air ($\varepsilon_{r1} = 1$) to dielectric ($\varepsilon_{r2} = 6$). We distinguish between parallel and perpendicular polarization. A zero angle of incidence ($\vartheta_i = 0$) represents normal incidence. In this case the coefficients for parallel and perpendicular polarization are equal $r_\parallel = r_\perp$ and $t_\parallel = t_\perp$. With increasing angles of incidence the values diverge from each other. For parallel polarization we see a Brewster angle with zero reflection $r_\parallel(\vartheta_{iB}) = 0$.

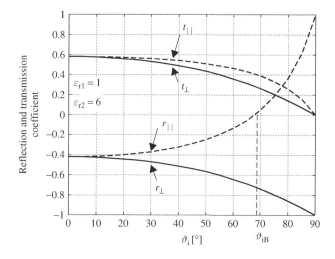

Figure 2.20 Reflection and transmission coefficients of a wave passing from air $\varepsilon_{r1} = 1$ to dielectric $\varepsilon_{r2} = 6$ for parallel ($\parallel$) and perpendicular ($\perp$) polarization.

2.5.4 Spherical Waves

In the previous section we discussed plane waves that exist in the entire space. Such a wave would carry infinite energy. Therefore it is not possible to excite such a wave. In order to fill up the entire space there must be substantial decay in amplitude as the wave propagates away from its source. This leads us to *spherical waves*.

Real sources (e.g. antennas) are always finite in their extension and can be viewed as point sources at very great distances. Therefore, far away from their source, waves tend to behave like spherical waves. A *scalar spherical wave* function is given by

$$G(r) = \frac{e^{-jkr}}{r} \qquad \text{(Scalar spherical wave)} \qquad (2.130)$$

where the phase is linear with distance r and the amplitude is inversely proportional to distance. Figure 2.21 visualizes the amplitude distribution of a spherical wave at a fixed instant in time.

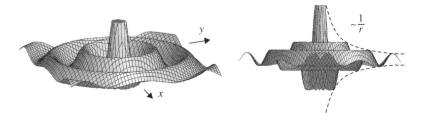

Figure 2.21 Amplitude distribution of a spherical wave at a fixed instant in time.

In the *farfield* (fields far away from a radiating source) the scalar spherical wave term is combined with a transversal electrical field strength that depends on the angles ϑ and φ.

$$\vec{E}\,(r, \vartheta, \varphi) = \vec{E}_{\mathrm{tr}}\,(\vartheta, \varphi)\,\frac{e^{-jkr}}{kr} \tag{2.131}$$

The electromagnetic wave propagates in a radial direction. Like plane waves spherical waves have electric field strength transversal to the direction of travel. For the magnetic field component we can write a similar term, whereas electric and magnetic field strength are perpendicular to each other.

$$\vec{H}\,(r, \vartheta, \varphi) = \vec{H}_{\mathrm{tr}}\,(\vartheta, \varphi)\,\frac{e^{-jkr}}{kr} \qquad \text{with} \qquad \vec{E}_{\mathrm{tr}} \perp \vec{H}_{\mathrm{tr}} \tag{2.132}$$

The complex Poynting vector points in the radial direction and is inversely proportional to the distance squared.

$$\vec{S} = \frac{1}{2}\vec{E} \times \vec{H}^* \sim \frac{1}{r^2}\vec{e}_r \tag{2.133}$$

The surface of a sphere is proportional to the distance squared $S = 4\pi r^2$. Hence, the average (radiated) power that passes through a closed surface around the source is constant [12].

$$P_{\mathrm{av}} = \frac{1}{2}\oiint_{A(V)} \mathrm{Re}\left\{\vec{E} \times \vec{H}^*\right\} \cdot d\vec{A} = \text{const.} \tag{2.134}$$

Spherical waves do not contradict the principle of finite power. Planes waves, although they cannot fill the entire space, are nonetheless an important concept when we look at electromagnetic fields far away from a source. Figure 2.22 depicts a wave that radiates

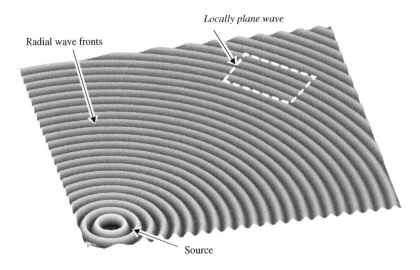

Figure 2.22   Spherical wave and locally plane wave.

radially outward from a source. At some distance the wave fronts become more and more plane. If we look at a spatial region that is *small in size* we can regard the wave as a *locally plane wave*.

2.6 Summary

In Section 2.1 we reviewed the basics of static electric and magnetic fields with an emphasis on physical understanding. A focus of our discussion was the interrelation between field quantities (like electric and magnetic field strength) and circuit parameters (like voltage, current, capacitance, inductance). Furthermore, we saw how differential vector operators relate to the sources of irrotational and solenoidal vector fields.

From basic field theory we stepped forward to Maxwell's equations in Section 2.2. These equations provide a mathematical description of the time-varying behaviour of electromagnetic phenomena in general. We primarily focused on physical meaning and visual interpretation of both the differential and the integral form of Maxwell' equations. Finally, we included constitutive relations and interface conditions to complete our set of relations.

It is difficult to identify solutions to Maxwell's equations in general. If we assume fields to be of static (time-invariant) or of quasi-statical (gradually time variant) nature, the complexity of Maxwell's equations will be reduced considerably; which in turn makes it much easier to determine the solution. Unfortunately for matters of high frequency technology these assumptions are insufficient in most cases and therefore the full set of Maxwell's equations needs to be considered simultaneously (see Section 2.3).

In this context, two aspects are of particular importance: the skin effect on the surface of good electrical conductors (Section 2.4) and the effect of propagating electromagnetic waves in space (Section 2.5) and on transmission lines. In Chapter 3 we will take a closer look at propagating electromagnetic waves on transmission lines and the resulting design implications for RF circuits. In Chapter 8 we will follow up the effect of propagating waves in space.

Today, the process of identifying solutions to practical electromagnetic field problems has been considerably facilitated by the use of modern three-dimensional EM simulation programs. After having set up the geometry on a CAD graphical user interface, the program will identify approximations to the solution by using suitable numerical algorithms. In Chapters 6 to 8 we will see various examples for the wide range of application of 3D-EM simulation software.

2.7 Problems

2.1 Consider a two wire transmission line, where two parallel thin wires have a distance of $d = 1$ cm. Each conductor carries a steady current of $I_0 = 10$ mA. The currents flow in opposite directions. The transmission line is located in the plane $z = 0$ and has an infinite extent in $\pm y$-direction.

a) Draw the distribution of the magnetic field strength $\vec{H}$ in a cross-sectional area (xz-plane).

b) Calculate the magnitude of the magnetic field strength H along the x-axis.

2.2 An air-filled coaxial transmission line has a solid cylindrical inner conductor of $R_1 = 2\,\mathrm{mm}$, an outer conductor with an inner radius of $R_2 = 4\,\mathrm{mm}$ and an outer radius $R_3 = 5\,\mathrm{mm}$. The inner and outer conductor carry steady currents I in opposite directions.

a) Determine the magnetic field strength by applying Faraday's law in integral form.
b) Calculate the inductance per unit length $L'_{DC} = L/\Delta\ell$ for DC conditions $(f = 0\,\mathrm{Hz})$.
c) Calculate the inductance per unit length $L'_{RF} = L/\Delta\ell$ for RF conditions. (Assume that the current flows in a thin layer at the surface of the conductors (*skin effect*).)

2.3 A plane wave propagates through a dielectric $(\varepsilon_r \neq 1,\ \mu_r = 1)$. The time-variant electric and magnetic field strengths are given by

$$\vec{E}(x, t) = E_0 \cos(\omega t - kx)\,\vec{e}_y \tag{2.135}$$

$$\vec{H}(x, t) = 4\frac{\mathrm{A}}{\mathrm{m}} \cos(\omega t - kx)\,\vec{e}_z \tag{2.136}$$

where $\omega = 2\pi 50\,\mathrm{MHz}$ is the angular frequency and $Z_F = 300\,\Omega$ is the characteristic wave impedance.

a) Calculate the amplitude of the electric field strength E_0.
b) What is the relative permittivity ε_r of the material?
c) Determine the wave number k.
d) At what speed does the wave propagate?

2.4 A long cylindrical coil is filled with magnetic material $\mu_r = 500$. The coil has a length of $\ell = 7$ cm, a diameter of $D = 0.2$ cm and consists of $n = 100$ turns. The current through the wire of the cylindrical coil is $I = 1$ mA.
Assume the following simplifications: inside the coil the magnetic field is constant $(H_i = \mathrm{const.})$ whereas outside the coil the magnetic field happens to be zero.

a) Calculate the magnetic field strength H_i inside the coil from Faraday's law in integral form.
b) Determine the magnetic energy stored by the coil.
c) What is the inductance of the coil?

2.5 A plane wave is given by

$$\vec{E}(x) = E_0 e^{-jkx}\vec{e}_z \tag{2.137}$$

a) Calculate the phasor of the magnetic field strength $\vec{H}$ using Ampere's law in differential form.
b) Determine the Poynting vector $\vec{S}$ and the average power P_{av} that flows through a surface of $A = 1\,\mathrm{m}^2$.

2.6 Prove that the phasor of the electric field strength $\vec{E}(z)$ in Equation 2.84 is a solution of Equation 2.83.

References

1. Balanis CA (1989) *Advanced Engineering Electromagnetics*. John Wiley & Sons.
2. Fleisch D (2008) *A Student's Guide to Maxwell's Equations*. Cambridge University Press.
3. Ida N (2004) *Engineering Electromagnetics*. Springer.
4. Jackson JD (1998) *Classical Electrodynamics*. John Wiley & Sons.
5. Kraus JD, Fleisch DA (1999) *Electromagnetics with Applications*. McGraw-Hill.
6. Paul CR (1998) *Introduction to Electromagnetic Fields*. McGraw-Hill.
7. Detlefsen J, Siart U (2009) *Grundlagen der Hochfrequenztechnik*. Oldenbourg.
8. Golio M, Golio J (2008) *The RF and Microwave Handbook*. CRC Press.
9. Meinke H, Gundlach FW (1992) *Taschenbuch der Hochfrequenztechnik*. Springer.
10. Glisson, Jr. TH (2011) *Introduction to Circuit Analysis and Design*. Springer.
11. Blume S (1988) *Theorie elektromagnetischer Felder*. Huethig.
12. Balanis CA (2005) *Antenna Theory*. John Wiley & Sons.

Further Reading

Furse C, Durney DH, Cristensen DA (2009) *Basic Introduction to Bioelectromagnetics*. CRC Press.
Gandhi OP (2002) Electromagnetic Fields: Human Safety Issues. *Annu. Rev. Biomed. Eng*. Vol. 4, pp. 211–34.
Griffiths DJ (2008) *Introduction to Electrodynamics*. Prentice Hall International.
ICNIRP (1998) *Guidelines For Limiting Exposure to time-varying Electric, Magnetic, and Electromagnetic Fields (up to 300 GHz)*. International Commission on Non-Ionizing Radiation Protection.
Kark K (2010) *Antennen und Strahlungsfelder*. Vieweg.
Leuchtmann P (2005) *Einfuehrung in die elektromagnetische Feldtheorie*. Pearson.
Schwab A (2002) *Begriffswelt der Feldtheorie*. Springer.
Schwartz M (1987) *Principles of Electrodynamics*. Dover Pubn Inc.

3

Transmission Line Theory and Transient Signals on Lines

In this chapter we derive the equations that describe the electrical behaviour of transmission lines. Transmission lines are key components to many passive circuits, like filters, power dividers, matching circuits and couplers. We will start our investigation into transmission lines with physical considerations and will apply straightforward mathematical methods. Furthermore, our approach will be illustrative, in order to fully understand the implications of the theory for the design of RF circuits.

Section 3.1 is about classical transmission line theory. A theory that allows an effective treatment of lines, especially in the case of sinusoidal (time-harmonic) signals. In Section 3.2 we will look at transient signals like pulse or step functions and study the effects that occur when such signals propagate along transmission lines.

3.1 Transmission Line Theory

3.1.1 Equivalent Circuit of a Line Segment

Let us consider the two-conductor transmission line in Figure 3.1a.

The two conductors are surrounded by a homogeneous material (for instance air). The currents I^+ and I^- are equal to each other but flow in opposite directions. The transmission line extends in $\pm z$-direction. The cross-section of the line is independent of the local coordinate z. The currents on the conductors produce a circulating magnetic field around the wires (see Figure 3.1b). Furthermore, there is a voltage between the conductors. Hence, we see an electric field starting at one conductor and ending at the other.

If we considered transient signals we would see a wave travelling down the line due to the finite propagation speed of the electromagnetic phenomena. The electric and magnetic fields on the transmission line are perpendicular to each other and perpendicular to the direction of the wave. Therefore, our two-conductor line carries a TEM wave, where TEM stands for *transversal electromagnetic*. A TEM wave represents the fundamental wave

---

RF and Microwave Engineering: Fundamentals of Wireless Communications, First Edition. Frank Gustrau.
© 2012 John Wiley & Sons, Ltd. Published 2012 by John Wiley & Sons, Ltd.

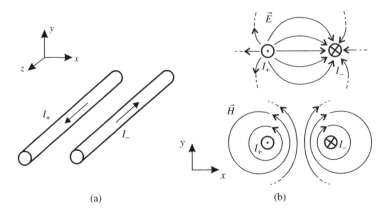

Figure 3.1 (a) Geometry of a two-wire line and (b) distribution of electric and magnetic field strength in a cross-section.

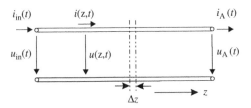

Figure 3.2 Definition of current and voltage on a transmission line.

mode that can propagate on a transmission line. The TEM mode[1] is able to propagate in the entire frequency spectrum with the same speed. The speed depends solely on the material parameters.

Figure 3.2 defines currents and voltages on a transmission line. The transmission line has two terminals. On the left side a generator is commonly connected that feeds a signal into the line. Therefore, we use index 'in' for generator. On the right side a load (e.g. an antenna) is usually connected. Hence, we use index 'A' for load or antenna. Furthermore, there are voltage $u(z, t)$ and current $i(z, t)$ at any point z along the line.

In the following we further move from arbitrary time-dependent signals $(u(z, t); i(z, t))$ to sinusoidal signals and use a phasor representation of voltage and current $(U(z); U(z))$. Furthermore, we consider a short section of the line with a length of Δz.

In order to derive an *equivalent circuit model* for a short transmission line section we investigate the physical electromagnetic field effects of a transmission line (Figure 3.1b). First, there is a magnetic field in the vicinity of the conductors. From Section 2.1.2 we know that magnetic energy is stored in the field which can be described by an inductance L. Second, an electronic field exists between the conductors. The electric energy which

---

[1] At this point we restrict ourself to TEM waves. In later chapters we will see that *higher order modes* may exist. These modes depend on geometry and material properties and they typically arise at higher frequencies, if the transverse dimensions of the line are on the order of the wavelength.

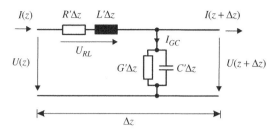

Figure 3.3   Equivalent circuit of a short transmission line segment.

is associated with this field can be described by a capacitance C. Third, in reality the conductors have finite conductivity, that is they are lossy. A lossy conductor is represented by an ohmic resistance R. Fourth, the isolator that separates the two conductors may have losses. These losses are represented by a resistance (or conductance) too. We choose a conductance G.

The equivalent circuit model in Figure 3.3 includes all effects of energy storing and power loss. It basically consists of a lossy coil (serial inductance and resistance) and a lossy capacitor (parallel capacitance and conductance). It is immediately obvious that all circuit elements (L, C, R, G) are proportional to the length of the line Δz. Therefore, we introduce circuit elements per unit length and use an apostrophe to identify the new quantities (primed quantities).

$$L' = \frac{L}{\Delta z} \qquad \text{(Inductance per unit length} \leftrightarrow \text{magnetic field energy)} \qquad (3.1)$$

$$C' = \frac{C}{\Delta z} \qquad \text{(Capacitance per unit length} \leftrightarrow \text{electric field energy)} \qquad (3.2)$$

$$R' = \frac{R}{\Delta z} \qquad \text{(Resistance per unit length} \leftrightarrow \text{ohmic losses in conductors)} \qquad (3.3)$$

$$G' = \frac{G}{\Delta z} \qquad \text{(Conductance per unit length} \leftrightarrow \text{dielectric losses in isolator)} \qquad (3.4)$$

3.1.2   Telegrapher's Equation

We will now apply *Kirchhoff's laws* on the equivalent circuit model in Figure 3.3. We will start with the loop or voltage rule. The total voltage around a closed loop is zero.

$$U(z) = \underbrace{U_{RL}(z)}_{I(z)\left(R'+j\omega L'\right)\Delta z} + U(z + \Delta z) \qquad (3.5)$$

We rearrange the equation to

$$\frac{U(z) - U(z + \Delta z)}{\Delta z} = I(z)\left(R' + j\omega L'\right) \qquad (3.6)$$

and consider the quotient on the left-hand side.

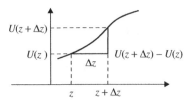

Figure 3.4 Difference quotient and differential quotient (derivative).

Figure 3.4 shows that the quotient of the two differences equals the negative slope of the function $U(z)$. If we let the length converge to zero ($\Delta z \to 0$) the difference quotient becomes the (negative) derivative of $U(z)$ with respect to z. So we get the following differential equation

$$-\frac{dU(z)}{dz} = I(z)\left(R' + j\omega L'\right) \tag{3.7}$$

Next we consider Kirchhoff's current law or node rule. At any node the sum of all the currents is zero. The equivalent circuit in Figure 3.3 gives us

$$I(z) = \underbrace{I_{GC}(z)}_{U(z+\Delta z)(G'+j\omega C')\Delta z} + I(z + \Delta z) \tag{3.8}$$

We rearrange the node equation to

$$\frac{I(z) - I(z + \Delta z)}{\Delta z} = \underbrace{U(z + \Delta z)}_{\to U(z)\ \text{for}\ \Delta z \to 0} \cdot \left(G' + j\omega C'\right) \tag{3.9}$$

and again let the length of the line segment approach zero ($\Delta z \to 0$). As before the difference quotient becomes the (negative) derivative with respect to the local coordinate. Now, we arrive at a second differential equation.

$$-\frac{dI(z)}{dz} = U(z)\left(G' + j\omega C'\right) \tag{3.10}$$

So Kirchhoff's laws give us two coupled first order differential equations in Equations 3.7 and 3.10 for current and voltage. Let us solve Equation 3.7 for the current $I(z)$

$$I(z) = -\frac{dU(z)}{dz}\frac{1}{(R' + j\omega L')} \tag{3.11}$$

and substitute $I(z)$ in Equation 3.10. We end up with a wave equation for the voltage $U(z)$ on the transmission line. The equation is also known as the *telegrapher's equation*.

$$\boxed{\frac{d^2U(z)}{dz^2} = U(z)\left(R' + j\omega L'\right)\left(G' + j\omega C'\right)} \qquad \text{(Telegrapher's equation)} \tag{3.12}$$

The bracket terms on the right-hand side include the four circuit elements L', C', R' and G' of the equivalent network. We introduce a new quantity

$$\gamma^2 = \left(R' + j\omega L'\right)\left(G' + j\omega C'\right) \tag{3.13}$$

where γ is the *propagation constant*. The propagation constant is complex-valued and an important parameter of a transmission line.

$$\boxed{\gamma = \alpha + j\beta = \sqrt{(R' + j\omega L')(G' + j\omega C')}} \qquad \text{(Propagation constant)} \qquad (3.14)$$

The real part α and the imaginary part β of the propagation constant have a physical meaning as we will see shortly.

$$\alpha = \operatorname{Re}\{\gamma\} \qquad \text{(Attenuation constant)} \tag{3.15}$$

$$\beta = \operatorname{Im}\{\gamma\} \qquad \text{(Phase constant)} \tag{3.16}$$

By using the propagation constant γ we can rewrite the telegrapher's equation as

$$\frac{\mathrm{d}^2 U(z)}{\mathrm{d}z^2} - \gamma^2 U(z) = 0 \tag{3.17}$$

The telegrapher's equation is a one-dimensional wave equation. We already know wave equations from Section 2.5.1 about plane waves in free space. Solutions of the telegrapher's equation are voltage waves that travel either in the positive or in the negative z-direction.

Mathematically we can express the superposition of forward and backward travelling voltage waves as

$$U(z) = U_f e^{-\gamma z} + U_r e^{\gamma z} \tag{3.18}$$

where the index 'f' indicates a forward travelling wave and the index 'r' indicates a backward travelling (or reflected) wave. The amplitudes U_f and U_r of the waves are complex-valued voltages

$$U_f = |U_f| e^{j\varphi_f} \qquad \text{and} \qquad U_r = |U_r| e^{j\varphi_r} \tag{3.19}$$

We can easily prove that the superposition of voltages waves in Equation 3.18 is a solution to the telegrapher's equation by substituting Equation 3.18 into Equation 3.17.

3.1.3 Voltage and Current Waves on Transmission Lines

Now, we want to take a closer look at the complex voltage waves in Equation 3.18. Let us start with the first summand

$$U(z) = U_f e^{-\gamma z} = U_f e^{-\alpha z} e^{-j\beta z} \tag{3.20}$$

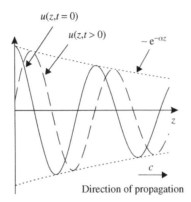

$u(z,t = 0)$

$u(z,t > 0)$

$\sim e^{-\alpha z}$

z

c

Direction of propagation

Figure 3.5 Instantaneous value of the voltage wave for $t = 0$ and $t \approx T/4$. (The phase angle φ_f is set to zero without loss of generality).

where α is the attenuation constant and β is the phase constant. In order to further investigate the term we will move back from the phasor domain to time domain $u(z,t)$. Hence, we multiply the phasor by $e^{j\omega t}$ and take the real part of the expression.

$$u(z, t) = \mathrm{Re}\left\{U(z)e^{j\omega t}\right\} = \mathrm{Re}\left\{\left|U_f\right| e^{j\varphi_f} e^{-\alpha z} e^{-j\beta z} e^{j\omega t}\right\} \tag{3.21}$$

The voltage as a function of space and time reads

$$u(z, t) = \left|U_f\right| \cdot e^{-\alpha z} \cdot \cos\left(\omega t - \beta z + \varphi_f\right) \tag{3.22}$$

with a constant factor $\left|U_f\right|$, an exponentially decaying term $e^{-\alpha z}$ and a harmonic (sinusoidal) term $\cos\left(\omega t - \beta z + \varphi_f\right)$. The three terms can be interpreted as follows

$$\left|U_f\right| \triangleq \text{Absolute value of amplitude} \tag{3.23}$$

$$e^{-\alpha z} \triangleq \text{Attenuation factor} \tag{3.24}$$

$$\cos\left(\omega t - \beta z + \varphi_f\right) \triangleq \text{Wave propagating in positive } z\text{-direction} \tag{3.25}$$

The angle φ_f indicates the *initial phase*. Figure 3.5 shows the voltage wave as a function of the local coordinates for two different instances in time. For $t = 0$ we see a cosine-type signal with an exponentially decaying amplitude $\sim e^{-\alpha z}$. Hence, the decay in amplitude is controlled by the attenuation constant[2] α. At a later time $t \approx T/4$ the wave has travelled a distance of a quarter of a wavelength ($\Delta z \approx \lambda/4$) in a positive z-direction.

We will recall the basic parameters of waves that we already know from our discussion of plane waves in Section 2.5.1. The angular frequency ω and the phase constant β are given by

$$\omega = 2\pi f = \frac{2\pi}{T} \quad \text{and} \quad \beta = \frac{2\pi}{\lambda} \tag{3.26}$$

---

[2] The unit of the attenuation constant α is 1/m or Neper/m. Neper is a pseudo-unit, meaning its value is one. Datasheets of technical transmission lines list α in dB/m (decibel per meter). We can easily convert the values by using the following equation: 1 Neper/m $\approx$ 8.686 dB/m.

where T is the time period and λ is the wavelength. The phase constant β is commonly used for waves on transmission lines whereas for plane waves we use wave number k. However, both parameters are the same $\beta = k = 2\pi/\lambda$.

We can calculate the propagation speed (*phase velocity*) of a sinusoidal wave by looking at the argument of the cosine function. The cosine function shows a maximum value if the argument equals zero.

$$\omega t - \beta z + \varphi_f = 0 \tag{3.27}$$

In order to investigate how this maximum moves with time we solve the equation for z as a function of time

$$z(t) = \frac{\omega t + \varphi_f}{\beta} \tag{3.28}$$

The derivative of $z(t)$ with respect to time yields the speed of propagation

$$v_{ph} = c = \frac{dz}{dt} = \frac{\omega}{\beta} = \frac{\lambda}{T} = \lambda f \tag{3.29}$$

The phase velocity $c = \lambda/T$ indicates that in a time interval of a period T the wave propagates a distance of a wavelength λ.

A transmission line has isolating material between the two conductors. The speed of propagation depends on the material parameters (relative permittivity ε_r and relative permeability μ_r) of that medium

$$c = \frac{c_0}{\sqrt{\varepsilon_r \mu_r}} \tag{3.30}$$

where c_0 is the light speed in vacuum.

$$c_0 = \frac{1}{\sqrt{\varepsilon_0 \mu_0}} \approx 3 \cdot 10^8 \frac{m}{s} \tag{3.31}$$

Our inspection of the term $U_f e^{-\gamma z}$ showed that it describes an exponentially decaying voltage wave travelling in a *positive* z-direction. In our solution (Equation 3.18) of the telegrapher's equation there was a second term given by $U_r e^{\gamma z}$. This second term has a positive sign in the exponent. Therefore, it describes a voltage wave travelling in the *negative* z-direction.

The solution of the telegrapher's equation in Equation 3.18 describes the superposition of two exponentially decaying *voltage waves* that travel in *opposite* directions, where U_f is the amplitude of the forward travelling wave and U_r is the amplitude of the backward travelling (reflected) wave. The *propagation constant* $\gamma = \alpha + j\beta$ controls the attenuation and wavelength and is therefore a *characteristic transmission line parameter*.

Next we will take a look at the *current* on the transmission line. Therefore, we substitute the general voltage solution of Equation 3.18 into the differential equation in Equation 3.11.

$$I(z) = -\frac{dU(z)}{dz} \cdot \frac{1}{R' + j\omega L'} = -\frac{1}{R' + j\omega L'} \cdot \frac{d}{dz}\left(U_f e^{-\gamma z} + U_r e^{\gamma z}\right) \qquad (3.32)$$

After the derivation with regard to the spatial coordinate we arrive at

$$I(z) = -\frac{1}{R' + j\omega L'}\left[(-\gamma)U_f e^{-\gamma z} + \gamma U_r e^{\gamma z}\right] = \frac{\gamma}{R' + j\omega L'}\left[U_f e^{-\gamma z} - U_r e^{\gamma z}\right] \quad (3.33)$$

Next, we use our definition of γ from Equation 3.14

$$\gamma = \alpha + j\beta = \sqrt{(R' + j\omega L')(G' + j\omega C')} \qquad (3.34)$$

and write

$$I(z) = \underbrace{\sqrt{\frac{G' + j\omega C'}{R' + j\omega L'}}}_{1/Z_0}\left[U_f e^{-\gamma z} - U_r e^{\gamma z}\right] \qquad (3.35)$$

On the right-hand side we see the square root of a quotient containing all four elements of our equivalent circuit model in Figure 3.3. The term has the unit of an admittance $(1/\Omega)$. Therefore, we define the *characteristic line impedance* Z_0 as the reciprocal value

$$\boxed{Z_0 = \sqrt{\frac{R' + j\omega L'}{G' + j\omega C'}}} \qquad \text{(Characteristic impedance)} \qquad (3.36)$$

By using the characteristic impedance Z_0 we arrive at two compact equations for voltage and current on a transmission line

$$U(z) = U_f e^{-\gamma z} + U_r e^{\gamma z} \qquad (3.37)$$

$$I(z)Z_0 = U_f e^{-\gamma z} - U_r e^{\gamma z} \qquad (3.38)$$

It is important to note that the characteristic impedance of the line is *not a circuit component* like, for instance, a resistor. It is defined by the four equivalent circuit elements L', C', R' and G'. However, Equations 3.37 and 3.38 give us an interesting interpretation of the characteristic impedance. Let us consider the special case where the amplitude of the backward travelling wave equals zero ($U_r = 0$). In this case the right-hand sides in Equations 3.37 and 3.38 are equal. Hence, the left-hand sides are equal

$$Z_0 = \frac{U(z)}{I(z)} \qquad \text{for} \quad U_r = 0 \qquad (3.39)$$

The *characteristic impedance* Z_0 is the *voltage-to-current-ratio of a wave* that travels *only in one direction*. The characteristic impedance is an important transmission line parameter.

So, we conclude that we have three parameters that are commonly used to describe the behaviour of a transmission line:

- complex propagation constant $\gamma = \alpha + j\beta$,
- characteristic impedance Z_0, and
- – most obviously – the (total) length of the line ℓ_t.

3.1.4 Load-Terminated Transmission Line

Generally on a transmission line voltage and current waves can propagate in opposite directions. Now we will investigate a transmission line that is terminated by a load impedance Z_A and determine the amplitudes of the forward and backward travelling wave, U_f and U_r respectively.

Figure 3.6 shows a line with a total length ℓ_t. The line is terminated by an impedance $Z_A = U_0/I_0$. Up to now we have not defined the origin of our spatial coordinate z. We define z to be zero at the end of the line ($z = 0$). So any location on the line is given by a negative value of z. Next, we define a new spatial variable $\ell = -z$, that starts at the end of the line too but runs in the opposite direction.

The voltage on the transmission line (see Equation 3.37) is given by the superposition of two waves

$$U(z) = U_f e^{-\gamma z} + U_r e^{\gamma z} \tag{3.40}$$

At the end of the line ($z = -\ell = 0$) we get

$$U_0 = U(0) = U_f + U_r \tag{3.41}$$

The current on the transmission line (see Equation 3.38) is given by

$$I(z)Z_0 = U_f e^{-\gamma z} - U_r e^{\gamma z} \tag{3.42}$$

So, the current at the end of the line is

$$I_0 Z_0 = I(0)Z_0 = U_f - U_r \tag{3.43}$$

Now we solve Equation 3.41 for U_f

$$U_f = U_0 - U_r \tag{3.44}$$

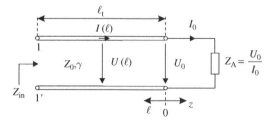

Figure 3.6 Transmission line with impedance Z_A at the end of the line.

and substitute U_f into Equation 3.43

$$I_0 Z_0 = U_0 - 2U_r \tag{3.45}$$

Hence, the amplitude of the backward travelling wave is

$$\boxed{U_r = \frac{1}{2}\left(U_0 - I_0 Z_0\right)} \quad \text{(Amplitude of backward travelling wave)} \tag{3.46}$$

If we substitute U_r into Equation 3.44 the amplitude of the forward travelling wave is

$$\boxed{U_f = \frac{1}{2}\left(U_0 + I_0 Z_0\right)} \quad \text{(Amplitude of forward propagating wave)} \tag{3.47}$$

Let us interpret our result by looking again at the transmission line and load impedance in Figure 3.6. If the load impedance represents, for instance, an antenna, the purpose of the transmission line would be to transfer a signal to that antenna. In a practical configuration it would be desirable that the antenna *fully absorbs* the forward propagating wave (and the power it carries) and radiates it into space. In other words there should be no reflected, backward travelling wave. Equation 3.46 gives us the conditions under which the reflected wave disappears.

$$U_0 - I_0 Z_0 = 0 \quad \rightarrow \quad Z_0 = \frac{U_0}{I_0} \stackrel{!}{=} Z_A \tag{3.48}$$

There is no reflected wave if the characteristic impedance Z_0 equals the ratio of voltage to current at the end of the line $Z_0 = U_0/I_0$. On the other hand this ratio equals the load impedance $Z_A = U_0/I_0$.

> If the load impedance equals the characteristic line impedance ($Z_0 = Z_A$), then only a forward travelling wave exists on the transmission line ($U_f = U_0 \neq 0$). All transmitted power is absorbed by the load impedance. There is no reflected wave ($U_r = 0$). This scenario is referred to as *matched* line termination (impedance matching).

Now we use Equation 3.40 and the results for U_f and U_r to calculate the voltage along the line as a function of the length coordinate $\ell = -z$.

$$U(\ell) = U_f e^{\gamma \ell} + U_r e^{-\gamma \ell} = \frac{1}{2}\left(U_0 + I_0 Z_0\right) e^{\gamma \ell} + \frac{1}{2}\left(U_0 - I_0 Z_0\right) e^{-\gamma \ell} \tag{3.49}$$

We rearrange the equation and get

$$U(\ell) = U_0 \underbrace{\frac{1}{2}\left(e^{\gamma \ell} + e^{-\gamma \ell}\right)}_{\cosh(\gamma \ell)} + I_0 Z_0 \underbrace{\frac{1}{2}\left(e^{\gamma \ell} - e^{-\gamma \ell}\right)}_{\sinh(\gamma \ell)} \tag{3.50}$$

$$= U_0 \cosh(\gamma \ell) + I_0 Z_0 \sinh(\gamma \ell) \tag{3.51}$$

Furthermore, we use Equation 3.42 and the results for U_f and U_r to calculate the current along the line as a function of the length coordinate $\ell = -z$.

$$I(\ell)Z_0 = U_f e^{\gamma \ell} - U_r e^{-\gamma \ell} = \frac{1}{2}(U_0 + I_0 Z_0) e^{\gamma \ell} - \frac{1}{2}(U_0 - I_0 Z_0) e^{-\gamma \ell} \qquad (3.52)$$

Again we rearrange the equation and get

$$I(\ell) = \frac{U_0}{Z_0} \underbrace{\frac{1}{2}(e^{\gamma \ell} - e^{-\gamma \ell})}_{\sinh(\gamma \ell)} + I_0 \underbrace{\frac{1}{2}(e^{\gamma \ell} + e^{-\gamma \ell})}_{\cosh(\gamma \ell)} \qquad (3.53)$$

$$= I_0 \cosh(\gamma \ell) + \frac{U_0}{Z_0} \sinh(\gamma \ell) \qquad (3.54)$$

Finally, we summarize our relations for the voltage and current on a trasmission line.

$$U(\ell) = U_0 \cosh(\gamma \ell) + I_0 Z_0 \sinh(\gamma \ell) \qquad (3.55)$$

$$I(\ell) = I_0 \cosh(\gamma \ell) + \frac{U_0}{Z_0} \sinh(\gamma \ell) \qquad (3.56)$$

We are now able to calculate voltage $U(\ell)$ and current $I(\ell)$ at any location ℓ as a function of the voltage and current at the end of the line, U_0 and I_0 respectively.

3.1.5 Input Impedance

If we would like to analyse a network that contains a transmission line an important quantity to know would be the *input impedance* of a line. Figure 3.6 shows that the input impedance Z_{in} is the impedance that occurs between the input terminals 1-1'. We can easily calculate the input impedance from Equations 3.55 and 3.56 as

$$Z_{in} = \frac{U(\ell_t)}{I(\ell_t)} = \frac{U_0 \cosh(\gamma \ell_t) + I_0 Z_0 \sinh(\gamma \ell_t)}{I_0 \cosh(\gamma \ell_t) + \dfrac{U_0}{Z_0} \sinh(\gamma \ell_t)} \qquad (3.57)$$

where ℓ_t is the total length of the line. The ratio of the voltage U_0 and the current I_0 at the end of the line is determined by the load impedance

$$Z_A = \frac{U_0}{I_0} \qquad (3.58)$$

In Equation 3.57 we factor out the term U_0/I_0 and replace it by Z_A.

$$Z_{in} = \underbrace{\frac{U_0}{I_0}}_{Z_A} \cdot \frac{\cosh(\gamma \ell_t) + \overbrace{\dfrac{I_0}{U_0}}^{1/Z_A} Z_0 \sinh(\gamma \ell_t)}{\cosh(\gamma \ell_t) + \underbrace{\dfrac{U_0}{I_0 Z_0}}_{Z_A/Z_0} \sinh(\gamma \ell_t)} \qquad (3.59)$$

So we end up with a compact and useful formula for the input impedance

$$Z_{\text{in}} = Z_A \cdot \frac{\cosh(\gamma \ell_t) + \dfrac{Z_0}{Z_A} \sinh(\gamma \ell_t)}{\cosh(\gamma \ell_t) + \dfrac{Z_A}{Z_0} \sinh(\gamma \ell_t)} \qquad \text{(Input impedance of} \atop \text{lossy transmission line)} \qquad (3.60)$$

The input impedance Z_{in} of a transmission line depends on the load impedance Z_A and the three transmission line parameters: propagation constant γ, characteristic impedance Z_0 and total length of line ℓ_t.

Example 3.1 *Transmission Line with Matched Termination*
The formula in Equation 3.60 describes the general case. In practical applications most often we choose $Z_A = Z_0$, at least approximately. This case is referred to as matched *termination. The formula in Equation 3.60 is simplified to*

$$Z_{\text{in}} = Z_A = Z_0 \qquad \text{(For arbitrary length of line } \ell_t) \qquad (3.61)$$

The input impedance equals the characteristic impedance. Interestingly the length of the line and the propagation constant do not affect the input impedance in the case of matched termination. This simplification is very welcome in practical designs: we consider an antenna that has an impedance of, let's say, $Z_A = 50\,\Omega$ mounted on a building. The connection cable has a characteristic impedance of $Z_0 = 50\,\Omega$ too. We assume that reconstruction work imposes a slightly greater line length. Fortunately, we can simply add additional $50\,\Omega$-transmission lines since the length of the line does not alter the input impedance.[3]

In the case of a *matched* line ($Z_A = Z_0$) the input impedance equals the characteristic impedance of the line ($Z_{\text{in}} = Z_0$). The input impedance is *independent* of the line length ℓ_t and propagation constant γ. This convenient situation is frequently used in RF applications.

Example 3.2 *Electrically Short Line*
An electrically short *line has a length that is small compared to wavelength ($\ell_t \ll \lambda$). At low frequencies most transmission lines can be considered electrically short, because of the huge size of wavelengths. For example, at a frequency $f = 100\,\text{KHz}$ the free space wavelength is $\lambda = c_0/f = 3\,000\,\text{m}$. For electrically short lines the argument $\gamma \ell_t$ in*

---

[3] Of course we have to ensure that the additional losses due to the increased length of line do not significantly degrade the transmitted signal.

Equation 3.60 approaches zero $(\gamma \ell_t \to 0)$. Hence, $\cosh(\gamma \ell_t) \to 1$ and $\sinh(\gamma \ell_t) \to 0$. Finally, the input impedance is

$$Z_{in} = Z_A \qquad \text{(For arbitrary } Z_0 \text{ and } \gamma\text{)} \tag{3.62}$$

If a transmission line is *electrically short*, that is, the length of the line is small compared to wavelength $(\ell_t \ll \lambda)$, we do not have to consider it a transmission line. The input impedance equals the load impedance independently of the transmission line parameters (propagation constant and characteristic impedance).

3.1.6 Loss-less Transmission Lines

Technical transmission lines are generally designed to have low-loss. In a first approach these lines can often be approximated by loss-less lines. The idealization is of practical importance because it simplifies mathematical considerations.

Let us look at our equivalent circuit model in Figure 3.3. Losses are associated with resistance and conductance. In the loss-less case we replace the resistance by a short circuit $(R'\Delta z = 0)$ and the conductance by an open circuit $(G'\Delta z = 0)$.

Under these assumptions the propagation constant will be a purely *imaginary* quantity

$$\gamma = \alpha + j\beta = \sqrt{(R' + j\omega L')(G' + j\omega C')} = j\omega\sqrt{L'C'} \tag{3.63}$$

The attenuation constant α and the phase constant β are given by

$$\alpha = 0 \qquad \text{and} \qquad \beta = \omega\sqrt{L'C'} \tag{3.64}$$

A loss-less line has an attenuation constant which is zero and a phase constant which is proportional to the frequency f (or angular frequency $\omega = 2\pi f$).

In the loss-less case the characteristic impedance is a purely *real* quantity.

$$Z_0 = \sqrt{\frac{R' + j\omega L'}{G' + j\omega C'}} = \sqrt{\frac{L'}{C'}} \tag{3.65}$$

A loss-less line has a characteristic impedance $Z_0 \in \mathbb{R}$ which is *real* and independent of the frequency $Z_0 \neq f(\omega)$. As the characteristic impedance represents the ratio of voltage to current for a forward travelling wave, we see that on a loss-less transmission line forward travelling waves have voltage and current that are *in phase*.

From the results in Equation 3.64 and Equation 3.65 we can derive a useful relation for the caculation (and measurement) of the characteristic impedance Z_0. Let us first rewrite Equation 3.64

$$\frac{\omega}{\beta} = \frac{1}{\sqrt{L'C'}} \tag{3.66}$$

From Equation 3.29 we know that ω/β equals the phase velocity v_{ph}. Hence, we arrive at the following equation

$$\frac{\omega}{\beta} = \frac{1}{\sqrt{L'C'}} = v_{\text{ph}} = \frac{c_0}{\sqrt{\varepsilon_r \mu_r}} \tag{3.67}$$

If we solve Equation 3.67 for L' and substitute it into Equation 3.65 we get

$$Z_0 = \sqrt{\frac{L'}{C'}} = \frac{\sqrt{\varepsilon_r \mu_r}}{c_0 C'} \tag{3.68}$$

The other way round, let us solve Equation 3.67 for C' and substitute it into Equation 3.65. Then we have

$$Z_0 = \sqrt{\frac{L'}{C'}} = \frac{c_0 L'}{\sqrt{\varepsilon_r \mu_r}} \tag{3.69}$$

The characteristic impedance Z_0 of a loss-less transmission line can be determined by a calculation (or measurement) of the capacitance per unit length C' *or* inductance per unit length L'. Furthermore we need to know the parameters ε_r and μ_r of the isolating material.

Next we will calculate the input impedance Z_{in} of a loss-less line. We already determined the input impedance of a *lossy* line as

$$Z_{\text{in}} = Z_A \cdot \frac{\cosh(\gamma \ell_t) + \dfrac{Z_0}{Z_A} \sinh(\gamma \ell_t)}{\cosh(\gamma \ell_t) + \dfrac{Z_A}{Z_0} \sinh(\gamma \ell_t)} \tag{3.70}$$

In the loss-less case we can replace the propagation constant by $\gamma = j\beta$. This gives us

$$\cosh(\gamma \ell) = \cos(\beta \ell) \tag{3.71}$$

$$\sinh(\gamma \ell) = j \sin(\beta \ell) \tag{3.72}$$

We can prove Equations 3.71 and 3.72 quite easily. We express the hyperbolic functions by exponential functions.

$$\cosh(x) = \frac{1}{2}\left(e^x + e^{-x}\right) \qquad \text{and} \qquad \sinh(x) = \frac{1}{2}\left(e^x - e^{-x}\right) \tag{3.73}$$

Furthermore we use *Euler's formula*

$$e^{j\varphi} = \cos(\varphi) + j\sin(\varphi) \tag{3.74}$$

By using Equation 3.73 the results are

$$\cosh(\gamma\ell) = \cosh(j\beta\ell) = \frac{1}{2}\left(e^{j\beta\ell} + e^{-j\beta\ell}\right) \tag{3.75}$$

$$= \frac{1}{2}\left[\cos(\beta\ell) + j\sin(\beta\ell) + \cos(\beta\ell) - j\sin(\beta\ell)\right] = \cos(\beta\ell)$$

and

$$\sinh(\gamma\ell) = \sinh(j\beta\ell) = \frac{1}{2}\left(e^{j\beta\ell} - e^{-j\beta\ell}\right) \tag{3.76}$$

$$= \frac{1}{2}\left[\cos(\beta\ell) + j\sin(\beta\ell) - \cos(\beta\ell) + j\sin(\beta\ell)\right] = j\sin(\beta\ell)$$

So, we have proved Equations 3.71 and 3.72.

Next, we substitute the relations in Equation 3.70 and get

$$Z_{in} = Z_A \cdot \frac{\cos(\beta\ell_t) + j\dfrac{Z_0}{Z_A}\sin(\beta\ell_t)}{\cos(\beta\ell_t) + j\dfrac{Z_A}{Z_0}\sin(\beta\ell_t)} \tag{3.77}$$

By using the tangent the input impedance is

$$\boxed{Z_{in} = Z_A \cdot \frac{1 + j\dfrac{Z_0}{Z_A}\tan(\beta\ell_t)}{1 + j\dfrac{Z_A}{Z_0}\tan(\beta\ell_t)}} \qquad \text{(Input impedance of lossless transmission line)} \tag{3.78}$$

where ℓ_t is the length of the line.

Equation 3.78 is a central equation when dealing with transmission lines. Unfortunately, it is quite inconvenient to handle, since it requires the manipulation of complex numbers. In Section 3.1.11 we will introduce a graphical tool (*Smith Chart*) that allows us a quick and easy evaluation of the interrelation between load impedance and input impedance of a line.

In the loss-less case voltage and current on the line are given by

$$U(\ell) = U_0\cos(\beta\ell) + jI_0Z_0\sin(\beta\ell) \tag{3.79}$$

$$I(\ell) = I_0\cos(\beta\ell) + j\frac{U_0}{Z_0}\sin(\beta\ell) \tag{3.80}$$

3.1.7 Low-loss Transmission Lines

Only quite short lines can be considered loss-less. On physical transmission lines of greater length the amplitude of the propagating wave decays significantly according to the representation in Figure 3.5. The decay results from the non-zero attenuation constant $\alpha \neq 0$. Therefore, we will now try to find a realistic approximation for the attenuation constant in the case of low-losses.

First, we have to define the term *low-loss*. Let us look at the equivalent circuit model in Figure 3.3. We can interpret the circuit as a lossy coil with an impedance of

$$Z_{\text{coil}} = (R + j\omega L) = (R' + j\omega L')\Delta z \tag{3.81}$$

and a lossy capacitor with an admittance of

$$Y_{\text{cap}} = (G + j\omega C) = (G' + j\omega C')\Delta z \tag{3.82}$$

The loss-less case (Section 3.1.6) was defined by $R' = G' = 0$. In the case of *low-losses* we allow the lossy components to be greater than zero, but we require the imaginary parts of the impedance and admittance to be much greater than the real parts

$$\left|R'\right| \ll \left|j\omega L'\right| \qquad \text{and} \qquad \left|G'\right| \ll \left|j\omega C'\right| \tag{3.83}$$

If we apply the conditions in Equations 3.83 to our general equation for the characteristic impedance we get

$$Z_0 = \sqrt{\frac{R' + j\omega L'}{G' + j\omega C'}} \approx \sqrt{\frac{L'}{C'}} \in \mathbb{R} \tag{3.84}$$

A low-loss transmission line has – as an approximation – the same characteristic impedance as a loss-less line. It is still *real-valued* and independent of frequency.

Now we evaluate the propagation constant. Since we are interested in the attenuation of the wave along the line we will focus particularly on the attenuation constant α. The propagation constant is given by

$$\gamma = \alpha + j\beta = \sqrt{(R' + j\omega L')(G' + j\omega C')}$$

$$= j\omega\sqrt{L'C'}\sqrt{\left(1 + \frac{G'}{j\omega C'}\right)\left(1 + \frac{R'}{j\omega L'}\right)} \tag{3.85}$$

$$= j\omega\sqrt{L'C'}\sqrt{1 + \frac{G'}{j\omega C'} + \frac{R'}{j\omega L'} + \frac{G'}{j\omega C'}\frac{R'}{j\omega L'}}$$

The quotients $G'/(j\omega C') \ll 1$ and $R'/(j\omega L') \ll 1$ are values much smaller than one. The last term under the square root is a product of small values and can therefore be omitted. This gives us the following approximation

$$\gamma = \alpha + j\beta \approx j\omega\sqrt{L'C'}\sqrt{1 + \frac{G'}{j\omega C'} + \frac{R'}{j\omega L'}} \tag{3.86}$$

In order to separate the real and imaginary parts of the propagation constant we use the Taylor series of $\sqrt{1 \pm x}$ at $x = 0$ for $|x| \ll 1$.

$$\sqrt{1 \pm x} = 1 \pm \frac{x}{2} - \frac{x^2}{8} \pm \frac{x^3}{16} - \ldots \approx 1 \pm \frac{x}{2} \qquad \text{for} \qquad |x| \ll 1 \qquad (3.87)$$

From Equation 3.86 we get

$$\gamma \approx j\omega\sqrt{L'C'}\left(1 + \frac{G'}{j2\omega C'} + \frac{R'}{j2\omega L'}\right) = j\underbrace{\omega\sqrt{L'C'}}_{\beta} + \underbrace{\frac{G'\sqrt{L'}}{2\sqrt{C'}} + \frac{R'\sqrt{C'}}{2\sqrt{L'}}}_{\alpha \,=\, \alpha_{\text{diel}} + \alpha_{\text{met}}} \qquad (3.88)$$

The phase constant of the low-loss case corresponds to the phase constant of the loss-less case.

$$\boxed{\beta \approx \omega\sqrt{L'C'}} \qquad (3.89)$$

By using the characteristic impedance in Equation 3.84 we write the attenuation constant as

$$\boxed{\alpha \approx \alpha_{\text{diel}} + \alpha_{\text{met}} \approx \frac{G'Z_0}{2} + \frac{R'}{2Z_0}} \qquad (3.90)$$

In the low-loss case the *characteristic impedance* Z_0 and the *phase constant* β do not deviate much from the loss-less case. So, the relations from the loss-less case represent reasonable approximations.

The *attenuation constant* α is given by the sum of two terms. One term corresponds to dielectric losses and the other term represents metallic (ohmic) losses. We can determine the attenuation constant from measured (or calculated) values of the equivalent circuit elements R' and G'.

3.1.8 Transmission Line with Different Terminations

Now we will look at three different terminations of a *loss-less* transmission line in detail: matched load, short circuited line and open circuited line. We are interested in input impedance Z_{in} as well as voltage distribution $U(\ell)$ and current distribution $I(\ell)$ along the line.

3.1.8.1 Matched Load

We begin with a loss-less transmission line which is terminated by a matched load $Z_A = Z_0$ (see Figure 3.7a).

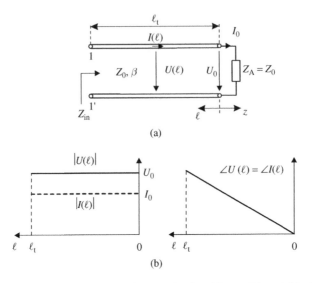

(a)

(b)

Figure 3.7 (a) Transmission line with matched termination ($Z_A = Z_0$) and (b) absolute value and phase angle of volt and current along the transmission line.

By using Equation 3.78 the input impedance reads

$$Z_{in} = Z_A \cdot \frac{1 + j\dfrac{Z_0}{Z_A}\tan(\beta\ell_t)}{1 + j\dfrac{Z_A}{Z_0}\tan(\beta\ell_t)} = Z_A = Z_0 \qquad \text{(For arbitrary line length } \ell_t) \qquad (3.91)$$

The input impedance equals the characteristic impedance of the transmission line and is independent of the line length ℓ_t and propagation constant $\gamma = j\beta$. (We have discussed this scenario of a *matched load* already in Example 3.1 for the more general case of lossy lines.)

Next we look at the voltage $U(\ell)$ and current $I(\ell)$ on the line. In Section 3.1.6 we derived formulas for voltage and current in Equations 3.79 and 3.80.

$$U(\ell) = U_0 \cos(\beta\ell) + j I_0 Z_0 \sin(\beta\ell) \qquad (3.92)$$

$$I(\ell) = I_0 \cos(\beta\ell) + j\frac{U_0}{Z_0}\sin(\beta\ell) \qquad (3.93)$$

where U_0 and I_0 are voltage and current at the end of the line. By using

$$Z_A = \frac{U_0}{I_0} \overset{!}{=} Z_0 \qquad (3.94)$$

we find for the voltage on the line

$$U(\ell) = U_0 \cos(\beta\ell) + j \underbrace{I_0 Z_0}_{U_0} \sin(\beta\ell) = U_0 \left[\cos(\beta\ell) + j\sin(\beta\ell)\right] \qquad (3.95)$$

$$= U_0 e^{j\beta\ell} \qquad (3.96)$$

We have seen such an exponential expression before, when we discussed solutions of the telegrapher's equation in Section 3.1.2. The term $U(\ell) = U_0 e^{j\beta\ell}$ describes a voltage wave that travels in a *negative* ℓ-direction (or positive z-direction), that is, the wave travels from the input terminals 1-1' to the load. (We notice that there is *no* backward travelling wave here.)

Likewise the current on the line becomes

$$I(\ell) = I_0 \cos(\beta\ell) + j \underbrace{\frac{U_0}{Z_0}}_{I_0} \sin(\beta\ell) = I_0 \left[\cos(\beta\ell) + j \sin(\beta\ell)\right] \qquad (3.97)$$

$$= I_0 e^{j\beta\ell} \qquad (3.98)$$

The current is described by a forward travelling wave too.

We will now plot the absolute values and phases of the voltage and current functions given in Equation 3.96 and Equation 3.98 as a function of ℓ (see Figure 3.7b). The absolute values of voltage $|U(\ell)|$ and current $|I(\ell)|$ are constant at any point on the line.

$$|U(\ell)| = U_0 = \text{const.} \qquad \text{and} \qquad |I(\ell)| = I_0 = \text{const.} \qquad (3.99)$$

The phases of voltage and current increase linearly with the coordinate ℓ and decrease linearly with the coordinate z.

$$\angle U(\ell) = \angle I(\ell) = \beta\ell = -\beta z \qquad (3.100)$$

Matched load termination ($Z_A = Z_0$) is the *most wanted (standard) case* in RF engineering and from a mathematical point of view the most simple case. The *input impedance* equals the characteristic impedance of the line ($Z_{in} = Z_0 = Z_A$) independent of the total line length ℓ_t. Voltage and current wave propagate *forward* from input terminals to the load. There is no backward travelling (reflected) wave. Therefore, all supplied power is transferred to the load. The *absolute values* of the voltage and current are *constant* on the line. The *phases* decrease *linearly* from the input terminal to the load. Voltage and current are *in phase*.

3.1.8.2 Short-Circuited Termination

Next we consider a short-circuited transmission line ($Z_A = 0$) according to Figure 3.8a. With Equation 3.78 we calculate the input impedance as

$$Z_{in} = Z_A \cdot \frac{1 + j\dfrac{Z_0}{Z_A}\tan(\beta\ell_t)}{1 + j\dfrac{Z_A}{Z_0}\tan(\beta\ell_t)} = j Z_0 \tan(\beta\ell_t) = j X_{in} \qquad (3.101)$$

The input impedance is reactive, it has only an *imaginary* part. Figure 3.8b shows the reactance X_{in} as a function of the length of line ℓ_t. The reactance runs periodically

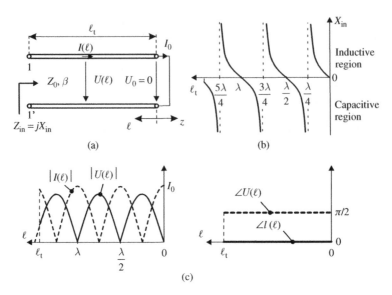

Figure 3.8 (a) Transmission line with short circuit termination ($Z_A = 0$), (b) imaginary part of input impedance and (c) absolute value and phase of voltage and current angle along the transmission line.

through regions of positive and negative values due to the periodicity of the tangent. If the reactance is *positive* we can interpret it as an *inductance* and write

$$Z_{in} = j X_{in} = j \omega L \tag{3.102}$$

On the other hand, if the reactance is *negative* we can interpret it as a *capacitance* and write

$$Z_{in} = j X_{in} = \frac{1}{j \omega C} = -j \frac{1}{\omega C} \tag{3.103}$$

The sign of the reactance changes periodically, so it moves through alternating regions of inductive and capacitive behaviour. For a line length of $0 < \ell_t < \lambda/4$ the input impedance is inductive; for a line length of $\lambda/4 < \ell_t < \lambda/2$ the input impedance is capacitive, and so forth.

An interesting line is a quarter-wave line, that is a line with a length of $\ell_t = \lambda/4$. In the case of a $\lambda/4$-line the short-circuit termination is transformed to an open-circuit input impedance.[4] If the line length is $\lambda/2$ the short-circuit load impedance is transformed to a short-circuit input impedance.

Finally, we investigate the voltage on the line. Due to the short-circuited line termination the voltage at the end of the line is zero ($U_0 = 0$).

$$U(\ell) = \underbrace{U_0}_{= 0} \cos(\beta \ell) + j I_0 Z_0 \sin(\beta \ell) = j I_0 Z_0 \sin(\beta \ell) \tag{3.104}$$

---

[4] This result may appear somewhat peculiar at a first glance: a short-circuited $\lambda/4$-line behaves like an open circuit. However, it has many practical applications. We will see throughout the whole book that $\lambda/4$-lines are of great use.

In order to interpret this result we rewrite the sine function by using exponential terms

$$\sin(x) = \frac{1}{2j} \left(e^{jx} - e^{-jx} \right) \tag{3.105}$$

Hence, the voltage on the line is

$$U(\ell) = \frac{1}{2} I_0 Z_0 \left(e^{j\beta\ell} - e^{-j\beta\ell} \right) \tag{3.106}$$

The voltage is a superposition of two voltage waves travelling in opposite directions. The forward and backward travelling waves have amplitudes of the same absolute value but opposite sign.

The current on the line then becomes

$$I(\ell) = I_0 \cos(\beta\ell) + j \underbrace{\frac{U_0}{Z_0}}_{= 0} \sin(\beta\ell) = I_0 \cos(\beta\ell) \tag{3.107}$$

In order to interpret this result we rewrite the cosine function by using exponential terms

$$\cos(x) = \frac{1}{2} \left(e^{jx} + e^{-jx} \right) \tag{3.108}$$

Hence, the current on the line is

$$I(\ell) = \frac{1}{2} I_0 \left(e^{j\beta\ell} + e^{-j\beta\ell} \right) \tag{3.109}$$

So the current is a superposition of two current waves travelling in opposite directions. The forward and backward travelling waves have equal amplitudes.

Voltage and current exhibit *standing wave* behaviour. The absolute values of voltage and current change sinusoidally. Both show stationary minima and maxima along the line. The distance between two adjacent maxima (or minima) is half a wavelength (see Figure 3.8c). At the end of the line the voltage is zero and the current is maximum according to the short circuit.

The input impedance of a short-circuited transmission line is reactive. Depending on the line length it can be inductive or capacitive. A short-circuited line starts with *inductive* behaviour for a short length ($\ell_t < \lambda/4$). This is easy to remember since the current I_0 that flows in the short circuit is accompanied by a magnetic field. From Section 2.1.2 we know that a component that stores magnetic energy has inductance.

Voltage and current on the line are *standing waves*. Their maxima (and minima) are displaced by $\lambda/4$. The phases of voltage and current are constant on the line. Voltage and current are out of phase by 90°, as is characteristic for reactive components.

In practice short-circuited lines may be used to mimic inductor or capacitor behaviour.

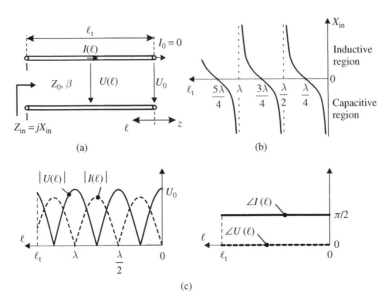

Figure 3.9 (a) Transmission line with open-circuit termination ($Z_A \rightarrow \infty$), (b) imaginary part of input impedance and (c) absolute value and phase angle of voltage and current along the transmission line.

3.1.8.3 Open-Circuited Termination

Let us now we consider an open-circuited transmission line ($Z_A \rightarrow \infty$) according to Figure 3.9a. With Equation 3.78 we calculate the input impedance as

$$Z_{in} = Z_A \cdot \frac{1 + j\dfrac{Z_0}{Z_A} \tan(\beta \ell_t)}{1 + j\dfrac{Z_A}{Z_0} \tan(\beta \ell_t)} = -jZ_0 \cot(\beta \ell_t) = jX_{in} \qquad (3.110)$$

The input impedance is reactive, it has only an *imaginary* part. Figure 3.9b shows the reactance X_{in} as a function of the length of line ℓ_t. The reactance runs periodically through regions of positive and negative values due to the periodicity of the co-tangent. For a line length of $0 < \ell_t < \lambda/4$ the input impedance is capacitive; for a line length of $\lambda/4 < \ell_t < \lambda/2$ the input impedance is inductive, and so forth. In the case of a $\lambda/4$-line the open-circuit termination is transformed to a short circuit input impedance. If the line length is $\lambda/2$ the open-circuit load impedance is transformed to an open-circuit input impedance.

Now, we investigate the voltage on the line. Due to the open-circuited line termination the current at the end of the line is zero ($I_0 = 0$).

$$U(\ell) = U_0 \cos(\beta \ell) + j \underbrace{I_0}_{=\,0} Z_0 \sin(\beta \ell) = U_0 \cos(\beta \ell) \qquad (3.111)$$

The current on the line then becomes

$$I(\ell) = \underbrace{I_0}_{= 0} \cos(\beta\ell) + j\frac{U_0}{Z_0}\sin(\beta\ell) = j\frac{U_0}{Z_0}\sin(\beta\ell) \qquad (3.112)$$

Voltage and current exhibit *standing wave* behaviour. The absolute values of voltage and current change sinusoidally. Both show stationary minima and maxima along the line. The distance between two adjacent maxima (or minima) is half a wavelength (see Figure 3.9c). At the end of the line the voltage is maximum and the current is zero according to the open circuit.

The input impedance of an open-circuited transmission line is reactive. Depending on the line length it is either inductive or capacitive. An open-circuited line starts with *capacitive* behaviour for short length ($\ell_t < \lambda/4$). This is easy to remember since the voltage U_0 that appears at the open end of the line is accompanied by an electric field. From Section 2.1.1 we know that a component that stores electric energy has capacitance.

Voltage and current on the line are *standing waves*. Their maxima (and minima) are displaced by $\lambda/4$. The phases of voltage and current are constant on the line. Voltage and current are *out of phase by* 90°, as is characteristic for reactive components.

In practice open-circuited lines can be used to mimic inductor or capacitor behaviour.

3.1.8.4 Arbitrary Line Termination

Lastly we consider an arbitrary line termination $Z_A = R_A + jX_A$ (see Figure 3.10a). The input impedance is given by the general relation in Equation 3.78 without any simplification. From previous discussions we know that forward and backward travelling waves of different amplitudes exist on a line, where U_f is the amplitude of the forward travelling wave and U_r is the amplitude of the backward travelling wave. As an example Figure 3.10b shows a typical voltage and current distribution on a line. The maximum voltage is given by

$$\left|U_{\text{max}}\right| = \max\left\{\left|U_f e^{j\beta\ell} + U_r e^{-j\beta\ell}\right|\right\} = \left|U_f\right| + \left|U_r\right| \qquad (3.113)$$

and the minimum voltage is

$$\left|U_{\text{min}}\right| = \min\left\{\left|U_f e^{j\beta\ell} + U_r e^{-j\beta\ell}\right|\right\} = \left|U_f\right| - \left|U_r\right| \qquad (3.114)$$

Locations of the maxima of the voltage correspond to locations of the minima of the current and vice versa. The distance between two maxima (or minima) is again half a wavelength.

The ratio of the maximum voltage U_{max} to the minimum voltage U_{min} on the line is the *voltage standing wave ratio* (VSWR). It equals the ratio of the maximum current I_{max}

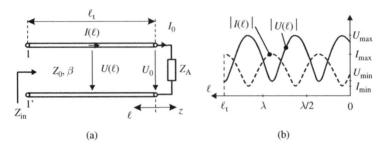

(a)   (b)

Figure 3.10 (a) Transmission line with general termination Z_A, (b) absolute values of voltage and current along the transmission line.

to the minimum current I_{min} on the line.[5]

$$\text{VSWR} = \left| \frac{U_{max}}{U_{min}} \right| = \left| \frac{I_{max}}{I_{min}} \right| \qquad (3.115)$$

The voltage standing wave ratio (VSWR) is a useful quantity to describe the *matching* of the load impedance Z_A to the characteristic impedance Z_0 of the line as we will see in the following example.

Example 3.3 *VSWR of Different Line Terminations*

We consider a transmission line with a line impedance of Z_0. First we terminate the line with a load impedance that equals the characteristic impedance $Z_A = Z_0$. In Figure 3.7 we see that $U_{max} = U_{min}$. Hence,

$$\text{VSWR}\left(Z_A = Z_0\right) = 1 \qquad (3.116)$$

Next, we consider a short-circuited line. In Figure 3.8 we see that U_{max} is finite and $U_{min} = 0$. So,

$$\text{VSWR}\left(Z_A = 0\right) \rightarrow \infty \qquad (3.117)$$

Finally, we consider an open-circuited line. In Figure 3.9 we see that $U_{max} = U_0$ and $U_{min} = 0$. So,

$$\text{VSWR}\left(Z_A = \infty\right) \rightarrow \infty \qquad (3.118)$$

From our discussions in Section 3.1.8 we conclude that a VSWR of 1 means no backward travelling (reflected) wave. A VSWR of infinite value means that the reflected wave equals the forward propagating wave on the line. Therefore, VSWR is a useful parameter to describe the amount of power that is delivered to the load. For matching purposes VSWR should be close to one.

---

[5] So, we can omit the letter V and refer to the ratio as *standing wave ratio* (SWR).

3.1.9   Impedance Transformation with Loss-less Lines

An interesting application of a loss-less line is a *matching circuit*, that is, a load impedance is transformed to a desired input impedance. Later in this book we will investigate matching circuits in detail (see Section 6.3). Here, we want to understand the concept of matching a *real* (ohmic) load to a *real* input impedance by a loss-less line. Transmission lines with a length of $\lambda/4$ and $\lambda/2$ provide simple relations between input and load resistance.

3.1.9.1   Quarter-Wave Transformer

First, we investigate the input impedance of a $\lambda/4$-line

$$Z_{in} = Z_A \cdot \frac{1 + j\dfrac{Z_0}{Z_A}\tan(\beta\ell_t)}{1 + j\dfrac{Z_A}{Z_0}\tan(\beta\ell_t)} \qquad \text{where} \qquad \beta\ell_t = \frac{2\pi}{\lambda}\cdot\frac{\lambda}{4} = \frac{\pi}{2} \qquad (3.119)$$

Since $\tan(\pi/2) \to \infty$, we find the following simple relation

$$\boxed{Z_{in} = \frac{Z_0^2}{Z_A}} \qquad \text{or} \qquad \boxed{Z_{in}Z_A = Z_0^2} \qquad (\lambda/4\text{-transformer}) \qquad (3.120)$$

Example 3.4 *Impedance Matching with Quarter-Wave Transmission Line*
   We want to transform a load impedance $Z_A = 200\,\Omega$ to an input impedance of $Z_{in} = 50\,\Omega$ by using a quarter-wave transformer. The frequency of our application is $f = 2.45\,GHz$. According to Equation 3.120 the characteristic impedance of the line must be

$$Z_0 = \sqrt{Z_{in}Z_A} = \sqrt{50\,\Omega \cdot 200\,\Omega} = 100\,\Omega \qquad (3.121)$$

Let us suppose that the transmission line is air-filled ($\varepsilon_r = 1$), that is, the propagation speed of waves equals $c_0 = 3\cdot 10^8$ m/s. Then our line length becomes

$$\ell_t = \frac{\lambda}{4} = \frac{c_0}{4f} = 3.06\,cm \qquad (3.122)$$

3.1.9.2   Half-Wave Transformer

Now we consider a line with $\ell_t = \lambda/2$. The input impedance is

$$Z_{in} = Z_A \cdot \frac{1 + j\frac{Z_0}{Z_A}\tan(\beta\ell_t)}{1 + j\dfrac{Z_A}{Z_0}\tan(\beta\ell_t)} \qquad \text{where} \qquad \beta\ell_t = \frac{2\pi}{\lambda}\cdot\frac{\lambda}{2} = \pi \qquad (3.123)$$

Since $\tan(\pi) = 0$, this simplifies to

$$\boxed{Z_{in} = Z_A} \qquad (\text{Half-wave transformer}) \qquad (3.124)$$

If the line length is half a wavelength, the input impedance equals the load impedance regardless of the the characteristic impedance of the line.

3.1.10 Reflection Coefficient

At this point we will introduce a key term of transmission line theory: the *reflection coefficient*. We derive this new term by looking at the voltage $U(\ell)$ on the line. It can be expressed as a superposition of waves travelling in opposite directions.

$$U(\ell) = \underbrace{U_f e^{j\beta\ell}}_{\substack{\text{forward propagating} \\ \text{wave}}} + \underbrace{U_r e^{-j\beta\ell}}_{\substack{\text{reflected} \\ \text{wave}}} \qquad (3.125)$$

where

$$U_f = \frac{1}{2}(U_0 + I_0 Z_0) \triangleq \text{Amplitude of forward propagating wave} \qquad (3.126)$$

$$U_r = \frac{1}{2}(U_0 - I_0 Z_0) \triangleq \text{Amplitude of reflected wave} \qquad (3.127)$$

$$Z_A = \frac{U_0}{I_0} \triangleq \text{Load impedance} \qquad (3.128)$$

We define the new quantity *reflection coefficient at point* ℓ as a ratio of reflected to forward propagating wave (see Figure 3.11).

$$\boxed{r(\ell) = \frac{U_r e^{-j\beta\ell}}{U_f e^{j\beta\ell}} = \frac{U_r}{U_f} e^{-j2\beta\ell}} \qquad \text{(Reflection coefficient)} \qquad (3.129)$$

We will now find the relation between the reflection coefficient and the load and characteristic impedance at the end of the line ($\ell = 0$). By using the definition from Equation 3.129 and the amplitudes U_f and U_r from Equations 3.126 and 3.127, respectively, we get

$$r_A = r(\ell = 0) = \frac{U_r}{U_f} = \frac{U_0 - I_0 Z_0}{U_0 + I_0 Z_0} = \frac{U_0/I_0 - Z_0}{U_0/I_0 + Z_0} \qquad (3.130)$$

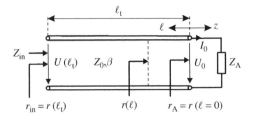

Figure 3.11 Reflection coefficient on a loss-less transmission line.

With Equation 3.128 the reflection coefficient at the end of the line is

$$\boxed{r_A = \frac{Z_A - Z_0}{Z_A + Z_0}}$$ (3.131)

From the reflection coefficient r_A at the end of the line we can now calculate the reflection coefficient at the input terminals r_{in}.

$$r_{in} = r(\ell_t) = \frac{U_r e^{-j\beta\ell_t}}{U_f e^{j\beta\ell_t}} = \frac{U_r}{U_f} e^{-j2\beta\ell_t} = r_A e^{-j2\beta\ell_t}$$ (3.132)

Let us summarize some important properties

- The *absolute value* of the reflection coefficient is constant on the line: $|r_A| = |r_{in}| = |r(\ell)| = \text{const.} \quad \forall \ell$
- The *phase* of the reflection coefficient decreases linearly with ℓ.

$$\varphi = -2\beta\ell = -4\pi \frac{\ell}{\lambda}$$ (3.133)

- For passive terminations ($\text{Re}\{Z_A\} \geq 0$) the absolute value of the reflection coefficient is equal to or less than one $|r(\ell)| \leq 1$

> The reflection coefficient is a complex quantity that can be plotted in the complex plane. Passive terminations correspond to reflection coefficients inside the unit circle ($U_r \leq U_f$). The curve of the reflection coefficient $r(\ell)$ along a line is a segment of a circle around the origin of the complex plane.

If we follow the reflection coefficient of a line *from the load impedance* to the input terminals we move along a circle in a *clockwise* direction (see Figure 3.12b). (If we went from input to load, the direction of rotation would be counter-clockwise, i.e. mathematically positive.)

In Section 3.1.9 we have seen that the line lengths of $\lambda/4$ and $\lambda/2$ showed simple relations between load and input impedance. Now we will see what these special lengths mean for the transformation of the reflection coefficient.

$$\ell = \lambda/4 \rightarrow 2\beta\ell = 2\frac{2\pi}{\lambda} \cdot \frac{\lambda}{4} = \pi = 180° \rightarrow r_{in} = -r_A$$ (3.134)

$$\ell = \lambda/2 \rightarrow 2\beta\ell = 2\frac{2\pi}{\lambda} \cdot \frac{\lambda}{2} = 2\pi = 360° \rightarrow r_{in} = r_A$$ (3.135)

A quarter-wave line rotates the reflection coefficient by an angle of 180°. This changes the sign of the reflection coefficient. A half-wave line rotates the reflection coefficient by an angle of 360°. Therefore, the refection coefficient at the input terminal equals the reflection coefficient at end of the line.

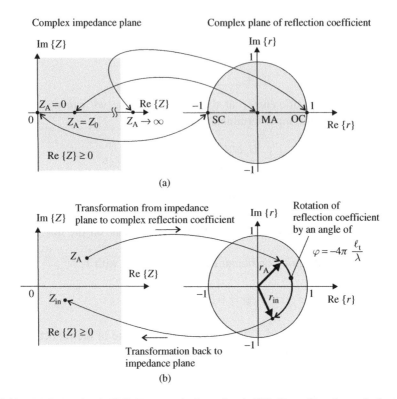

Figure 3.12 (a) Open circuit (OC, $Z_A \to \infty$), short circuit (SC, $Z_A = 0$) and matched termination (MA, $Z_A = Z_0$) in the complex impedance plane and in the complex plane of reflection coefficient; (b) transformation of load impedance Z_A into the complex plane of the reflection coefficient gives r_A; clockwise rotation in the plane of the reflection coefficient gives r_{in}; transformation back into the impedance plane gives the input impedance Z_{in}.

In order to get a visual understanding of how reflection coefficients correspond to impedances we investigate the short circuit, open circuit and matched termination load ($Z_A = Z_0$).

$$\text{Short circuit}(Z_A = 0) \to r_A = \frac{Z_A - Z_L}{Z_A + Z_L} = -1 \to \text{SC} \qquad (3.136)$$

$$\text{Open circuit}(Z_A \to \infty) \to \qquad r_A = 1 \qquad \to \text{OC} \qquad (3.137)$$

$$\text{Matched termination}(Z_A = Z_0) \to \qquad r_A = 0 \qquad \to \text{MA} \qquad (3.138)$$

The locations in the complex impedance plane and in the complex plane of the reflection coefficient are shown in Figure 3.12a.

In Section 3.1.8 we introduced the voltage standing wave ratio VSWR. It is a measure of the quality of the matching of the load impedance to the characteristic impedance of the line. We can easily relate VSWR to our new quantity reflection coefficient.

VSWR (see Equation 3.115) is given by

$$\text{VSWR} = \left| \frac{U_{\max}}{U_{\min}} \right| \tag{3.139}$$

where $|U_{\max}| = |U_f| + |U_r|$ (see Equation 3.113) and $|U_{\min}| = |U_f| - |U_r|$ (see Equation 3.114).

If we substitute Equation 3.113 and Equation 3.114 in Equation 3.139 we get

$$\text{VSWR} = \left| \frac{U_{\max}}{U_{\min}} \right| = \frac{|U_f| + |U_r|}{|U_f| - |U_r|} = \frac{1 + \dfrac{|U_r|}{|U_f|}}{1 - \dfrac{|U_r|}{|U_f|}} \tag{3.140}$$

With our definition of the reflection coefficient in Equation 3.129 we write

$$\boxed{\text{VSWR} = \frac{1 + |r|}{1 - |r|}} \quad \leftrightarrow \quad \boxed{|r| = \frac{\text{VSWR} - 1}{\text{VSWR} + 1}} \tag{3.141}$$

The voltage standing wave ratio VSWR *and* the reflection coefficient r are commonly used to describe the quality of matching. In order to convert the values between both measures we note these helpful formulas here.

3.1.11   Smith Chart

A loss-less transmission line can be considered a two-port network, that transforms a load impedance Z_A into an input impedances Z_{in}. The relation depends on the three transmission line parameters (characteristic impedance Z_0, phase constant β and length of line ℓ_t) and is given by a rather inconvenient equation as

$$Z_{in} = Z_A \cdot \frac{1 + j\dfrac{Z_0}{Z_A} \tan(\beta \ell_t)}{1 + j\dfrac{Z_A}{Z_0} \tan(\beta \ell_t)} \tag{3.142}$$

In the previous section we saw that the reflection coefficient is easier to handle on a line. The line simply contributes to the phase of the reflection coefficient.

$$r_{in} = r_A e^{-j2\beta \ell_t} \tag{3.143}$$

In order to employ this rather simple relation, we introduce a graphical tool that allows us to convert impedances to reflection coefficients and manipulate the phase of the reflection coefficient. This diagram is known as the *Smith chart*. The Smith chart shows the complex plane of the reflection coefficient. We remember that passive load impedances ($\text{Re}\{Z_A\} \geq 0$) result in reflection coefficients $|r| \leq 1$. So the outer circle in Figure 3.13a corresponds to the unit circle. The centre point (MA) is the origin of the complex plane.

Inside the unit circle we see solid-line circles and dashed-line segments of circles. These lines represent normalized impedances $z = Z/Z_0$. So the Smith chart shows us the

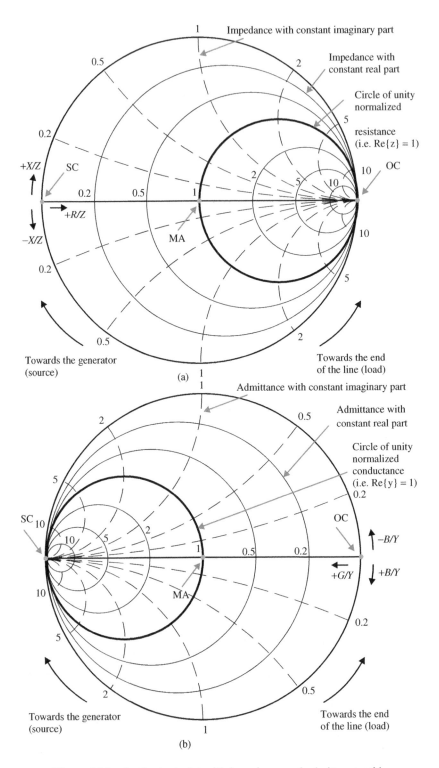

Figure 3.13 Smith chart plot with impedance and admittance grid.

reflection coefficient and normalized impedance in one diagram. Figure 3.13a shows the so-called *impedance form* of the Smith chart.

$$r = \frac{Z - Z_0}{Z + Z_0} = \frac{z - 1}{z + 1} \qquad \text{where} \qquad z = \frac{Z}{Z_0} \qquad (3.144)$$

Dashed lines and the real axis indicate normalized impedances with a constant imaginary part (constant normalized reactance) and solid lines indicate normalized impedances with a constant real part (constant normalized resistance).

The *Smith diagram* simultaneously represents *normalized impedances* and *reflection coefficients*. It allows easy conversion between the two and is useful in order to determine the reflection coefficient along the line. Since a transmission line only changes the phase of the reflection coefficient, we simply rotate the reflection coefficient around the centre point of the diagram.

If we plot normalized admittances $y = Y/Y_0 = YZ_0$ into the complex plane of the reflection coefficient we get the *admittance form* of the Smith chart. The admittance form is shown in Figure 3.13b. In order to work practically with the diagram it is often useful to have both forms in one diagram. Coloured Smith chart plots that overlay impedance and admittance forms are available on the Web.

The Smith chart is a graphical tool that was historically used to replace complex calculations by graphical methods when dealing with transmission lines and passive circuits. Today, it is still useful in the design of transmission lines and matching circuits. Furthermore, the Smith chart is used to display measurement results (e.g. on a Vector Network Analyser) or simulation results (from a circuit or EM simulator). Nowadays, computerized Smith chart tools are available as freeware or shareware software (e.g. [1]) that now supersede paper and pencil work.

In the following examples we will study generic transmission line problems on a Smith chart to get a basic understanding of how to work with the diagram.

Example 3.5 *Quarter-Wave Transformer*
We consider the $\lambda/4$-transformer from Example 3.4. A load impedance $Z_{A1} = 200\,\Omega$ was matched to an input impedance $Z_{in1} = 50\,\Omega$ by a $\lambda/4$-line with a characteristic impedance of $Z_{01} = \sqrt{50\,\Omega \cdot 200\,\Omega} = 100\,\Omega$. Now we will display this transformation using the Smith chart.

As a first step we normalize the load impedance $Z_{A1} = 200\,\Omega$ to the characteristic impedance of the line $Z_{01} = 100\,\Omega$. The normalized load impedance is then $z_{A1} = Z_{A1}/Z_{01} = 2$. We plot the normalized impedance as point (1) in the Smith chart in Figure 3.14. From Equation 3.134 we know that a $\lambda/4$-line corresponds to a clockwise rotation of $180°$. This gives us point (2), that is a normalized impedance of $z_{in1} = 0.5$. In order to calculate the input impedance we renormalize this value and get $Z_{in1} = z_{in1} \cdot Z_{01} = 0.5 \cdot 100\,\Omega = 50\,\Omega$.

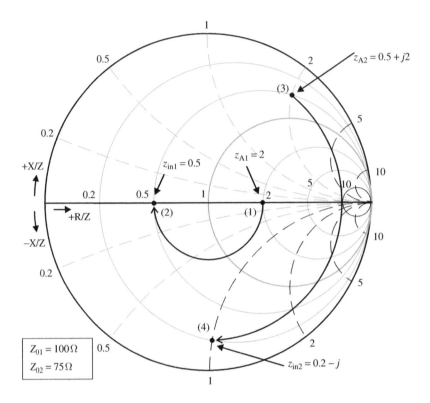

Figure 3.14   Solutions for Examples 3.5 and 3.6.

Example 3.6 *General Impedance Transformation by a Transmission Line*

Let us consider a transmission line with a characteristic impedance of $Z_{02} = 75\,\Omega$ and an electrical length of $\ell_t/\lambda = 0.194$. The load impedance is $Z_{A2} = (37.5 + j150)\Omega$. We search for the input impedance Z_{in2}.

In order to apply the Smith chart, we normalize the load impedance: $z_{A2} = Z_{A2}/Z_{02} = 0.5 + j2$ and plot point (3) on the diagram (see Figure 3.14). According to Equation 3.133 the angle for clockwise rotation is $|\varphi| = 4\pi\ell_t/\lambda = 2.4378\ldots \approx 139.68°$. This gives us point (4), that is a normalized input impedance of $z_{in2} = 0.2 - j$. We denormalize z_{in2} and get $Z_{in2} = z_{in2} \cdot Z_{02} = (0.2 - j) \cdot 75\,\Omega = (15 - j75)\Omega$.

Example 3.7 *Impedance Matching with Line and Series Reactance*

As a last example we will match a load $Z_A = (25 - j25)\Omega$ to an input impedance of $Z_{in} = 50\,\Omega$. The matching circuit consists of a $50\,\Omega$-line and a series reactance X_S. We will determine the necessary length of the line ℓ_t and the value of the reactance. The frequency of operation is $f = 1\,GHz$. The line is filled with PTFE ($\varepsilon_r = 2.1$).

First, we normalize the load impedance with respect to the characteristic impedance $z_A = 0.5 - j0.5$ and plot it on Figure 3.15 as point (1). In order to match the input impedance we have to move this point to the centre point ($r = 0$) of the diagram. There are two possible ways to proceed:

- *First solution: We rotate point (1) in a clockwise direction around the centre point until it intersects with the unity (normalized) resistance circle at point (2). From the angle of rotation of $\varphi = 180°$ we determine an electrical line length of $\lambda/4$ or a physical line length of $\ell_t = c_0/(4f\sqrt{\varepsilon_r}) \approx 5.18$ cm. At point (2) our normalized impedance is $z = 1 + j$. In order to reach the matching point (4) we need a series impedance of $z_S = -j$, or denormalized $Z_S = -j50\,\Omega$. A negative reactance can be realized with a capacitor: $Z_S = 1/(j\omega C) = -j(1/\omega C)$. So, the series reactance is a capacitor with a capacitance of $C = 1/(50\,\Omega\,2\pi f) \approx 3.2$ pF.*
- *Alternative solution: We rotate point (1) in a clockwise direction around the centre point until it intersects with the unity (normalized) resistance circle at point (3). The angle of rotation is now $\varphi = 307°$. So we calculate an electrical line length of 0.426λ or a physical line length of $\ell_t = 0.426c_0/(f\sqrt{\varepsilon_r}) \approx 8.8$ cm. At point (3) our normalized impedance is $z = 1 - j$. In order to reach the matching point (4) we need a series impedance of $z_S = +j$, or denormalized $Z_S = +j50\,\Omega$. A positive reactance can be realized with an inductor: $Z_S = j\omega L$. So, the series reactance is an inductor with an inductance of $L = 50\,\Omega/(2\pi f) \approx 8$ nH.*

We will take a closer look at matching circuits in Section 6.3.

3.2 Transient Signals on Transmission Lines

In Section 3.1 we considered sinusoidal signals $u(z, t) = \text{Re}\left\{Ue^{j\omega t}\right\}$. These continuous wave (CW) signals represent the steady state. Continuous wave signals are monofrequent and described by their complex amplitude U. In the case of *periodic* but non-sinusoidal signals we can develop the periodic signal in a series of sinusoidal signals (*Fourier series*). If the transient signal is *non-periodic* we apply the Fourier transform instead of the Fourier series. In the case of a linear system we can consider the individual sinusoidal components and use the principle of superposition to calculate the resulting signal.

An important class of signals are *digital base band signals* that represent symbols by rapidly switching between two logical states, for example 5 V for a logical '1' and 0 V for a logical '0'. In order to understand the implications of reflection and propagation directly in the time domain, we can benefit from the results in Section 3.1 where we defined transmission line parameters and the reflection coefficient. The fundamental aspects can be derived by looking at step and rectangular pulse functions.

3.2.1 Step Function

Let us first look at the step function $s(t)$

$$\boxed{s(t) = \begin{cases} 0 & \text{for } t < 0 \\ 1 & \text{for } t \geq 0 \end{cases}} \tag{3.145}$$

In a graphical representation we use a vertical line to indicate the voltage step. In praxis such an immediate transition is not possible. However, it is a reasonable approximation in basic considerations.[6]

---

[6] We look at more realistic trapezoid signals in Section 3.3.

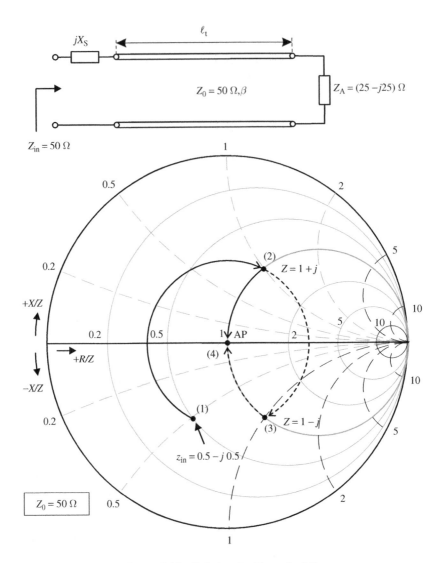

Figure 3.15 Solution for Example 3.7.

3.2.1.1 Matched Source and Load

We consider a source with internal resistance R_I, a loss-less transmission line with characteristic impedance Z_0 and a load resistance R_A. The generator voltage is given by a step function $u_G = U_G s(t)$. We start with the *ideal* case of matched load and generator $R_I = Z_0 = R_A$ (see Figure 3.17a).

We are interested in the voltages at the beginning and end of the line $u_{in}(t)$ and $u_A(t)$, respectively. For $t < 0$ all voltages are zero. At $t = 0$ the generator voltage jumps to u_0 and remains on that value forever (see Figure 3.17b).

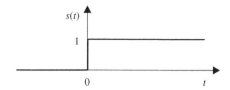

Figure 3.16   Step function $s(t)$.

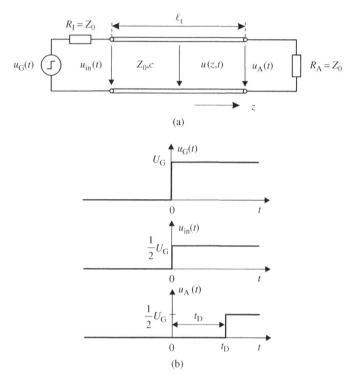

Figure 3.17   Source with step function $u_G(t) = U_G s(t)$ (Matched source and load ($R_I = Z_0$ $= R_A$)).

The input impedance at $t = 0$ equals the characteristic impedance of the line $Z_{in} = Z_0$ independent of the load resistance R_A due to the finite propagation speed of electro-magnetic waves. In other words, at $t = 0$ the generator starts a forward propagating wave and the ratio of voltage and current of that wave equals the characteristic line impedance (see Section 3.1.3).

In Section 3.1 we looked at the steady state case (sinusoidal signals). Due to the superposition of forward and backward travelling waves the impedance – in general – depends on the load impedance.

We apply the voltage divider rule to calculate the voltage at the input terminals u_{in} for $t = 0$ as

$$u_{in}(0) = \frac{Z_0}{R_I + Z_0} u_G(0) = \frac{1}{2} u_G(0) = \frac{1}{2} U_G \tag{3.146}$$

The output voltage $u_A(t)$ remains zero until the signal has propagated to the end of the line. We calculate the propagation time (or delay time) as

$$t_D = \frac{\ell_t}{c} = \frac{\ell_t}{c_0} \sqrt{\varepsilon_r} \tag{3.147}$$

where c is the speed of the wave on the line and ℓ_t is the length of the line. On a practical transmission line the speed of light may be reduced due to dielectric material (relative permittivity $\varepsilon_r \geq 1$). When the step has arrived at the end of the line *no reflection* occurs since load and line are matched $R_A = Z_0$.

We conclude that with matched load and source ($R_A = Z_0 = R_I$) the only effect of the line is *propagation delay* due to *finite speed* of propagation.

3.2.1.2   Matched Source and Resistive Load

Next we consider a mismatched load ($R_A \neq Z_0$) in Figure 3.18a.

From Section 3.1 we know that a signal that arrives at the end of the line is reflected back. The reflection coefficient that relates the reflected and forward propagating step is given by

$$r_A = \frac{R_A - Z_0}{R_A + Z_0} \tag{3.148}$$

For $R_A > Z_0$ the reflection coefficient is positive; for $R_A < Z_0$ the reflection coefficient is negative.

At the end of the line the forward propagating and reflected wave add up to

$$u_A(t_D) = \frac{1}{2} U_G \left(1 + r_A\right) \tag{3.149}$$

Figure 3.18b shows the time plots. The reflected step arrives at $t = 2t_D$ at the input terminals. Since the generator is matched to the line ($R_I = Z_0$) no further reflection occurs and the voltage is

$$u_{in}(2t_D) = \frac{1}{2} U_G \left(1 + r_A\right) = \frac{1}{2} U_G \left(1 + \frac{R_A - Z_0}{R_A + Z_0}\right)$$

$$= \frac{1}{2} U_G \left(\frac{2R_A}{R_A + Z_0}\right) = U_G \frac{R_A}{R_I + R_A} \tag{3.150}$$

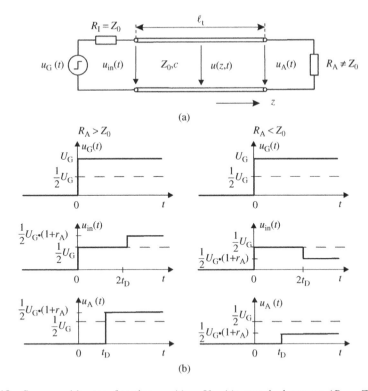

(a)

(b)

Figure 3.18 Source with step function $u_G(t) = U_G s(t)$: matched source ($R_I = Z_0$) and mismatched resistive load ($R_A \neq Z_0$).

Interestingly, this result equals the DC-case (time-invariant current). From the voltage divider rule we get

$$u_{in}(t \to \infty) = \frac{R_A}{R_I + R_A} U_G \qquad (3.151)$$

Example 3.8 *Short-Circuit and Open-Circuit Line Termination*

We will look now at two special cases: short-circuit termination ($R_A = 0$) and open-circuit termination ($R_A \to \infty$).

Let us start with the short-circuit load $R_A = 0$. From Equation 3.148 we calculate the reflection coefficient $r_{A,SC} = -1$. So with Equation 3.149 we see that the voltage at the end of the line $u_A(t)$ is zero for any time. This is the expected behaviour for a short circuit. The input voltage $u_{in}(t)$ steps to $U_G/2$ (voltage divider rule) at $t = 0$ and falls to 0 V again after the reflected step arrives at the input terminals at $t = 2t_D$ (see Figure 3.19a).

Next we consider the open circuit $R_A \to \infty$. From Equation 3.148 we calculate the reflection coefficient $r_{A,OC} = +1$. So with Equation 3.149 we see that the voltage at the end of the line $u_A(t)$ steps to U_G at $t = t_D$. The input voltage $u_{in}(t)$ steps to $U_G/2$ (voltage divider rule) at $t = 0$ and to U_G at $t = 2t_D$ (see Figure 3.19b).

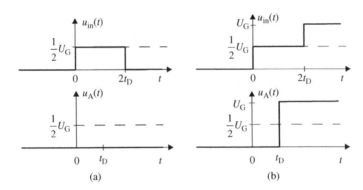

Figure 3.19 Source with step function $u_G(t) = U_G s(t)$: matched source ($R_I = Z_0$) and (a) short-circuit and (b) open-circuit termination (see Example 3.8).

3.2.1.3 Matched Source and Reactive Load

Now we will terminate the transmission line with a reactive element (capacitor or inductor). At the input terminals the line is still matched to the source ($R_I = Z_0$). The circuit is shown in Figure 3.20a.

Capacitive Load
First, we look at a capacitive load. The time plots of the voltages are shown in Figure 3.20b. At $t = 0$ the voltage source steps to its maximum value U_G. According to the voltage divider rule the input voltage steps to $U_G/2$ as in our previous discussions. It takes the time interval t_D until the signal change arrives at the end of the line. The capacitor at the end of the line is *uncharged* ($U_C = 0$) since the circuit has been inactive for all times $t < 0$.

From basic circuit theory we know that the voltage across a capacitor *cannot change instantaneously*. Hence, the voltage is a continuous function of time [2]. At $t = t_D$ the voltage step arrives at the capacitor. The time course of the voltage across the capacitor follows an exponential function

$$u_A(t) = \begin{cases} 0 & \text{for } t < t_D \\ U_G\left(1 - e^{-\frac{t-t_D}{\tau_{RC}}}\right) & \text{for } t \geq t_D \end{cases} \tag{3.152}$$

where the rate of change is controlled by the time constant τ_{RC}

$$\tau_{RC} = Z_0 C \tag{3.153}$$

Figure 3.20b shows the time plots. The time needed to charge the capacitor increases with capacitance C and characteristic impedance Z_0. For $t \to \infty$ the voltage across the capacitor $u_A(t \to \infty)$ approaches U_G.

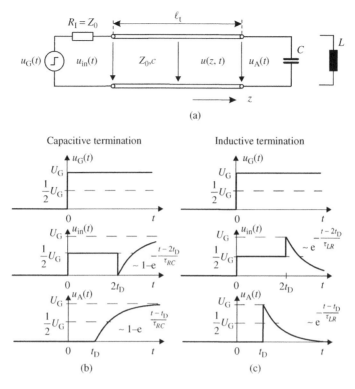

(a)

(b)

(c)

Figure 3.20 Source with step function $u_0(t) = U_0 s(t)$: (a) schematic with matched source ($R_I = Z_0$) and reactive load. Time plots for (b) capacitive and (c) inductive termination.

At the input terminals we observe $u_{in}(t) = U_G/2$ in the time interval $[0 \ldots 2t_D]$. For $t \geq 2t_D$ we observe an increase due to an exponential function.

$$
u_{in}(t) = \begin{cases} 0 & \text{for } t < 0 \\[2mm] \dfrac{1}{2}U_G & \text{for } 0 \leq t < 2t_D \\[3mm] U_G \left(1 - e^{-\frac{t-2t_D}{\tau_{RC}}}\right) & \text{for } t \geq 2t_D \end{cases} \tag{3.154}
$$

If we compare the voltage across a capacitor with the short- and open-circuit termination in Example 3.8 we see the following behaviour: when the step arrives at the end of the line the capacitor behaves as a short circuit ($u_A(t_D) = 0$). For $t \rightarrow \infty$ the capacitance behaves like an open circuit ($u_A(t \rightarrow \infty) = U_G$).

Inductive Load

Next we consider an inductive termination (see Figure 3.20a). The time plots of the voltages are shown in Figure 3.20c.

From basic circuit theory we know that the current through an inductor *cannot change instantaneously*. At $t = t_D$ the voltage step arrives at the end of the line. Since the current through the inductor is a continuous function of time there is no current at that instant in time ($t = t_D$) through the inductor. The inductor behaves like an open circuit. Hence, $u_A(t_D) = U_G$. For $t > t_D$ the current through the inductor rises due to an exponential function and the voltage decreases. For $t \to \infty$ the voltage is zero. The voltage at the end of the line is given by

$$u_A(t) = \begin{cases} 0 & \text{for } t < t_D \\ U_G e^{-\dfrac{t - t_D}{\tau_{LR}}} & \text{for } t \geq t_D \end{cases} \tag{3.155}$$

where τ_{LR} is the time constant

$$\tau_{LR} = \frac{L}{Z_0} \tag{3.156}$$

The voltage at the input terminals is given by

$$u_{in}(t) = \begin{cases} 0 & \text{for } t < 0 \\ \dfrac{1}{2} U_G & \text{for } 0 \leq t < 2t_D \\ U_G e^{-\dfrac{t - 2t_D}{\tau_{LR}}} & \text{for } t \geq 2t_D \end{cases} \tag{3.157}$$

If we compare the voltage across an inductor with the short- and open-circuit termination in Example 3.8 we see the following behaviour: when the step arrives at the end of the line the inductor behaves as an open circuit ($u_A(t_D) = U_G$). For $t \to \infty$ the inductor behaves like a short circuit ($u_A(t \to \infty) = 0$).

3.2.1.4   Mismatched Source and Load

In our previous discussion there was only *one* reflection at the end of the line due to a mismatched load ($R_A \neq Z_0$). Now we consider a mismatched source too ($R_I \neq Z_0$). The circuit is shown in Figure 3.21a. This leads to additional reflections of backwards propagating waves at the beginning of the transmission line. We will limit our investigation to an ohmic internal resistance R_I. The reflection coefficient for backwards propagating waves at the input terminals then becomes

$$r_I = \frac{R_I - Z_0}{R_I + Z_0} \tag{3.158}$$

Since there are now reflections on *both* ends of the line we expect waves to propagate back and forth on the line. In order to keep track of these multiple reflections we introduce the

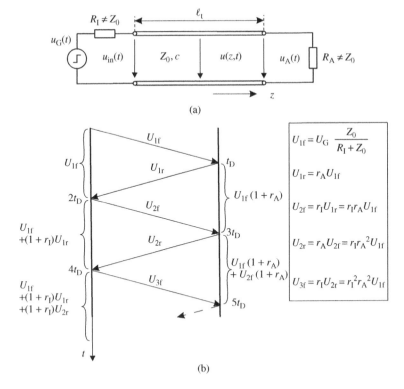

Figure 3.21 (a) Source with step function $u_G(t) = U_G s(t)$: mismatched source ($R_I \neq Z_0$) and mismatched load ($R_A \neq Z_0$). (b) Bounce diagram to track reflections.

bounce diagram or *lattice diagram* in Figure 3.21b. The time axis is vertically oriented. U_{1f} is the amplitude of the forward propagating step function. We can calculate it from the voltage divider rule as

$$U_{1f} = U_G \frac{Z_0}{R_I + Z_0} \tag{3.159}$$

At $t = t_D$ the step arrives at the end of the line and is reflected with a coefficient r_A (see Equation 3.148). At the end of the line the forward propagating voltage step U_{1f} and the reflected signal

$$U_{1r} = r_A U_{1f} \tag{3.160}$$

superimpose. Therefore, the voltage at the end of the line is now

$$u_A(t_D) = U_{1f}(1 + r_A) \tag{3.161}$$

At $t = 2t_D$ the reflected signal is back at the input terminal. It superimposes with the initial step. In the time interval $[2t_D \ldots 4t_D]$ we see a voltage of $U_{1f} + (1 + r_I)U_{1r}$ and so forth. In the bounce diagram in Figure 3.21b we see how the signal contributions add up over time.

For $t \to \infty$ the voltage converges to the steady state (DC) value that can be calculated from the voltage divider rule as

$$U_\infty = U_{in}(t \to \infty) = U_A(t \to \infty) = U_G \frac{R_A}{R_1 + R_A} \qquad (3.162)$$

Let us look at an example to illustrate the time course of the voltage at the input terminals and the load.

Example 3.9 *Propagation of Signals on a Line with Mismatched Source and Load*
We consider a transmission line with a characteristic impedance of $Z_0 = 50\,\Omega$ and a length of $\ell_t = 0.2\,\text{m}$. The line is filled with dielectric material ($\varepsilon_r = 2.25$). The amplitude of the step function is $U_G = 1\,\text{V}$. The source has an internal resistance of $R_1 = 150\,\Omega$. The line is terminated by a load of $R_A = 200\,\Omega$.

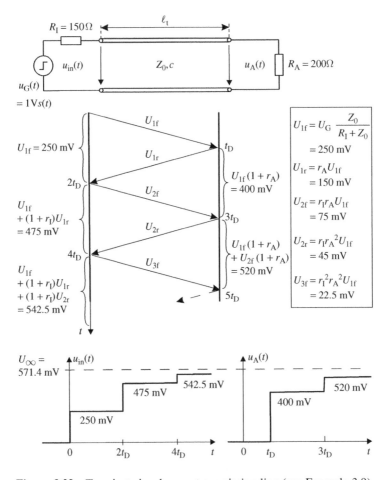

Figure 3.22 Transient signals on a transmission line (see Example 3.9).

With Equation 3.147 we have a propagation time of $t_D = 1$ ns from the input to the load. The reflection coefficients at the input and the load are $r_I = 0.5$ and $r_A = 0.6$, respectively. Figure 3.22 shows the voltage at the input terminals and at the end of the line. For $t \to \infty$ the voltage approaches a value of $U_\infty = U_G R_A/(R_I + R_A) = 571.4$ mV.

3.2.2 Rectangular Function

Next we consider rectangular pulses. We write the rectangular function as the difference of two-step functions (see Figure 3.23) by

$$\text{rect}(t) = s(t) - s(t - t_P) = \begin{cases} 1 & \text{for } 0 \leq t < t_P \\ 0 & \text{otherwise} \end{cases} \tag{3.163}$$

where t_P is the length of the pulse.

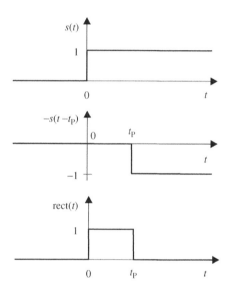

Figure 3.23 Rectangular (rect) function.

Since transmission lines are linear we can use the principle of superimposition and apply the bounce diagram (Figure 3.21) on both step functions. The situation becomes most simple when the pulses that run back and forth on the line do not overlap at the ends of the line, that is when the pulse duration t_P is shorter than the doubled propagation delay $t_P < 2t_D$.

Let us look at an example to illustrate the bouncing pulses on the line.

Example 3.10 *Rectangular Pulse on a Transmission Line with Mismatched Terminations*
We consider a transmission line with a characteristic impedance of $Z_0 = 50\,\Omega$ and a length of $\ell_t = 0.2$ m. The line is filled with dielectric material ($\varepsilon_r = 2.25$). The amplitude of the rectangular pulse is $U_G = 1$ V with a pulse duration of $t_P = 0.5$ ns. The source has an internal resistance of $R_I = 150\,\Omega$. The line is terminated by a load of $R_A = 200\,\Omega$.

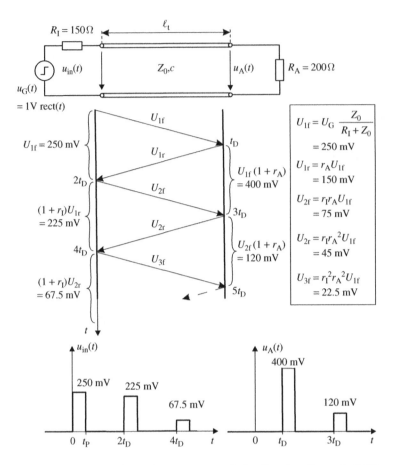

Figure 3.24   Rectangular pulses on a transmission line (see Example 3.10).

With Equation 3.147 we have a propagation time of $t_D = 1$ ns from the input to the load. The reflection coefficients at the input and the load are $r_I = 0.5$ and $r_A = 0.6$, respectively. Figure 3.24 shows the voltage at the input terminals and at the end of the line.

Mismatched terminations of transmission lines lead to unwanted reflections of pulses. The bouncing pulses may cause severe problems in practical applications since (parasitic) reflected pulses interfere with pulses that are sent at a later instant in time. Therefore, proper (matched) termination $R_I = Z_0 = R_A$ is required to overcome this problem.

3.3   Eye Diagram

In Section 3.2 we investigated the propagation of transient signals (voltage step and pulse) on transmission lines. In digital base band circuitry, information is transmitted by a series

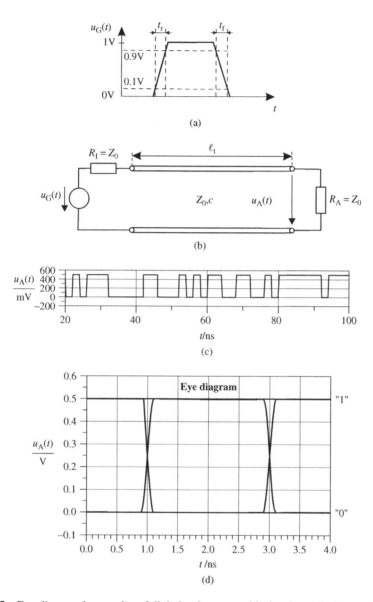

Figure 3.25 Eye diagram for a series of digital pulses on an ideal and matched transmission line.

of pulses. In practice a pulse has a more trapezoid than rectangular shape due to lower bandwidth.[7]

Commonly, the transition time from 10% to 90% of a signal amplitude change is referred to as *rise time* t_r and the transition time from 90% to 10% is called *fall time* t_f (see Figure 3.25a).

---

[7] More rapidly changing signals show less convenient frequency content (higher bandwidth). High frequency components can cause severe coupling and radiation problems.

If we transmit such a pulse signal over a matched, loss-less, non-dispersive transmission line the only effect we observe is time delay due to the finite speed of signals on the line (see Figure 3.25b).

$$u_A = \frac{1}{2} u_G(t - t_D) \tag{3.164}$$

where the factor $1/2$ is due to the voltage divider rule (at the input terminals we see half of the source voltage). We illustrate the effect by selecting a voltage source that generates a random sequence of logical 'ones' (higher voltage output) and logical 'zeros' (lower voltage output). Figure 3.25c shows as an example the unperturbed digital output voltage that equals (except for time delay and voltage drop) the source voltage.

A practical graphical tool to access the quality of a digital signal after it has experienced transmission over a (non-ideal) path is the *eye diagram*. The transient signal is segmented in time intervals of equal length (like a repetitively sampled time signal on an oscilloscope). The length of the interval is chosen according to the data rate. The centre of the diagram corresponds to the centre of the pulses. In Figure 3.25d the time interval corresponds to twice the pulse duration. Due to the ideal conditions of the transmission path (loss-less, matched, non-dispersive) the transitions from one logical state to the other are undistorted and follow the input signal. The centre of the diagram–between the cross-over points–looks like an eye, hence the name *eye diagram*.

To demonstrate the influence of a non-ideal transmission path we consider two lines with different characteristic impedances Z_{01} and Z_{02} (see Figure 3.26a). Both lines have mismatched terminations $Z_{01} \neq R_I$ and $Z_{02} \neq R_A$. This mismatch will lead to multiple reflections on the lines. Furthermore, we have included a capacitance, representing a capacitive load at the centre of the circuit. A capacitor is a frequency-dependent element. Figure 3.26b shows the distorted output signal and Figure 3.26c shows the corresponding eye diagram.

The reflections on the line overlay the pulses sent by the generator. The reflections can be positive or negative and tend to fan out the lines in the diagram. The capacitor limits the slope of the transitions (we remember the exponential time course from the previous section). These effects reduce the vertical and horizontal opening of the eye. In practical designs we have to ensure a defined minimum *opening of the eye* so that the receiver can distinguish reliably between the two logical states.

3.4 Summary

In Section 3.1 we provided a description of transmission line effects for harmonic signals. On a transmission line, voltage and current waves can propagate in opposite directions. These waves determine the electric behaviour at the input and load terminals. In order to describe a transmission line as a circuit component we need the three transmission line parameters: length of line ℓ_t, characteristic impedance Z_0 and propagation constant γ. The characteristic impedance Z_0 is defined as a ratio between voltage and current on a forward propagating wave. In the case of a loss-less line the characteristic impedance is real and voltage and current are in-phase. For a low-loss line this is still a good approximation.

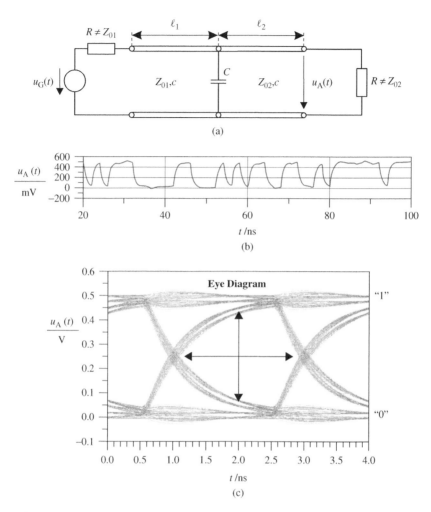

Figure 3.26 Eye diagram for a series of digital pulses on a mismatched and frequency dependent transmission path.

The propagation constant $\gamma = \alpha + j\beta$ is a complex valued with a real part (attenuation constant) and an imaginary part (phase constant). The attenuation constant α describes the exponential decay of a wave that travels along the line. The phase constant β depends on the wavelength and is – for a given frequency – a measure for the propagation velocity c.

In Section 3.2 and 3.3 we looked at transient signals on lines. In order to avoid reflections and bouncing signals transmission lines have to be matched at their terminals. The eye diagram represents a valuable tool to access the quality of digital signals that are propagated along a non-ideal transmission path. With the eye diagram we can visualize the effects of reflections and frequency-dependent transmission channels.

3.5 Problems

3.1 Determine the input impedance in Example 3.6 from Equation 3.78.

3.2 Reproduce the example shown in Figure 3.24 for $R_I = 15\,\Omega$ and $R_A = 10\,\Omega$. Plot $u_{in}(t)$ and $u_A(t)$ as functions of time. Compare your results with simulation results from a circuit simulator.

3.3 Match the load impedance $Z_A = (120 - j80)\,\Omega$ to an input impedance of $Z_{in} = 50\,\Omega$ using the Smith chart diagram. As a matching circuit consider a serial line and a short-circuited stub line with characteristic impedances of $Z_0 = 50\,\Omega$. Determine the lengths of the air-filled lines ($\varepsilon_r = 1$) for a frequency of $f = 1\,\text{GHz}$.

3.4 In order to determine the characteristic impedance of a loss-less transmission line we measure the capacitance of a short line segment as $C = 1.5\,\text{pF}$. The length of the line segment is $\ell = 12\,\text{mm}$. The line is filled with a homogeneous dielectric material ($\varepsilon_r = 1.44$).

1. Calculate the propagation velocity c of an electromagnetic wave on the line.
2. Determine the characteristic impedance Z_0 of the line.
3. Compute the inductance per unit length L'.

   In the following the frequency is give as $f = 250\,\text{MHz}$.

4. Specify the propagation constant γ.
5. Determine the input impedance Z_{in} of an open-circuited line ($\ell_t = 25\,\text{cm}$). (**Hint:** The light speed in vacuum is $c_0 = 3 \cdot 10^8$ m/s).

3.5 Consider an air-filled line ($\varepsilon_r = 1$) with a characteristic impedance of $Z_0 = 25\,\Omega$ (Figure 3.27). The line is teminated by a capacitance C and two resistors R_{AS} and R_{AP}. The generator constists of an internal resistance of $R_{in} = 25\,\Omega$ and a voltage

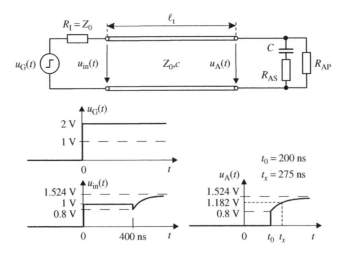

Figure 3.27 Circuit and time plots for Problem 3.5.

source with a maximum of $U_G = 2$ V. Figure 3.27 shows the input voltage $u_{in}(t)$ as a function of time. Determine

1. the length of the line ℓ_t.
2. the capacitance C and the resistances R_{AS} and R_{AP}.

References

1. Dellsperger F (2010) Smith Chart Tool http://www.fritz.dellsperger.net/downloads.htm.
2. Glisson TH (2011) *Introduction to Circuit Analysis and Design*. Springer.

Further Reading

Bowick C (2008) *RF Circuit Design*. Newnes.
Gustrau F, Manteuffel D (2006) *EM Modeling of Antennas and RF Components for Wireless Communication Systems*. Springer.
Hagen, JB (2009) *Radio-Frequency Electronics: Circuits and Applications*. Cambridge University Press.
IMST GmbH (2010) *Empire: Users Guide*. IMST GmbH.
Kark K (2010) *Antennen und Strahlungsfelder*. Vieweg.
Ludwig R, Bogdanov G (2008) *RF Circuit Design: Theory and Applications*. Prentice Hall.
Meinke H, Gundlach FW (1992) *Taschenbuch der Hochfrequenztechnik*. Springer.
Pozar DM (2005) *Microwave Engineering*. John Wiley & Sons.
Zinke O, Brunswig H (2000) *Hochfrequenztechnik 1*. Springer

4

Transmission Lines and Waveguides

This chapter provides an overview of technically important types of transmission lines. We will look in detail at the coaxial line and derive the transmission line parameters (characteristic impedance and propagation constant) mathematically based on fundamental physical considerations. For all other transmission line types we will pull together the most important facts in an illustrative manner.

4.1 Overview

Transmission lines may be used in various ways and applications. First, transmission lines can carry signals from one circuit component (e.g. the output of an amplifier) to another circuit component (e.g. an antenna). Ideally, in this area of application the transmission line is loss-less and without reflection at both ends. In order to make practical lines low-loss, good conductors (e.g. copper) and low-loss dielectrics are used to construct the lines. Reflections at both ends of a line are reduced by matching the load and internal source resistance to the characteristic impedance of the line.

Another area of application for transmission lines is the design of passive circuits. Transmission line segments can be arranged to – for instance – build filters, power dividers, couplers and matching circuits. In the case of transmission line circuits, reflections are useful for obtaining the desired transfer behaviour.

Figure 4.1 shows some technically important types of transmission lines. Coaxial lines, two-wire lines and optical waveguides are used to bridge larger distances. Planar transmission lines (microstrip, striplines) are suitable for the design of passive circuits. Rectangular waveguides are used especially in high power applications (e.g. radar).

A *coaxial line* (Figure 4.1a) consists of concentric cylindrical conductors separated by a homogeneous low-loss dielectric material. This type of line is the standard line for high-frequency measurement and for the interconnection of laboratory equipment. The *enclosed construction* is an advantageous property: the electromagnetic fields exist only between the outer and inner conductor and in very thin superficial layers of the conductors (skin

RF and Microwave Engineering: Fundamentals of Wireless Communications, First Edition. Frank Gustrau.
© 2012 John Wiley & Sons, Ltd. Published 2012 by John Wiley & Sons, Ltd.

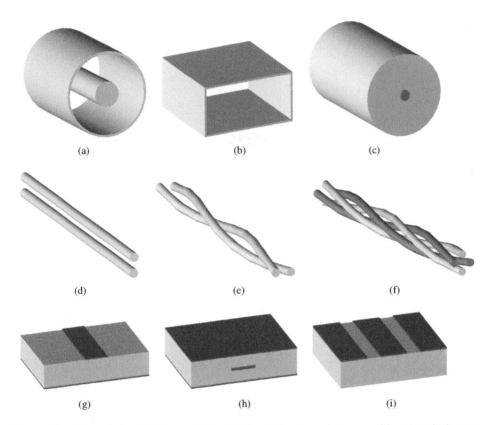

Figure 4.1 Transmission line types: (a) Coaxial line, (b) rectangular waveguide, (c) optical waveguide, (d) two-wire line, (e) twisted two-wire line, (f) star-quad line, (g) microstrip line, (h) stripline and (i) coplanar line.

effect). There is *no external field* in the volume beyond the outer conductor. On the other hand external fields cannot couple into the transmission line due to the shielding effect of the outer conductor. Due to the homogeneous dielectric material, the fundamental wave mode of a coaxial line is *transversal electromagnetic* (TEM). The electric and magnetic fields between the conductors have only transverse components with respect to the axis of the coaxial line.

Transmission lines in general are categorized as *balanced* or *unbalanced*. A transmission line is said to be *balanced* if it consists of two conductors of the same type that have equal impedances to a common ground. It is said to be unbalanced if the inner and outer conductor of a coaxial line are of different extensions and it is intuitively clear that the conductors have different capacitances to ground. Hence, the coaxial line is an *unbalanced* line.

A *two-wire line* (Figure 4.1d) consists of two straight, parallel, cylindrical conductors. This type of transmission line has an *open* structure: the electric and magnetic fields extend into the surrounding space. Therefore, the two-wire line is susceptible to external fields that couple into the line. The two-wire line is commonly driven by *differential signals* (see Section 4.9). Since the two conductors are of equal shape and size the two-wire line is a *balanced* line. The fundamental wave mode is *transversal electromagnetic* (TEM).

From the basic concept of a balanced two-wire line two practical types of cable are developed: *twisted-pair* cable and *star-quad* cable. A twisted-pair cable consists of a bundle of two-wire lines (shielded multiple twisted pair). Each two-wire line in that bundle is stranded (twisted) as shown in the Figure 4.1e. Twisting the line can significantly reduce the interference from external fields. Each pair has a different length of twist to minimize coupling between adjacent pairs. In a star-quad configuration (see Figure 4.1f) two pairs are stranded to form a cable. Opposite conductors form a two-wire line. The pairs are arranged orthogonally so as to minimize coupling. In practical cables the conductors shown in Figure 4.1d−f are surrounded and separated by dielectric isolating material to achieve mechanical stability and avoid shunt faults. (The isolation material is omitted in the graphical representations.)

In planar circuits the most common types of transmission lines are *microstrip* lines, *striplines* (or *triplate* transmission lines) and *coplanar* lines. They are used to realize frequency dependent passive circuits like filters or couplers.

The *microstrip* line in Figure 4.1g consists of a dielectric substrate layer with conductors on each side. On the bottom side we find a broad ground plane metallization. On top of the substrate there is a metallic strip. The electromagnetic wave that travels along a microstrip line exists in the substrate materials as well as in air above the metallic strip. We will see in Section 4.3 that this leads to a *quasi-TEM* wave.[1] The microstrip line is an open and unbalanced line. It can be miniaturized to be used in microwave integrated circuits. The open construction is ideal for hybrid circuits that combine transmission lines and lumped elements (like SMD-components or ICs). Two coupled microstrip lines can be used to form a balanced line (see Section 4.9).

The *stripline* in Figure 4.1h is an unbalanced transmission line, where the centre conductor lies between two ground planes. The electromagnetic wave that propagates on such a line is restricted to the area inside the substrate. Hence, the stripline is an enclosed line. Striplines can be found in multilayer boards when traces in deeper layers run between ground planes. The fundamental wave mode is *transversal electromagnetic* (TEM) due to the homogeneous dielectric material used.

In a *coplanar* line (Figure 4.1i) the metallic strip runs between the ground planes on the *same* side of the substrate. Strip and ground are located in the *same* plane, hence, the name *co*planar. An electromagnetic wave would be present in air and substrate, so the wave type is *quasi-TEM*.

Until now we have discussed *two-conductor lines* that carry TEM or quasi-TEM waves as fundamental wave modes. In particular, two conductor lines are able to carry steady (DC) current. A *rectangular waveguide* deviates from this concept, since it has only *one* conductor. Figure 4.1b shows the metallic boundary of a rectangular waveguide. It is intuitively clear that such a structure cannot carry a steady (DC) current. A rectangular waveguide can guide electromagnetic waves only for frequencies above a so-called *cut-off frequency*. Above the cut-off frequency electromagnetic waves are reflected between the side walls and produce a forward propagating wave [1]. Rectangular waveguides are capable of carrying comparatively high power and show low-loss.

---

[1] Unlike a TEM wave that has only transversal field components a quasi-TEM wave has (small) longitudinal field components.

Finally, Figure 4.1c shows an optical waveguide that consists only of non-conducting, dielectric material. The fundamental design has a thin cylindrically dielectric core and thicker cylindrically dielectric cladding, where the core has a slightly higher permittivity than the outer material. Electromagnetic waves in the core are reflected at the core/cladding interface due to total internal reflection (see Section 2.5.3). Optical waveguides have low attenuation, which is what makes them suitable for bridging great distances in telecommunications. Due to their lack of conducting materials, optical waveguides are unsusceptible to interference from external electromagnetic fields.

4.2   Coaxial Line

A coaxial line consists of two concentric cylindrical conductors (see Figure 4.2), where R_i and R_o are the inner and outer radius, respectively. The two conductors are separated by a homogeneous dielectric material with a relative permittivity of ε_r. The voltage U between the inner and outer conductor corresponds to a radial electric field strength E_R and the current I in the conductors produces a circulating magnetic field strength H_φ.

In the following sections we will derive the transmission line parameters of a coaxial line. The cylindrical symmetry of the problem allows a mathematically quite simple treatment using cylindrical coordinates. In our considerations we employ our knowledge of electromagnetic theory from Chapter 2 and the definitions from the transmission line theory in Chapter 3.

4.2.1   Specific Inductance and Characteristic Impedance

Let us consider a loss-less transmission line. We can calculate the characteristic impedance Z_0 from the inductance per unit length L', the material parameters ε_r and μ_r and the light speed in vacuum c_0 (see Equation 3.69).

$$Z_0 = \sqrt{\frac{L'}{C'}} = \frac{c_0 L'}{\sqrt{\varepsilon_r \mu_r}} \tag{4.1}$$

Technical transmission lines commonly use dielectric, non-magnetic materials. Hence, we set $\mu_r = 1$. Figure 4.3a shows the geometry and the spherical coordinate system for the

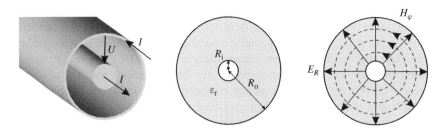

Figure 4.2   Geometry of a coaxial line and electric and magnetic field distribution of a TEM wave.

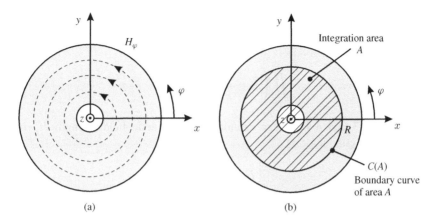

Figure 4.3 Distribution of magnetic field strength and integration area of Equation 4.2.

following calculations. The z-axis corresponds to the axis of the coaxial line. We start by applying Ampere's law in the integral form (see Equation 2.51 and Problem 2.2).

$$\oint_{C(A)} \vec{H} \cdot d\vec{s} = \iint_{A} \left(\vec{J} + \frac{\partial \vec{D}}{\partial t} \right) \cdot d\vec{A} \tag{4.2}$$

We are interested in results for high frequencies. From our discussion of the skin effect in Chapter 2 we know that the current flows only superficially in good conductors. Hence, the conductors are free of electromagnetic fields. So, we calculate the magnetic field only in the region between the conductors ($R_i \leq R \leq R_0$). Let us further assume that the current I in the inner conductor flows in a positive z-direction.

Based on our understanding of electromagnetic field theory we expect that the magnetic field runs in circles around the inner conductor (Figure 4.3a). Hence, there is only a φ-component of the magnetic field strength ($H_\varphi \neq 0$). The radial and z-component of the magnetic field strength are zero ($H_z = H_R = 0$). Furthermore, we assume the magnetic field to be constant on concentric circles around the inner conductor ($H_\varphi(R = \text{const.}) = \text{const.}$). Hence, the integrals in Equation 4.2 are most easy to solve if we choose concentric circles around the z-axis for our integration as shown in Figure 4.3b.

Let us begin with the left-hand side of Equation 4.2. We integrate over the closed contour line $C(A)$ of area A. The boundary curve is a concentric circle around the z-axis. We conveniently evaluate the integral in cylindrical coordinates. The line element $d\vec{s}$ and the magnetic field strength $\vec{H}$ point in the same direction (φ-direction), so the scalar product becomes the product of the absolute values. The differential line element along the circular boundary curve ds_φ can be expressed in cylindrical coordinates as $ds_\varphi = Rd\varphi$ (see Appendix A.1). Furthermore, we can take the radius R and the magnetic field H_φ outside the integral, since they are constant along the circular boundary curve $C(A)$.

$$\oint_{C(A)} \vec{H} \cdot d\vec{s} = \int_0^{2\pi} H_\varphi ds_\varphi = \int_0^{2\pi} H_\varphi Rd\varphi = H_\varphi R \int_0^{2\pi} d\varphi = 2\pi R H_\varphi \tag{4.3}$$

On the right-hand side of Equation 4.2 the total current density $\vec{J} + \partial \vec{D}/\partial t$ is integrated over the circle area A. The total current that flows through area A is the current I on the inner conductor.

$$\iint_A \left(\vec{J} + \frac{\partial \vec{D}}{\partial t} \right) \cdot \mathrm{d}\vec{A} = I \tag{4.4}$$

From Equations 4.3 and 4.4 we can calculate the magnetic field strength H_φ between the inner and outer conductor as

$$\boxed{H_\varphi = \frac{I}{2\pi R}} \tag{4.5}$$

Next, we determine the inductance L of a line segment with a length of ℓ. According to Equation 2.32 the inductance is associated with the magnetic field energy W_m stored in that line segment.

$$W_m = \frac{1}{2} \iiint_V \vec{B} \cdot \vec{H} \, \mathrm{d}v = \frac{1}{2} L I^2 \qquad \text{where} \qquad \vec{B} = \mu \vec{H} = \mu_0 \mu_r \vec{H} \tag{4.6}$$

We will calculate the magnetic energy in cylindrical coordinates (see Appendix A.1) as

$$\begin{aligned}
W_m &= \frac{1}{2} \int_{R_i}^{R_o} \int_0^{2\pi} \int_0^{\ell} \mu_0 \left(\frac{I}{2\pi R} \right)^2 \mathrm{d}z \, \mathrm{d}\varphi \, R\mathrm{d}R \\[2mm]
&= \frac{1}{2} \mu_0 \left(\frac{I}{2\pi} \right)^2 \underbrace{\int_0^{\ell} \mathrm{d}z}_{\ell} \underbrace{\int_0^{2\pi} \mathrm{d}\varphi}_{2\pi} \underbrace{\int_{R_i}^{R_o} \frac{1}{R^2} R\mathrm{d}R}_{\ln R |_{R_i}^{R_o} = \ln \left(\frac{R_o}{R_i} \right)}
\end{aligned} \tag{4.7}$$

Hence, the magnetic energy is

$$W_m = \frac{1}{2} \mu_0 \ell \frac{I^2}{2\pi} \ln \left(\frac{R_o}{R_i} \right) = \frac{1}{2} L I^2 \tag{4.8}$$

In order to determine the characteristic impedance Z_0 (Equation 4.1) we need the inductance per unit length which is given by

$$\boxed{L' = \frac{L}{\ell} = \frac{\mu_0}{2\pi} \ln \left(\frac{R_o}{R_i} \right)} \tag{4.9}$$

From Equation 4.1 we find the characteristic impedance of the line as

$$Z_0 = \sqrt{\frac{L'}{C'}} = \frac{c_0 L'}{\sqrt{\varepsilon_r}} = \underbrace{\frac{1}{\sqrt{\varepsilon_0 \mu_0}}}_{c_0} \frac{1}{\sqrt{\varepsilon_r}} \frac{\mu_0}{2\pi} \ln \left(\frac{R_o}{R_i} \right) = \underbrace{\sqrt{\frac{\mu_0}{\varepsilon_0}}}_{Z_{F0}} \frac{1}{\sqrt{\varepsilon_r}} \frac{1}{2\pi} \ln \left(\frac{R_o}{R_i} \right) \tag{4.10}$$

where Z_{F0} is the characteristic impedance of free space

$$Z_{F0} = \sqrt{\frac{\mu_0}{\varepsilon_0}} = 120\pi \; \Omega \approx 377 \; \Omega \tag{4.11}$$

So, the characteristic impedance of a coaxial line filled with dielectric material is given by

$$\boxed{Z_0 = \frac{60 \; \Omega}{\sqrt{\varepsilon_r}} \ln\left(\frac{R_o}{R_i}\right)} \qquad \text{(Characteristic impedance of a coaxial line)} \tag{4.12}$$

Example 4.1 *Characteristic Impedance of Coaxial Lines*
  An air-filled ($\varepsilon_r = 1$) coaxial line with a radius ratio of $R_o/R_i = 2.3$ has a characteristic impedance of $Z_0 = 50 \; \Omega$. If a coaxial line is filled with dielectric PTFE material ($\varepsilon_r = 2.1$) the relationship between the radii has to be $R_o/R_i = 3.35$.

4.2.2 Attenuation of Low-loss Transmission Lines

If a practical transmission line has a significant length, waves that propagate along the line exhibit a reduction in amplitude due to attenuation. According to Equation 3.90 we can calculate the attenuation constant as

$$\alpha \approx \alpha_{\text{diel}} + \alpha_{\text{met}} = \frac{G' Z_0}{2} + \frac{R'}{2 Z_0} \tag{4.13}$$

where R' represents conductor (metallic) losses and G' represents losses associated with the dielectric material.

4.2.2.1 Conductor Losses

Conductor losses are represented by the resistance per unit length R'. Due to the skin effect at higher frequencies the current flows in superficial regions of conductors. In order to calculate the resistance we assume that the current flows homogeneously in a layer with a thickness that equals the skin depth δ.

$$\delta = \sqrt{\frac{2}{\mu \sigma \omega}} \tag{4.14}$$

In order to calculate the resistance we treat the inner and outer conductor separately. The current flows in the inner conductor in one direction and comes back in the outer conductor. Hence, the resistance of the inner (metallic) conductor R_{mi} and the resistance of the outer (metallic) conductor R_{mo} form a series circuit and the total resistance R_{m} is the sum of the two.

$$R_{\text{m}} = R_{\text{mi}} + R_{\text{mo}} \tag{4.15}$$

The resistances of the inner and outer conductors can be calculated by the formula for RF resistance (see Equation 2.87).

$$R_{\text{mi}} = \frac{\ell}{\sigma A_i} \qquad \text{and} \qquad R_{\text{mo}} = \frac{\ell}{\sigma A_o} \tag{4.16}$$

where A_i and A_o are the areas that are assumed to carry a homogeneous current.

$$A_i = 2\pi R_i \delta \quad \text{and} \quad A_o = 2\pi R_o \delta \tag{4.17}$$

With this relation the resistance per unit length becomes

$$\boxed{R' = \frac{R_m}{\ell} = \frac{1}{2\pi\sigma\delta}\left(\frac{1}{R_i} + \frac{1}{R_o}\right) \sim \sqrt{f}} \qquad \text{(Resistance per unit length)} \tag{4.18}$$

Since the skin depth δ is a function of frequency, the resistance per unit length increases with the square root of frequency. Equation 4.18 is valid for homogeneous conducting material. In practical coaxial lines the use of braided outer conductors or foils can increase the resistance.

4.2.2.2 Dielectric Losses

Dielectric materials can absorb energy due to the oscillating microscopic dipoles (polarization losses, Section 2.1.1). The loss tangent $\tan\delta_\varepsilon$ indicates how lossy a material is. A real (lossy) capacitor is commonly represented by a parallel circuit of an ideal capacitance C and a conductance G. The losses of the capacitor are then described by the loss tangent of the capacitor.

$$\tan\delta_C = \frac{G}{\omega C} \tag{4.19}$$

If the volume between the electrodes of the capacitor is completely filled with a homogeneous dielectric material, the loss tangent of the capacitor equals the loss tangent of the material $\tan\delta_\varepsilon = \tan\delta_C$. Hence, we can link our transmission line circuit element G' with the capacitance per unit length C' by

$$\tan\delta_\varepsilon = \frac{G'}{\omega C'} \tag{4.20}$$

The conductance per unit length G' then becomes

$$\boxed{G' = \omega C' \tan\delta_\varepsilon \sim f} \qquad \text{(Conductance per unit length)} \tag{4.21}$$

The dielectric losses increase linearly with frequency, if we assume the loss tangent to be constant in a certain frequency range.

Example 4.2 *Electrical Characteristics of a Coaxial Line*

Let us consider a coaxial line with conductors made of solid copper (electrical conductivity $\sigma = 5.7 \cdot 10^7$ S/m) and calculate the most important transmission line parameters for a frequency of $f = 2$ GHz. The coaxial line is defined by following parameters: radius of inner conductor $R_i = 2.4$ mm, radius of outer conductor $R_o = 6.2$ mm, relative propagation velocity of a TEM wave travelling along the line $c = 0.88c_0$ (88% of vacuum light speed) and loss tangent of dielectric material $\tan\delta_\varepsilon = 5 \cdot 10^{-5}$.

First, we calculate the relative permittivity from the reduction of the propagation velocity as $\varepsilon_r = (c_0/c)^2 = (1/0.88)^2 = 1.291$. We see that the percentage factor 88% equals the *reciprocal value of the square-root of the relative permittivity* $1/\sqrt{\varepsilon_r} = 0.88$. *This is very convenient since the term* $1/\sqrt{\varepsilon_r}$ *occurs in several important equations. So we can directly use the percentage factor in our calculations.*

From Equation 4.12 we find the characteristic impedance $Z_0 = 50.1\,\Omega$. According to Equation 4.9 the inductance per unit length is $L' = 0.190\,\mu\text{H/m}$. By using Equation 4.1 the capacitance per unit length becomes $C' = 75.7\,\text{pF/m}$. In order to determine the attenuation coefficient we first calculate the skin depth by using Equation 4.14 as $\delta = 1.468\,\mu\text{m}$. Equation 4.13 and Equation 4.18 give us an attenuation coefficient of $\alpha_{met} = 0.01066\,1/\text{m} = 0.0926\,\text{dB/m}$ for the conductor losses. According to Equation 4.13 and Equation 4.21 the attenuation constant for the dielectric losses is $\alpha_{diel} = 0.001194\,1/\text{m} = 0.0104\,\text{dB/m}$. Hence, the attenuation coefficient of the coaxial line is $\alpha = \alpha_{met} + \alpha_{diel} = 10.3\,\text{dB/100}\,\text{m}$.

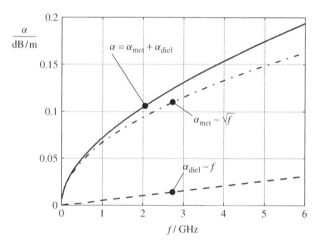

Figure 4.4 Frequency dependence of attenuation constant α of a low-loss coaxial line (see Example 4.2).

Figure 4.4 shows the results in the frequency range from 0 Hz to 6 GHz. Metallic losses are dominant. With increasing frequency the dielectric losses become more and more significant.

4.2.3 Technical Frequency Range

On a coaxial line TEM waves can propagate in the frequency range from DC to infinity. Above a critical frequency higher order modes may propagate. These modes differ from the TEM wave in propagation velocity (showing significant frequency dependence) and characteristic impedance. Propagation of higher order modes is generally unwanted since these modes lead to signal distortion. In order to have well-defined pure TEM mode wave propagation, the technical frequency range is limited by the higher order mode with the lowest cut-off frequency.

On a coaxial line the first non-TEM mode is the TE_{11} mode shown in Figure 4.5b. The TE_{11} mode has a distinct field pattern. Unlike the TEM mode the TE_{11} mode varies in

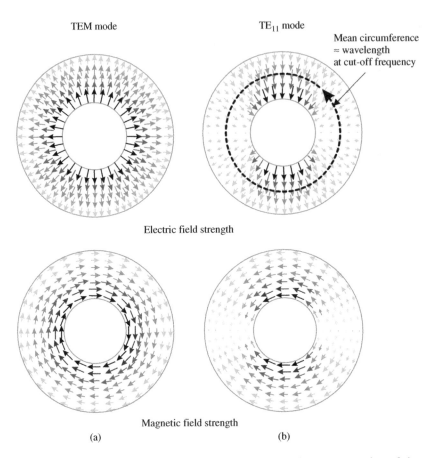

TEM mode

TE$_{11}$ mode

Mean circumference
≈ wavelength
at cut-off frequency

Electric field strength

Magnetic field strength

(a)   (b)

Figure 4.5 Distribution of electric and magnetic field strength in a cross-section of the coaxial line: (a) fundamental mode (TEM) and (b) higher order mode (TE$_{11}$).

the circumferential direction. It has two maxima (with opposite sign) and two zero points. This reminds us of a sine wave propagating in circumferential direction. Hence, the TE$_{11}$ mode may exist when the circumference U is on the order of a wavelength (or greater) $U \approx \lambda$. The mean circumference of our coaxial line is

$$U = 2\pi R = 2\pi \frac{R_\mathrm{o} + R_\mathrm{i}}{2} \approx \lambda = \frac{c}{f} = \frac{c_0}{f\sqrt{\varepsilon_\mathrm{r}}} \qquad (4.22)$$

From these considerations we estimate the cut-off frequency f_c of the coaxial line as

$$f_\mathrm{c} = \frac{c}{\pi\left(R_\mathrm{o} + R_\mathrm{i}\right)} = \frac{c_0}{\pi\left(R_\mathrm{o} + R_\mathrm{i}\right)\sqrt{\varepsilon_\mathrm{r}}} \qquad (4.23)$$

Coaxial lines should only be used below this cut-off frequency f_c. Similar formulas can be found in [2, 3].

Example 4.3 *Technical Frequency Range of a Coaxial Line*

A practical coaxial line for cellular communication systems is specified by the following parameters: inner radius $R_i = 8.85$ mm, outer radius $R_o = 21.5$ mm, propagation velocity of a TEM wave relative to vacuum light speed $c = 0.89\,c_0$. According to Equation 4.23 the cut-off frequency becomes $f_c \approx 2.8$ GHz.

4.2.4 Areas of Application

Coaxial lines are the standard lines in RF measurement. They are used to connect laboratory equipment like synthesizers, amplifiers, antennas and so on. Coaxial lines have several benefits.

- Coaxial lines are flexible, especially when the outer conductor is made of braided wires and the inner conductor is a stranded wire.
- TEM-wave propagation leads to low signal distortions.
- Low-loss coaxial line types may be used to bridge greater distances (on the order of ~100 m and more).
- The shielded construction reduces coupling to the electromagnetic environment (low interference, low radiation).

A drawback of the coaxial line is that – due to their closed (shielded) construction – coaxial lines can hardly be combined with lumped circuit elements (e.g. inductors, capacitors) to build circuits. Here open lines like microstrip lines, which will be discussed in the next section, are more advantageous.

4.3 Microstrip Line

4.3.1 Characteristic Impedance and Effective Permittivity

Figure 4.6 shows the cross-section of a *microstrip* line. A microstrip line consists of a dielectric (or semiconductor) substrate with a continuous ground plane on one side and a metallic strip on the other side. An electromagnetic wave that propagates on a microstrip line exists in two different media: air (above the metallic trace) and substrate (between strip and ground plane). At the air–substrate interface the fields have to meet electromagnetic boundary conditions. These boundary conditions lead to (small) longitudinal electric and magnetic field components. Since the longitudial components are small compared to the transversal field components the fundamental wave mode is referred to as *quasi-TEM* .

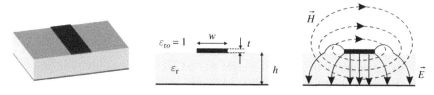

Figure 4.6   Geometry of a microstrip line and electric and magnetic field distribution of a quasi-TEM wave.

Table 4.1 Properties of typical substrate materials for planar circuits and microstrip lines

Material	Relative permittivity ε_r	Loss tangent $\tan \delta$	Typical thickness	Typical thickness of metallization
Polytetrafluoro-ethylene (PTFE)	2.1	0.0002	1.6 mm (63 mil)	35 µm
Glass reinforced epoxy laminate (FR4)	4.4...5	0.03	1.6 mm (63 mil)	35 µm
Rogers Duroid 6006	6.15	0.0027	0.635 mm (25 mil)	5 µm
Dupont 951	7.8	0.006	0.1 mm (4 mil)	5 µm
Alumina (Al$_2$O$_3$)	9.8	0.0001	0.635 mm (25 mil)	10 µm
Gallium arsenide (GaAs)	12.9	0.0004	0.1 mm (4 mil)	5 µm

Electromagnetic waves in air (with a relative permittivity of one) and in substrate (with a relative permittivity of ε_r) have different phase velocities. The resultant velocity c on a microstrip line is given somewhere in between these two velocities and can be expressed by the definition of an *effective relative permittivity*.

$$\varepsilon_{r,eff} = \left(\frac{c_0}{c}\right)^2 = \left(\frac{\lambda_0}{\lambda}\right)^2 \qquad \text{(Effective relative permittivity)} \qquad (4.24)$$

The effective relative permittivity is in the range $1 \leq \varepsilon_{r,eff} \leq \varepsilon_r$ and describes a virtual *homogeneous* material that has the same phase velocity as a quasi-TEM wave on the microstrip line.

Different commercial substrate materials are available for microstrip lines. Table 4.1 shows typical materials, electrical properties, substrate heights and metallization thicknesses [4–6]. Glass reinforced epoxy laminate (FR4) is a low-cost material. Due to relatively high losses it is only used at lower frequencies. Substrate heights are given in mm and mils.[2]

There are different manufacturing processes for the construction of planar circuits. *Monolithic Microwave Integrated Circuits* MMIC use semiconductor materials (like Gallium arsenide) to realize passive distributed structures (transmission lines, filters and couplers), passive concentrated components (capacitors, inductors) as well as active, non-linear circuits (transistors, diodes). This technique allows compact designs of microwave modules including power amplifiers, mixers and low-noise amplifiers for wireless communication systems. *Hybrid circuits* combine different techniques. Transmission lines on planar dielectric substrates interconnect concentrated elements (SMD-components), ICs and semiconductor modules that are attached to the surface of the substrate.

Although the geometry of a microstrip line looks quite simple, the transmission line parameters cannot be calculated analytically. Investigations show that the characteristic impedance Z_0 and the effective relative permittivity $\varepsilon_{r,eff}$ are primary functions of the ratio w/h (trace width w to substrate height h) and relative permittivity of the substrate ε_r. A more in-depth analysis also reveals additional influencing factors like frequency f and thickness of metallization t.

---

[2] 1 mil = 1/1000 inch = 0.0254 mm

In the literature (e.g. [7]) we find numerous approximate formulas and graphical representations. As an example we reproduce formulas for low frequencies from [7].

$$Z_0 = \frac{Z_{F0}}{\pi\sqrt{8(\varepsilon_r+1)}} \ln\left(1 + \frac{4h}{w'}\left[\frac{14+8/\varepsilon_r}{11}\cdot\frac{4h}{w'}\right.\right.$$
$$\left.\left.+\sqrt{\left(\frac{14+8/\varepsilon_r}{11}\cdot\frac{4h}{w'}\right)^2 + \frac{1+1/\varepsilon_r}{2}\pi^2}\right]\right) \tag{4.25}$$

where w' is the effective width

$$w' = w + \frac{t}{\pi}\ln\left(4e\left[\sqrt{\left(\frac{t}{h}\right)^2 + \left(\frac{1/\pi}{w/t+1.1}\right)^2}\right]^{-1}\right) \tag{4.26}$$

The effective relative permittivity can be approximated by

$$\varepsilon_{r,\text{eff}} = \frac{\varepsilon_r+1}{2} + \frac{\varepsilon_r-1}{2}\left[\left(1+\frac{12h}{w}\right)^{-\frac{1}{2}} + 0.04\left(1-\frac{w}{h}\right)^2\right] \quad \text{for} \quad w/h \le 1 \tag{4.27}$$

and

$$\varepsilon_{r,\text{eff}} = \frac{\varepsilon_r+1}{2} + \frac{\varepsilon_r-1}{2}\left(1+\frac{12h}{w}\right)^{-\frac{1}{2}} \quad \text{for} \quad w/h \ge 1 \tag{4.28}$$

Example 4.4 *Line Impedance and Effective Relative Permittivity*

Figure 4.7 shows plots of the characteristic impedance Z_0 and the effective relative permittivity $\varepsilon_{r,\text{eff}}$ as a function of w/h (ratio of trace width to substrate height) at low frequencies (1 GHz). The curves represent three different substrate materials: PTFE ($\varepsilon_r = 2.1$), FR4 ($\varepsilon_r = 4.4$) and alumina ($\varepsilon_r = 9.8$). From the figures we can find the

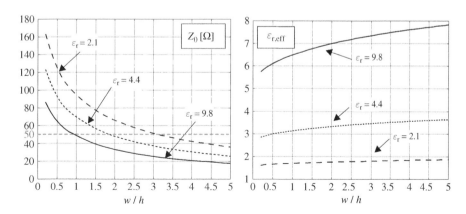

Figure 4.7 Characteristic line impedance Z_0 and effective relative permittivity $\varepsilon_{r,\text{eff}}$ as a function of w/h (w = width of trace; h = height of substrate).

following rule of thumb for a 50 Ω-line: on PTFE-substrate the trace is three times as wide as the substrate height (w/h ≈ 3), on FR4 the trace is twice as wide as the substrate height (w/h ≈ 2), on alumina the trace equals the substrate height (w/h ≈ 1).

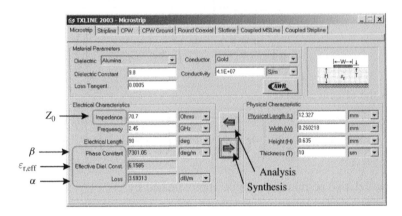

Figure 4.8 TX-Line software [8] for numerical calculation of transmission line parameters.

Unfortunately, Equations 4.25 to 4.28 cannot be solved directly for trace width from given characteristic impedance. Furthermore frequency effects are not included. More accurate values of the transmission line parameters of microstrip lines can be obtained by using commercial or freeware software tools (e.g. TX-Line [8]). Figure 4.8 shows – as an example – the user interface of a transmission line tool. With such a tool we can easily analyse and synthesize microstrip transmission lines as well as further commonly used transmission line types like striplines and coplanar lines. Some tools are capable of calculating coupled lines and some can even handle the effect of metallic side walls or ground planes above the microstrip structures, as this is important when circuits are put in metallic housings in order to minimize coupling effects to the electromagnetic environment.

Example 4.5 *Quarter-wave Transformer with Microstrip Line*

A microstrip line quarter-wave transformer is used to match a load impedance of $Z_A = 100\,\Omega$ to an input impedance of $Z_{in} = 50\,\Omega$ at a frequency of $f = 2.45\,\text{GHz}$. The substrate parameters are substrate height $h = 635\,\mu m$ and relative permittivity $\varepsilon_r = 9.8$. From transmission line theory we know that the following formula applies.

$$Z_{in}Z_A = Z_0^2 \tag{4.29}$$

Hence, the characteristic impedance of the microstrip line is

$$Z_0 = \sqrt{Z_{in}Z_A} = 70.7\,\Omega \tag{4.30}$$

By using the software tool TX-Line [8] we estimate the trace width as $w = 260\,\mu m$ for a characteristic impedance of $Z_0 = 70.7\,\Omega$. The effective relative permittivity is

$\varepsilon_{\text{r,eff}} = 6.16$. *Herewith we calculate the line length (quarter wavelength) as $\ell = c/ (4 f \sqrt{\varepsilon_{\text{r,eff}}}) = 12.3 \, \text{mm}$). By way of comparison we use Equations 4.25 to 4.28 to recalculate the transmission line parameters from our previously estimated geometric data. For the characteristic impedance we get $Z_0 = 71.7 \, \Omega$ and for the effective relative permittivity we calculate $\varepsilon_{\text{r,eff}} = 6.26$. The values from these simple formulas are in satisfactory agreement with our numerical results for most practical applications.*

4.3.2 Dispersion and Technical Frequency Range

As discussed earlier the transmission line parameters of a microstrip line are frequency dependent (dispersive). If the frequency increases more and more the electric field concentrates in the region between trace and ground. Hence, the characteristic impedance as well as the effective relative permittivity increase with frequency [7, 9]. A detailed analysis can be performed using transmission line tools or electromagnetic field simulators.

Due to dispersion, signal components with different frequencies propagate with different velocities c.

$$c(\omega) = \frac{c_0}{\sqrt{\varepsilon_{\text{r,eff}}(\omega)}} \tag{4.31}$$

With increasing length of line, broadband signals (like short pulses) will show distortions.

Like a coaxial line a microstrip line can guide *higher order wave modes* above a certain *cut-off frequency* f_c. Above this frequency a surface wave may propagate along the dielectric–air interface (*surface wave mode*). According to [10] we can estimate the higher order mode with the lowest cut-off frequency by the following relation.

$$f_c = \frac{75 \, \text{GHz}}{(h/1 \, \text{mm}) \sqrt{\varepsilon_r - 1}} \tag{4.32}$$

The cut-off frequency is a function of the substrate height h and the relative permittivity of the dielectric material ε_r. We can derive this formula from the cut-off frequency of a *dielectric slab waveguide*, that is a dielectric layer with a thickness $a = 2h$ between two air-filled half-spaces [11]. The factor of two results from the ground plane in the microstrip configuration which represents a plane of symmetry in the slab waveguide configuration. In a dielectric slab, waves can propagate in a zigzag course resulting from total reflection at the dielectric–air interfaces. So we can write for the cut-off frequency of the lowest non-quasi-TEM mode

$$\boxed{f_c = \frac{c_0}{4h\sqrt{\varepsilon_r - 1}}} \qquad \text{(Cut-off Frequency of a microstrip line)} \tag{4.33}$$

Example 4.6 *Technical Frequency Range of a Microstrip Line*

A microstrip line on an alumina substrate ($\varepsilon_r = 9.8$) with a height of $h = 0.635 \, \mu\text{m}$ has a cut-off frequency of approximately $f_c \approx 40 \, \text{GHz}$. Above this frequency higher order modes can propagate and cause unwanted effects.

4.3.3 Areas of Application

Microstrip lines are the standard transmission lines in planar circuits. Due to their open structure microstrip lines can easily be combined with concentrated elements like SMD-components. Furthermore, microstrip lines are low-cost, easy to manufacture and can be miniaturized. Microstrip transmission lines are used to realize compact filters, couplers, matching circuits and power dividers.

Due to their open structure microstrip lines show significant radiation losses at higher frequencies. A metallic enclosure can improve the situation; however, in a metallic enclosure box (housing) resonances can cause additional problems.[3] Unlike coaxial lines microstrip lines are not flexible due to the *rigid* substrate material. Losses are generally higher as with coaxial lines. Furthermore, dispersion can lead to signal distortions.

4.4 Stripline

A stripline (triplate) transmission line consists of a flat metallic trace centred between two infinite ground planes. The trace width is w and the distance between the ground planes is b (see Figure 4.9). The entire space between the ground planes is filled with a homogeneous dielectric material with a relative permittivity of ε_r. In real striplines the substrate width b_s is finite and the centre strip conductor has a thickness t. The fundamental wave mode is a TEM wave. The transversal magnetic field circles around the centre strip conductor. The electric field points from the strip conductor to ground.

4.4.1 Characteristic Impedance

In the literature we find approximate formulas for the characteristic impedance of striplines. A compact formula from [12] gives us

$$Z_0 = \frac{30\,\Omega}{\sqrt{\varepsilon_r}} \ln\left(1 + \frac{8h}{\pi w'}\left[\frac{16h}{\pi w'} + \sqrt{\left(\frac{8h}{\pi w'}\right)^2 + 6.27}\right]\right) \qquad (4.34)$$

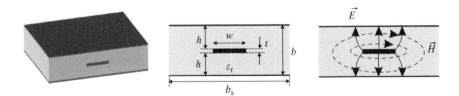

Figure 4.9 Geometry of a stripline and electric and magnetic field distribution of a TEM wave.

---

[3] We will investigate such resonances in Section 4.6.6 on cavity resonators.

with an effective trace width w' of

$$w' = w + \frac{t}{\pi} \ln \left(e \left[\sqrt{\left(\frac{1}{4(h/t) + 1} \right)^2 + \left(\frac{1/(4\pi)}{w/t + 1.1} \right)^m} \right]^{-1} \right) \quad \text{where} \quad m = \frac{6}{3 + t/h}$$

(4.35)

Example 4.7 *Line Impedance of a Stripline*

A stripline is given with the following parameters: substrate height $b = 2h + t = 1.6$ mm, relative permittivity $\varepsilon_r = 2.5$, trace width $w = 1.17$ mm and strip thickness $t = 10 \, \mu m$. For the auxiliary parameters we get $m = 1.9916$ and $w' = 1.191$ mm. According to Equation 4.34 the characteristic impedance of the stripline is $Z_0 = 50.0 \, \Omega$. The software tool TX-Line gives us the same value.

4.4.2 Technical Frequency Range

The technical frequency range is limited by the cut-off frequency of the first higher order mode. For a stripline with laterally infinite substrate ($b_s \rightarrow \infty$) the cut-off frequency of a stripline is given by the cut-off frequency of a *parallel-plate waveguide* [11] as

$$f_c = \frac{c_0}{2b\sqrt{\varepsilon_r}} \qquad \text{(Cut-off Frequency of a stripline)} \qquad (4.36)$$

In the case of a finite substrate (substrate width b_s) with electric side walls the closed metallic boundaries form a rectangular waveguide (see Section 4.6). Now, waveguide modes can propagate with a cut-off frequency of

$$f_c = \frac{c_0}{2b_s\sqrt{\varepsilon_r}} \qquad \text{(Cut-off Frequency of a stripline with side walls)} \qquad (4.37)$$

Example 4.8 *Cut-off Frequency of a Stripline*

We consider the stripline from Example 4.7 with the following parameters: substrate height $b = 1.6$ mm and relative permittivity $\varepsilon_r = 2.5$. The substrate has a width of $b_s = 10$ mm and is bounded by metallic side walls. According to Equation 4.37 the cut-off frequency is $f_c = 9.5$ GHz. For a laterally infinite substrate the cut-off frequency would be $f_c = 59.3$ GHz (Equation 4.36).

4.4.2.1 Areas of Application

Planar circuits with multilayer boards have microstrip lines on the top layer and striplines between ground planes in deeper layers. Figure 4.10 shows an example of a stripline that connects to two microstrip lines at both ends. The microstrip-to-stripline transition is realized by *vias* (vertical contacts through the dielectric substrates). Stripline transmission

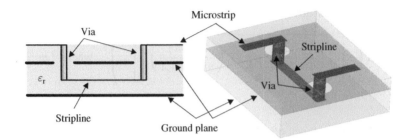

Figure 4.10 Multilayer-board with stripline and microstrip transmission lines.

lines can be used to realize compact filters, couplers, matching circuits and power dividers. Due to their closed construction and homogeneous dielectric material they do not show unintentional radiation and dispersion.

4.5 Coplanar Line

In a *coplanar* line a metallic strip (width w) runs between infinite ground planes separated by slots (slot width s) on the upper surface of a substrate. In real circuits the ground planes have a finite width $w_g > w$. Strip and ground are located in the *same* plane, hence, the name *co*planar. Figure 4.11a shows the geometry. The trace represents the signal conductor and the ground planes represent the return conductors. The distance d between the ground planes is

$$d = w + 2s \tag{4.38}$$

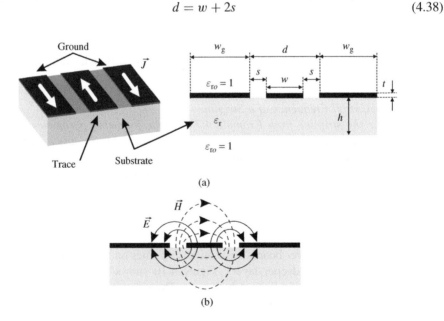

Figure 4.11 (a) Geometry of a coplanar transmission line and (b) electric and magnetic field distribution of a quasi-TEM wave.

An electromagnetic wave is present in air and substrate (height h, relative permittivity ε_r), so the fundamental wave type is *quasi-TEM* (see Figure 4.11b), where the magnetic fields runs in closed loops around the centre conductor and the electric fields point from centre conductor to the ground planes.

4.5.1 Characteristic Impedance and Effective Permittivity

In the literature [7, 11, 13] and [14] we find formulas for the characteristic impedance that involves elliptic integrals of the first kind that cannot be solved directly. In order to have less accurate but more easily evaluated expressions we reproduce here two equations from [11] and restrict our discussion to thick substrates and low frequencies.[4]

First, let us assume that we have *low frequencies* defined by a distance d between the ground planes. This distance shall be small compared to a quarter wavelength of an electromagnetic wave inside the substrate

$$d \ll \frac{\lambda_0}{4\sqrt{\varepsilon_r}} \tag{4.39}$$

where λ_0 is the free space wavelength.

For *thick substrates* (substrate height h much greater than total line width d) the electromagnetic wave is distributed over air and substrate in equal shares. Hence, the effective relative permittivity is

$$\varepsilon_{r,eff,thick} = \frac{\varepsilon_r + 1}{2} \tag{4.40}$$

If the substrate becomes thinner, a greater part of the wave is in air, hence the effective relative permittivity gets smaller.

Approximate expressions for the characteristic impedance Z_0 [11] are

$$Z_0 = \frac{120\,\Omega}{\sqrt{\varepsilon_{r,eff,thick}}} \cdot \ln\left(2\sqrt{\frac{d}{w}}\right) \qquad \text{for} \quad \frac{w}{d} \leq 0.17 \tag{4.41}$$

and

$$Z_0 = \frac{30\pi^2\,\Omega}{\sqrt{\varepsilon_{r,eff,thick}}} \left[\ln\left(2\frac{1 + \sqrt{w/d}}{1 - \sqrt{w/d}}\right)\right]^{-1} \qquad \text{for} \quad \frac{w}{d} \geq 0.17 \tag{4.42}$$

For higher frequencies the propagation constant and the characteristic impedance vary with frequency. Hence, the coplanar line is dispersive.

Example 4.9 *Characteristic Impedance of a Coplanar Line*

A coplanar waveguide on a semiconductor substrate is given with the following parameters: substrate height $h = 450\,\mu m$, relative permittivity $\varepsilon_r = 12.9$ (GaAs), trace width $w = 80\,\mu m$ and slot width $s = 55\,\mu m$. The substrate height $h = 450\,\mu m$ is significantly greater

---

[4] If we are interested in more accurate results for thin substrates we can apply expressions from literature or use a transmission line calculator (like TX-Line [8]). Alternatively, we could get into detailed analysis by using electromagnetic field simulation software.

than the ground plane distance $d = w + 2s = 190\,\mu m$. Therefore, we assume that the formula for the effective relative permittivity is a good approximation. From Equation 4.40 we get $\varepsilon_{r,eff,thick} = 6.95$. The trace-width-to-ground-distance-ratio is $w/d = 0.421$, hence we use Equation 4.42. The characteristic impedance is $Z_0 = 50.14\,\Omega$.

In order to compare our result with a transmission line calculator [8] we must assume additional parameters and choose the thickness of metallization $t = 3\,\mu m$ and frequency $f = 1\,GHz$. The calculated effective relative permittivity is slightly smaller $\varepsilon_{r,eff,thick} = 6.82$ than our approximation due to the air below the substrate. The characteristic impedance $Z_0 = 50.27\,\Omega$ is virtually identical to our approximation.

4.5.2 Coplanar Waveguide over Ground

A variation on the coplanar waveguide is the *coplanar waveguide over ground*. This type of line has an additional ground plane on the bottom side of the substrate (see Figure 4.12).

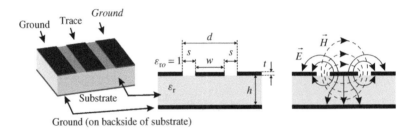

Figure 4.12 Geometry of a coplanar waveguide *over ground* and electric and magnetic field distribution of a quasi-TEM wave.

If the height of the substrate h is much greater than the total line width $d = w + 2s$ (thick substrate), there is little influence of the ground plane. The electromagnetic field is concentrated in the trace-slot-region on the upper side of the substrate. In order to support a *coplanar wave mode* the height of the substrate should be much greater than the total line width $h \gg d$.

If the height of the substrate decreases and the slots become larger the field pattern approaches the *microstrip wave mode* with strong coupling between trace and bottom ground. In this case the line behaves like a microstrip line with additional grounds on both sides of the top trace.

Example 4.10 *Characteristic Impedance of a Coplanar Waveguide Over Ground*

If we use the coplanar line from Example 4.9 and add a ground plane at the bottom of the substrate the influence on the characteristic impedance is quite moderate (-2%) due to the thick substrate ($h = 450\,\mu m \gg b = 190\,\mu m$).

$$Z_0 = 50.27\,\Omega \quad \text{(Without ground)} \quad \rightarrow \quad Z_0 = 49.45\,\Omega \quad \text{(With ground)} \qquad (4.43)$$

If we reduce the substrate height from $h = 450\,\mu m$ *to* $h = 100\,\mu m$ *the ground effect becomes more significant (−20%).*

$$Z_0 = 52.76\,\Omega \quad \text{(Without ground)} \quad \rightarrow \quad Z_0 = 42.42\,\Omega \quad \text{(With ground)} \qquad (4.44)$$

The calculations have been performed with a transmission line calculator [8].

4.5.3   Coplanar Waveguides and Air Bridges

A coplanar line is categorically a three-conductor line. Such a structure can support two fundamental modes of propagation.

Figure 4.13a shows the previously discussed fundamental mode of propagation which is known as *even mode* or *coplanar waveguide mode*. The current in the ground planes flows in one direction (*even*) and the current in the centre trace flows in the opposite direction. At a particular coordinate z along the transmission line both ground planes have equal electric potential. This fundamental mode of propagation is the desired propagation mode for a coplanar line.

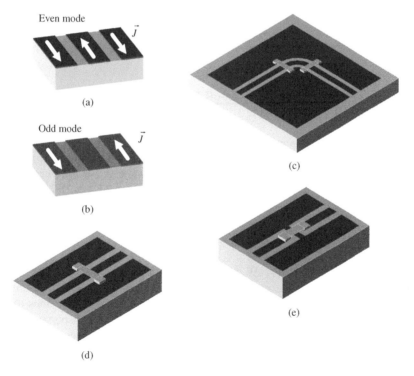

Figure 4.13   (a) Even mode and (b) odd mode current flow on a coplanar line, (c) coplanar bend with two air bridges, (d)(e) different types of coplanar air bridges.

Unfortunately, there is a second fundamental propagation mode: Figure 4.13b shows the *odd mode* or *slotline mode*. The currents in the ground planes flow in the opposite direction (*odd*). This mode is unwanted and must be suppressed.

There is no problem with the desired mode of propagation as long as the coplanar waveguide follows a straight line. However, if the coplanar line is bent, the current paths in the ground planes have different length. In order to ensure equal potentials on the ground plane *air bridges* are set between the ground planes [13, 14]. Figure 4.13c shows a 90°-bend with two air bridges. There are different types of air bridges. Figure 4.13d shows an air bridge that connects the ground planes on both sides by a strip above the centre conductor. In Figure 4.13e the ground plane connection is made by a perpendicular strip in the same plane. In this case the centre trace of the coplanar line has to go a more indirect route by bridging the ground connection.

An air bridge represents a discontinuity of a coplanar line and thus locally changes the characteristic impedance of the line. A change in the characteristic impedance gives rise to reflections and degrades the performance of the transmission line. In order to minimize reflections air bridges should be small and carefully designed.

4.5.4 Technical Frequency Range

Like a microstrip line a coplanar waveguide supports *higher order wave modes* above a certain *cut-off frequency* f_c. Above this frequency a surface wave [14] can propagate along the dielectric–air interface (*surface wave mode*). The cut-off frequency is given by

$$\boxed{f_c = \frac{c_0}{4h\sqrt{\varepsilon_r - 1}}} \qquad \text{(Cut-off frequency of a coplanar line)} \qquad (4.45)$$

Example 4.11 *Cut-off Frequency of a Coplanar Waveguide*

We consider a coplanar line with a substrate height of $h = 410\,\mu m$ and a relative permittivity of $\varepsilon_r = 12.9$. By using Equation 4.45 we calculate a cut-off frequency of the surface wave mode as $f_c = 53\,GHz$. So below this frequency only quasi-TEM modes can propagate.

4.5.5 Areas of Application

Coplanar lines allow high levels of integration and are especially suitable for hybrid circuits and integrated microwave circuits (MMIC). Since ground and trace are located at the upper surface of the substrate it is quite easy to connect coplanar lines to lumped components. Compared to microstrip lines coplanar lines are less dispersive. Losses of coplanar lines are comparable to those of microstrip lines. Air-bridges that suppress the the unwanted odd-mode must be considered in the design and fabrication process [14].

4.6 Rectangular Waveguide

All previously discussed types of transmission line consist of two distinct conductors where the current flows in opposite directions. In a microstrip line the strip on top of

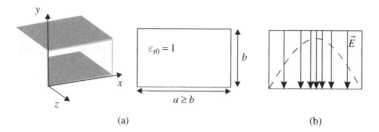

(a)   (b)

Figure 4.14 (a) Construction of a rectangular waveguide and (b) distribution of the electric field (Fundamental mode TE_{10}).

the substrate is the signal conductor and the ground plane is the return conductor. Hence, such transmission lines can carry DC current and support TEM (or quasi-TEM) waves as a fundamental wave mode.

A rectangular waveguide consist of only *one* conductor. The metallic conductor encloses a rectangular volume (see Figure 4.14a). The width of the volume is a and the height is b. Usually the width is greater than the height ($a \geq b$). A commonly used ratio is $a = 2b$. Table 4.2 shows some examples of commonly used dimensions of rectangular waveguide cross-sections. It is intuitively clear that a one-conductor line cannot support DC current. The first propagating wave type is a non-TEM wave type called TE_{10} (or H_{10}). This mode has a transversal electric field but magnetic field components $\vec{H}$ in the direction of propagation. Figure 4.14b shows the orientation of the electric field $\vec{E}$ in a cross-section of the rectangular waveguide. In the next section we will take a closer look at the wave type and derive the cut-off frequency that is associated with that wave mode.

4.6.1 Electromagnetic Waves between Electric Side Walls

In Chapter 2 we described plane waves that propagate in free space. A plane wave is a transversal electromagnetic wave (TEM) with a polarization defined by the direction of the electric field vector. Let us assume that such a plane wave propagates in a negative x-direction and is polarized in a y-direction as shown in Figure 4.15.

Table 4.2 Some examples of rectangular waveguide dimensions

Type	Width a	Height b	Cut-off frequency
R22	109.22 mm	54.61 mm	1.372 GHz
R40	58.17 mm	29.083 mm	2.577 GHz
R100	22.86 mm	10.16 mm	6.56 GHz
R120	19.050 mm	9.575 mm	7.87 GHz
R140	15.799 mm	7.899 mm	9.49 GHz
R500	4.775 mm	2.388 mm	31.39 GHz
R620	3.759 mm	1.880 mm	39.88 GHz
R1400	1.2954 mm	0.6477 mm	90.79 GHz

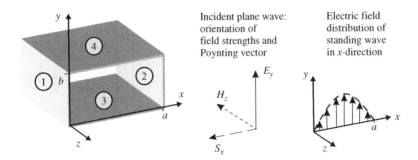

Figure 4.15 Construction of a transversal standing wave inside a rectangular waveguide.

The electric field of the plane wave is given by the following phasor

$$\vec{E}(x) = E_0 e^{+jkx} \vec{e}_y \tag{4.46}$$

Now we define an ideal electric wall at $x = 0$ (yz-plane, see ① in Figure 4.15). The ideal electric wall forces the tangential electric field to be zero. This is achieved by the generation of a reflected wave that travels in the opposite direction (positive x-direction). The reflection coefficient is $r = -1$. So – for positive values of x – we see a superposition of two waves of equal amplitude propagating in opposite directions resulting in a standing wave.

$$\vec{E}(x) = E_0 \left(e^{+jkx} + re^{-jkx}\right)\vec{e}_y = E_0 \left(e^{+jkx} - e^{-jkx}\right)\vec{e}_y = 2jE_0\sin(kx)\vec{e}_y \tag{4.47}$$

At $x = 0$ the electric field is zero. From our discussion of standing waves on transmission lines (Section 3.1.8), we know that further zeros (minima) can be found at integer multiples of half of the wavelength $x_{zero} = n\lambda_0/2$, where n is an integer value and λ_0 is the free space wavelength. So, if we step from $x = 0$ to $x = \lambda_0/2$ we find the next zero of the electric field distribution. If the field is zero, we can position another electric wall at $x = \lambda_0/2$ (see ② in Figure 4.15). A standing wave can exist between these two walls with a distance of half a (free space) wavelength λ_0. In a rectangular waveguide we have two side walls with a distance of a. If the width a equals half of the wavelength

$$a = \frac{\lambda_0}{2} = \frac{c_0}{2f} \tag{4.48}$$

there can exist a standing wave between the side walls with a frequency of

$$f_c = \frac{c_0}{2a} \tag{4.49}$$

The frequency f_c is called *cut-off* frequency and represents the lowest frequency where electromagnetic waves can propagate between the side walls.

In order to build a rectangular waveguide we need a bottom ③ and a top ④ metallic plate to limit the volume in the vertical direction. The previously defined standing wave is polarized in y-direction and is therefore perpendicular to these horizontal metallic planes.

Hence, we can define horizontal planes at any location ($y = \text{const.}$).[5] If we assume that the distance b between the horizontal planes ③ and ④ is smaller than or equal to the distance a ($b \leq a$) the distance a defines the cut-frequency of the lowest propagation mode.

The standing wave that can exist inside the waveguide at a frequency that equals the cut-off frequency $f_c = c_0/(2a)$ does not carry any effective power along the line. It is a reactive field mode. The electric field at any point along the line has the same phase. We do not see a wave propagating along the line. If we interpreted the field mode as a propagating mode we would say that the wavelength (minimum non-zero distance between two points with the same phase) is infinity. Hence, the phase velocity is infinity two. The group velocity (the velocity at which energy and information propagate) is zero.

Let us now consider a frequency that is greater than the cut-off frequency ($f > f_c$). Now the distance between the side walls is greater than half a wavelength. The wave no longer satisfies the boundary conditions at $x = 0$ and $x = a$. We can satisfy the boundary conditions when we consider a wave that impinges under a certain angle on plane ①. Due to reflections on both side walls we see an electromagnetic wave that propagates in a zigzag-course down the line. With increasing distance between the side walls the group or energy velocity of the wave increases and the phase velocity decreases. Figure 4.16 shows the instantaneous values of the electric field strength of two plane waves that superimpose under a certain angle. We see stationary nulls where we can place the side walls of a rectangular waveguide. Furthermore we see nulls that propagate with the phase velocity down the line. The distance between adjacent propagating nulls is half of the waveguide wavelength λ_W. The waveguide wavelength is greater than the free space wavelength $\lambda_W > \lambda_0$.

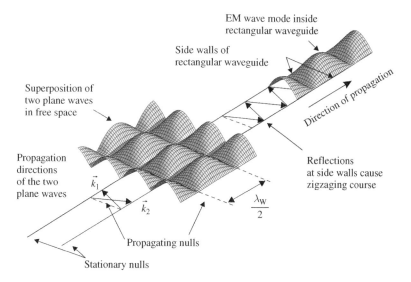

Figure 4.16 Superposition of two plane waves (instantaneous values of the electric field strength) and wave mode inside a rectangular waveguide.

---

[5] For the sake of simplicity we consider only the electric field in our discussion. Of course, the magnetic field also conforms to its boundary conditions at all four metallic planes.

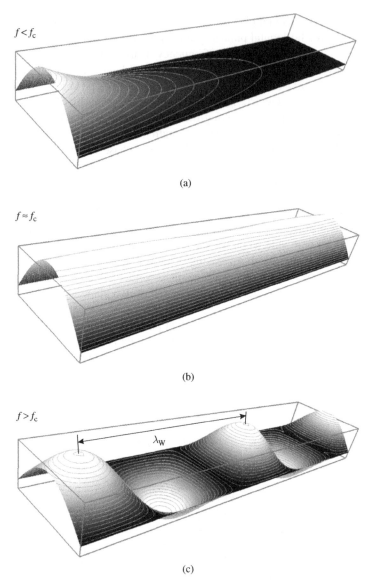

$f < f_c$

(a)

$f \approx f_c$

(b)

$f > f_c$

λ_W

(c)

Figure 4.17 Distribution of electric field strengths in a rectangular waveguide (a) below, (b) at, and (c) above cut-off frequency.

Figure 4.17 shows the electric field distribution in a R100 rectangular waveguide ($a = 22.86\,\text{mm}$, $b = 10.16\,\text{mm}$). According to Equation 4.49 the cut-off frequency is $f_c = 6.56\,\text{GHz}$. At a frequency of $f = 6\,\text{GHz}$ ($f < f_c$) (see Figure 4.17a) no wave propagation occurs. The field is reactive and decays exponentially along the line. If the frequency equals the cut-off frequency $f = f_c$ a standing wave between the side walls occurs inside the waveguide (see Figure 4.17b). All electric field components oscillate with the same phase. At a frequency of $f = 8.25\,\text{GHz}$ ($f > f_c$) we see wave propagation along the line with a waveguide wavelength of λ_W (see Figure 4.17c). The wave velocity and the wavelength can be calculated based on the previous discussion.

In the next section we will look at the technically important fundamental wave mode.

4.6.2 Dominant Mode (TE10)

Rectangular waveguide modes are separated into two categories: TE-waves (or H-waves) and TM-waves (or E-waves). Furthermore, two indices m and n are added to describe the modes.

TE-waves TE_{mn} (or H-waves H_{mn}) are *transversal electric* waves. The electric field in the direction of propagation is zero ($E_z = 0$). The magnetic field in the direction of propagation is non-zero ($H_z \neq 0$), hence the name H-wave.

TM-waves TM_{mn} (or E-waves E_{mn}) are *transversal magnetic* waves. The magnetic field in the direction of propagation is zero ($H_z = 0$). The electric field in the direction of propagation is non-zero ($E_z \neq 0$), hence the name E-wave.

The indices m and n indicate the number of field maxima in the different transversal directions. Index m is associated with the horizontal direction (where the waveguide has a side length of a). Index n is associated with the vertical direction (where the waveguide has a side length of $b \leq a$). For TE-waves the index specifies the number of electric field maxima. For TM-waves the index specifies the number of magnetic field maxima.

Following our previous discussion wave propagation starts at a cut-off frequency of

$$\boxed{f_c = \frac{c_0}{2a}} \qquad \text{(Cut-off frequency of dominant mode)} \qquad (4.50)$$

Below cut-off the wave exhibits an exponential decaying behaviour (Figure 4.17a).

The wave mode with the lowest cut-off frequency (fundamental wave mode) is the TE_{10} mode. The electric field distribution has one maximum in the horizontal direction ($m = 1$) and no maximum (i.e. the field is constant) in the vertical direction ($n = 0$). The distribution is shown in Figure 4.14b and Figure 4.18a.

The electric field strength in Figure 4.14b has only a (transversal) component in the vertical direction (y-direction). The magnetic field strength has a (transversal) component in the horizontal direction (x-direction) and an additional component in the longitutinal direction (z-direction). According to [11] we can write the magnetic field components as

$$H_z = H_{z0} \cos\left(\pi \frac{x}{a}\right) e^{-j \frac{2\pi}{\lambda_W} z} \qquad (4.51)$$

$$H_y = 0 \qquad (4.52)$$

$$H_x = H_{z0} \frac{2\pi}{\lambda_W k_c^2} \frac{\pi}{a} \sin\left(\pi \frac{x}{a}\right) e^{-j\left(\frac{2\pi}{\lambda_W} z - \frac{\pi}{2}\right)} \qquad (4.53)$$

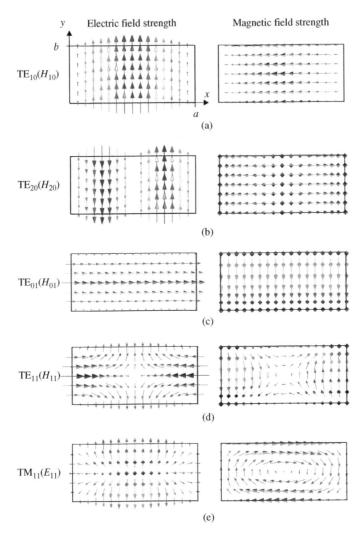

Figure 4.18 Modes in a R100 rectangular waveguide: distribution of electric and magnetic field in a cross-sectional area.

The electric field components are given as

$$E_x = E_z = 0 \tag{4.54}$$

$$E_y = H_{z0} Z_{F0} \frac{k}{k_c^2} \frac{\pi}{a} \sin\left(\pi \frac{x}{a}\right) e^{-j\left(\frac{2\pi}{\lambda_W} z + \frac{\pi}{2}\right)} \tag{4.55}$$

where k_c is the cut-off wave number of the TE_{10} wave mode

$$k_c = \beta_c = \frac{2\pi}{\lambda_c} = \frac{2\pi f_c}{c_0} = \frac{\pi}{a} \tag{4.56}$$

and λ_W is the wavelength of the guided wave

$$\lambda_W = \frac{\lambda_0}{\sqrt{1 - \left(\frac{f_c}{f}\right)^2}} > \lambda_0 \qquad \text{(Wavelength of guided wave)} \qquad (4.57)$$

The wavelength inside the waveguide is a function of frequency. The wavelength λ_W decays with increasing frequency and is always greater than the wavelength in free space λ_0.

The velocity of the wave is a function of frequency too. In the previous section we discussed the zigzag-course of a wave that is reflected between the side walls of a rectangular waveguide. This zigzag-course leads to an energy (or group) velocity v_{gr} that is always smaller than the speed of light in vacuum c_0.

$$v_{gr} = c_0 \frac{\lambda_0}{\lambda_W} = c_0 \sqrt{1 - \left(\frac{f_c}{f}\right)^2} < c_0 \qquad \text{(Group or energy velocity)} \qquad (4.58)$$

If we look at the field distribution inside the rectangular waveguide we see that it is periodic with the wavelength of the guided wave λ_W. If we look at instantaneous values of the electric and magnetic fields, we see nulls and maxima propagating along the line. The velocity of propagation is called phase velocity v_{ph} and given as

$$v_{ph} = c_0 \frac{\lambda_W}{\lambda_0} = \frac{c_0}{\sqrt{1 - \left(\frac{f_c}{f}\right)^2}} > c_0 \qquad \text{(Phase velocity)} \qquad (4.59)$$

The phase velocity is always greater than the speed of light in vacuum c_0. We remember that the field pattern results from a superposition of plane waves that propagate in different directions given by the wave vectors $\vec{k}_1$ and $\vec{k}_1$ (see Figure 4.16). Each plane wave propagates with the vacuum speed of light. Since points with equal phase (e.g. nulls or extrema) are located on planes we visualize plane waves in 2D representations as straight lines that move with the vacuum speed of light. These lines cross at a certain angle (like a pair of scissors) and define the propagation speed of the nulls from the guided wave.[6]

Group and phase velocity are related by the following equation.

$$v_{gr} v_{ph} = c_0^2 \qquad (4.60)$$

Since the phase velocity exhibits a strong frequency dependence just above the cut-off frequency, the technical frequency range of usage is commonly defined by a lower limit of $f_{min} = 1.25 f_c$. For a common cross-section with $a = 2b$ the cut-off frequency of the next mode is two times the cut-off frequency of the fundamental mode $2f_c$. It is intuitively clear that a standing wave with double frequency can exist between the side walls. This standing wave has two maxima in the x-direction and a null in the centre of

---

[6] Imagine cutting a piece of paper with a pair of scissors. The crossover point on the pair of scissors may move faster than the scissors themselves.

the rectangular waveguide (see Figure 4.18b). In order to have clearly defined conditions on a transmission line, only the fundamental mode should be able to propagate. In order to avoid excitation of the next mode (with a cut-off frequency of $2f_c$) the upper limit of the technical frequency range of usage is therefore given by $f_{max} = 1.9f_c$ [2].

$$\boxed{1.25f_c \leq f \leq 1.9f_c}$$ (Technical frequency range)   (4.61)

Based on current and power inside the rectangular waveguide we can define a characteristic line impedance Z_0 for the fundamental wave TE_{10} [11] as

$$\boxed{Z_0^{TE_{10}} = \frac{\pi^2 b}{8a} \frac{Z_{F0}}{\sqrt{1 - \left(\dfrac{f_c}{f}\right)^2}}}$$ (Characteristic line impedance of dominant mode)   (4.62)

where $Z_{F0} = 377\,\Omega$ is the characteristic wave impedance of free space.

The characteristic wave impedance [7] is given by

$$Z_F^{TE_{10}} = \frac{Z_{F0}}{\sqrt{1 - \left(\dfrac{f_c}{f}\right)^2}}$$ (Characteristic wave impedance of dominant mode)   (4.63)

The characteristic line impedance and characteristic wave impedance exhibit a strong frequency dependence.

Example 4.12 *Dominant Mode of a R100 Rectangular Waveguide*

A R100 rectangular waveguide has a cross-section with a width of a = 22.86 mm and a height of b = 10.16 mm. The cut-off frequency of the dominant mode is $f_c = 6.56$ GHz. The technical frequency range of usage extends from $f_{min} = 1.25 f_c = 8.2$ GHz to $f_{max} = 1.9 f_c = 12.5$ GHz. At a frequency of f = 8.5 GHz the waveguide wavelength is $\lambda_W = 5.55$ cm and the characteristic line impedance of the dominant mode is $Z_0^{TE_{10}} = 325\,\Omega$.[7]

4.6.3 Higher Order Modes

In Section 4.6.1 we have seen that at cut-off frequency $f_c = c_0/(2a)$ a standing wave in the rectangular waveguide satisfies the electric boundary conditions at the sidewalls: the electric field strength of the standing wave has nulls at $x = 0$ and $x = a$ and a maximum at $x = a/2$ (see Figure 4.18a). It is rather obvious that the electric field strength of a standing wave at *twice the cut-off frequency* has nulls at $x = 0$, $x = a/2$ and $x = a$ and maxima at $x = a/4$ and $x = 3a/4$ (see Figure 4.18b). Hence, at the positions of the side walls the boundary conditions are also satisfied and this second wave mode can exist.

---

[7] These values can be easily evaluated with EM simulation software. These software tools commonly calculate tree different characteristic impedances: the current-voltage characteristic impedance Z_{UI}, the power-voltage characteristic impedance Z_{PV} and the power-current characteristic impedance Z_{PI}. The values are equal for TEM waves and can be quite different for non-TEM waves [15, 16].

Accordingly, we can construct further higher order modes that satisfy the boundary conditions above corresponding cut-off frequencies. The cut-off frequencies of the higher order modes are given by the following general equation.

$$\boxed{f_{c,mn} = \frac{c_0}{2}\sqrt{\left(\frac{m}{a}\right)^2 + \left(\frac{n}{b}\right)^2}} \qquad \text{(Cut-off frequencies)} \qquad (4.64)$$

If $m = 0$ or $n = 0$ only TE modes exist. The electric field is vertical ($n = 0$) or horizontal ($m = 0$). Figure 4.18a–c show examples. If both indices are non-zero ($m \neq 0$ and $n \neq 0$) TE and TM modes have equal cut-off frequencies $f_{c,mn}$. TE_{mn} and TM_{mn} modes with equal cut-off frequencies are referred to as *degenerated modes* [17].

Figure 4.18 shows distributions of the electric and magnetic field strength in a cross-section of an R100 rectangular waveguide for higher order modes. On the website that accompanies this book animated plots of waveguide modes are shown.

Example 4.13 *Higher Order Modes of a R100 Rectangular Waveguide*

A R100 rectangular waveguide has a width of $a = 22.86$ mm and a height of $b = 10.16$ mm. Table 4.3 lists the cut-off frequencies of wave modes that can propagate below $f = 20$ GHz. Field plots of the first five modes are illustrated in Figure 4.18.

Table 4.3 Cut-off frequencies $f_{c,mn}$ in a R100 waveguide ($a = 22.86$ mm, $b = 10.16$ mm) (see Example 4.13)

Mode	TE_{10}	TE_{20}	TE_{01}	$\text{TE}_{11}, \text{TM}_{11}$	TE_{30}	$\text{TE}_{21}, \text{TM}_{21}$
Cut-off freq.	6.56 GHz	13.12 GHz	14.76 GHz	16.16 GHz	19.69 GHz	19.75 GHz

4.6.4 Areas of Application

Rectangular waveguides have high power handling capabilities and exhibit low-loss per length. This makes them suitable for high power applications (like Radar) and low power applications (like radio astronomy).[8] Rectangular waveguides are used to feed horn antennas (see Section 7.2) and to design filters with high quality-factors.

A drawback of a rectangular waveguide is the small bandwidth, that is the frequency range where only the fundamental wave mode can propagate. Furthermore, rectangular waveguides are highly dispersive. The propagation speed of a wave depends on frequency. Finally, rectangular waveguides can hardly be combined with concentrated components (like resistors, circuits) and require quite complex transitions when combined with other transmission line types. We will look at such a transition in the subsequent section. Due to the solid metallic walls, rectangular waveguides are rigid structures that are interconnected by flanges.

A further commonly used waveguide geometry is the circular waveguide (see Section 4.7). Other less used waveguide geometries may be found in [2] and [7].

---

[8] Detection of radio waves from distant astronomical objects.

4.6.5 Excitation of Waveguide Modes

Common laboratory equipment – like network analysers, generators, amplifiers, filters and antennas – usually has coaxial connectors (SMA (SubMiniature Type-A), Type N for frequencies up to 18 GHz, Precision 2.4 mm-, 2.92 mm- and 3.5 mm connectors for frequency ranges exceeding 18 GHz [18]).

In order to connect waveguide components to the coaxial world we need *coax-to-waveguide transitions*. Such a device is a two-port network with a coaxial port and a waveguide port. A coax-to-waveguide transition converts a TEM-wave (in the coaxial line section) into a fundamental waveguide mode (in the waveguide section) and vice versa. Figure 4.19a shows a transition from a coaxial 50 Ω-line to a R100 waveguide [19]. The component consists of a coaxial line with an elongated inner conductor that extends into an air-filled waveguide section. On top of the inner conductor a metallic cylinder is placed. The cylinder has a radius of $R = 2.05$ mm and a height of $h = 2.6$ mm. The extension of the inner conductor is $\Delta\ell = 2.3$ mm. The coaxial line is defined by an inner radius of $R_i = 0.65$ mm, an outer radius of $R_o = 2.05$ mm and a dielectric material with a relative permittivity of $\varepsilon_r = 1.9$. All other dimensions are given in the figure.

Between the top surface of the cylinder and the ceiling of the waveguide a vertical electric field exists that excites the fundamental waveguide mode. The dimensions are optimized to result in minimum reflection in the frequency range of operation. According to Equation 4.61 a R100 waveguide is used in the frequency range 8.2 GHz $\leq f \leq$ 12.5 GHz. The reflection coefficient is given in Figure 4.19b in logarithmic[9] scale $r_1^\ell/\mathrm{dB} = 20\lg |r_1|$. A value of $r_1^\ell = -20$ dB corresponds to a linear value of $r_1 = 0.1$ (i.e. reflected power of 1%). Therefore, the reflection coefficient is less than 0.1 in the frequency range of usage that is sufficient for most applications.

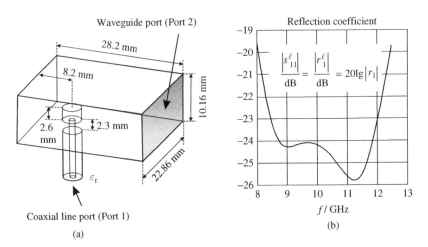

Figure 4.19 Transition from coaxial line to R100 waveguide: (a) Geometry and (b) reflection coefficient r_1 in a commonly used logarithmic scale.

---

[9] We look at logarithmic representation of reflection coefficients and scattering parameters in general in Chapter 5.

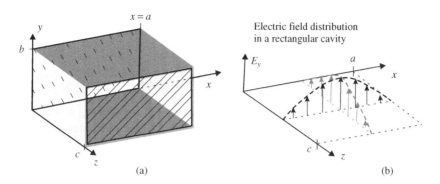

Figure 4.20 Electric field distribution (H_{101} mode) in a cavity resonator.

4.6.6 Cavity Resonators

Let us consider a rectangular waveguide section with a length $\ell = c$. If we close the open ends of the waveguide with metallic walls we get a rectangular cavity with a cuboid air volume of size $a \times b \times c$ (width$\times$height$\times$length). Figure 4.20a shows the geometry and coordinate system. Electric and magnetic fields inside the cavity must now fulfil their boundary conditions at all six metallic walls. At certain frequencies the cavity can resonate (box resonance).

We assume that the length of the waveguide section c is greater than the width a, which itself is greater than the height b ($c > a > b$). Then, the resonance mode with the lowest resonance frequency has the following field distribution [3].

$$E_y(x, y, z) = E_{y0} \sin\left(\frac{\pi x}{a}\right) \sin\left(\frac{\pi z}{c}\right) \tag{4.65}$$

$$H_x(x, y, z) = H_{x0} \sin\left(\frac{\pi x}{a}\right) \cos\left(\frac{\pi z}{c}\right) \tag{4.66}$$

$$H_z(x, y, z) = H_{z0} \cos\left(\frac{\pi x}{a}\right) \cos\left(\frac{\pi z}{c}\right) \tag{4.67}$$

All field components are independent of the coordinate y. There is no field variation in the vertical direction. The electric field exhibits only a transversal y-component with respect to the original propagating wave mode in the open waveguide, which is the z-direction. The magnetic field has a transversal component and a component in the original direction of propagation. The resonance mode with the lowest frequency is therefore a TE mode. If we look at the electric field distribution in Figure 4.20b we see that the electric field has one maximum in the x-direction and one maximum in the z-direction. Hence, the resonance mode (for $c > a > b$) is called TE$_{101}$ mode with a resonance frequency of

$$f_{R,101} = \frac{c_0}{2}\sqrt{\left(\frac{1}{a}\right)^2 + \left(\frac{1}{c}\right)^2} \tag{4.68}$$

At higher frequencies we find resonant modes with more complex field distributions. Equations can be found for instance in [3]. The resonance frequencies of all resonance modes are given by

$$f_{R,mnp} = \frac{c_0}{2}\sqrt{\left(\frac{m}{a}\right)^2 + \left(\frac{n}{b}\right)^2 + \left(\frac{p}{c}\right)^2} \tag{4.69}$$

where m, n and p are integer numbers with at least two of them being non-zero.

In practical applications circuits are often built into shielded enclosures to avoid electromagnetic interference with adjacent circuits. A metallic enclosure represents a cavity with specific resonances depending on the geometry of the casing. So, cavity resonances may be excited unintentionally by the circuit, causing unexpected behaviour of the circuit, for example instability of amplifiers and extra passbands in filters. So, the size of the metallic enclosure must be chosen carefully. Later, in Section 6.5.4 we will investigate a filter in a metallic enclosure and see how a shielded enclosure may affect the transmission coefficient of the circuit.

Example 4.14 *Resonant Modes of a Cavity*
Let us consider a rectangular cavity with width $a = 24$ mm, height $b = 10$ mm and length $c = 40$ mm. The resonance frequencies $f_{R,mnp}$ are given in Table 4.4. Figure 4.21 shows the distribution of the instantaneous electric field (TE_{101}-mode and TE_{102}-mode) in a horizontal cut plane. Both resonant modes are TE modes, the electric field is transversal to the original direction of propagation. The TE_{101} mode in Figure 4.21a has one maximum lengthwise, one maximum in the lateral direction, and shows no field variation in the vertical direction. The TE_{102} mode in Figure 4.21b has two maxima lengthwise, one maximum in the lateral direction, and shows no field variation in the vertical direction.

Table 4.4 Resonance frequencies $f_{R,mnp}$ in a rectangular cavity (see Example 4.14)

m	n	p	$f_{R,mnp}$
1	0	1	7.289 GHz
0	1	1	15.462 GHz
1	1	0	16.250 GHz
1	1	1	16.677 GHz
2	0	1	13.050 GHz
1	0	2	9.763 GHz
0	1	2	16.771 GHz
0	2	1	30.233 GHz
1	2	0	30.644 GHz
2	1	0	19.526 GHz
2	1	1	19.882 GHz
1	2	1	30.873 GHz
1	1	2	17.897 GHz

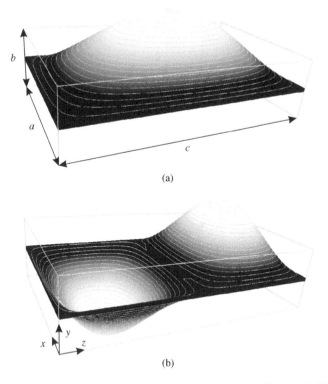

(a)

(b)

Figure 4.21   Instantaneous electric field distribution of (a) TE_{101} mode and (b) TE_{102} mode in a cavity resonator (see Example 4.14).

4.7   Circular Waveguide

A circular waveguide consists of a hollow metallic cylinder with an inner radius R (see Figure 4.22a). In the inner air-filled volume of the cylinder electromagnetic waves can propagate above mode-specific cut-off frequencies $f_{c,mn}$. Solutions of Maxwell's equations can be found using cylindrical coordinates and involve Bessel functions [1]. As in the case of a rectangular waveguide the solutions can be divided into TE and TM modes. The different modes are identified by integer numbers m and n.

The wave mode with the lowest cut-off frequency (fundamental mode) is TE_{11}. The electric field distribution in a cross-section area is shown in Figure 4.22b-c. Since the cross-section is a circle, there is no preferential direction for the orientation of the electric field [1, 2]. We see a vertical field orientation in Figure 4.22b and a horizontal field orientation in Figure 4.22c. In practical operation the orientation depends on the excitation of the mode. We can use orthogonal polarized waves of the fundamental wave mode to transmit two independent data streams simultaneously.

Table 4.5 lists some examples of practical waveguide dimensions as well as cut-off frequencies for the fundamental wave mode TE_{11}. The cut-off frequencies can be

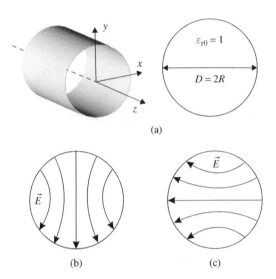

(a)

(b)   (c)

Figure 4.22 (a) Geometry of a circular waveguide and (b) horizontal and vertical orientation of the electric field (Fundamental mode TE$_{11}$).

Table 4.5 Some examples of circular waveguide dimensions

Type	Diameter $D = 2R$	Cut-off frequency
C22	97.87 mm	1.795 GHz
C40	44.45 mm	3.379 GHz
C104	20.244 mm	8.679 GHz
C120	17.475 mm	10.054 GHz
C140	15.088 mm	11.645 GHz
C495	4.369 mm	40.215 GHz
C660	3.166 mm	55.495 GHz

calculated using roots of the Bessel function and their derivative. The cut-off frequency of the fundamental (or dominant) wave mode is given by

$$f_{c,TE_{11}} = \frac{c_0}{2R \cdot x} \quad \text{where} \quad x = 1.7063 \qquad \text{(Cut-off frequency of TE}_{11}\text{-mode)} \qquad (4.70)$$

In the literature [1, 3] we find formulas to calculate cut-off frequencies for higher order modes. Table 4.6 lists values for the factor x to calculate the cut-off frequencies by using Equation 4.70. From the values of $f_c/f_{c,TE_{11}}$ we see that a circular waveguide exhibits less bandwidth ($f_{c,TM_{01}}/f_{c,TE_{11}} = 1.3061$) than a rectangular waveguide ($f_{c,20}/f_{c,10} = 2$ for commonly used rectangular cross-sections with $a \geq 2b$).

Field distributions of different circular waveguide modes are given in Figure 4.23 and Figure 4.24. Figure 4.23 shows the electric and magnetic field strength in a cross-section of the circular waveguide for the TE$_{11}$, TM$_{01}$ and TE$_{21}$ mode. TE$_{01}$ and TM$_{11}$ mode are displayed in Figure 4.24.

Table 4.6 Factor x for calculation of cut-off frequencies in circular waveguides by using Equation 4.70

Mode	TE_{11}	TM_{01}	TE_{21}	TM_{11}, TE_{01}	TE_{31}	TM_{21}
Factor x	1.7063	1.3064	1.0286	0.8199	0.7478	0.6117
$f_c/f_{c,TE_{11}}$	1	1.3061	1.6589	2.081	2.282	2.789

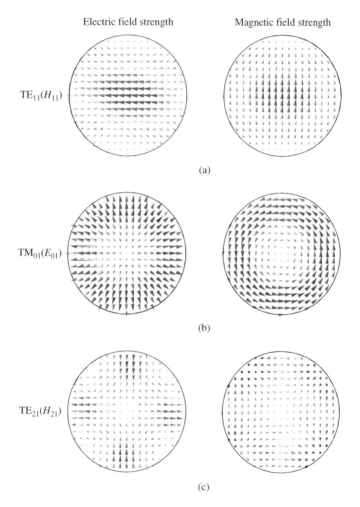

Electric field strength   Magnetic field strength

$TE_{11}(H_{11})$

(a)

$TM_{01}(E_{01})$

(b)

$TE_{21}(H_{21})$

(c)

Figure 4.23 Modes in a circular waveguide: distribution of electric and magnetic field in a cross-section area.

Electric field strength   Magnetic field strength

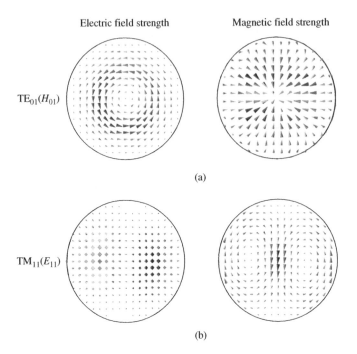

$TE_{01}(H_{01})$

(a)

$TM_{11}(E_{11})$

(b)

Figure 4.24 Modes in a circular waveguide (continued): distribution of electric and magnetic field in a cross-section area.

Interestingly, the TE_{01} mode is the one with minimum losses and therefore most suitable for transmission over large distances. Unfortunately, the low-loss mode is not the dominant mode. If the low-loss mode encounters a disturbance in the transmission line cross-section, power is converted into modes that can also propagate (mode conversion) [1].

As in the rectangular waveguide, the velocity of propagation and wavelength in the circular waveguide are functions of frequency. The wavelength inside the circular waveguide λ_W is greater than the free space wavelength λ_0 and given by

$$\lambda_W = \frac{\lambda_0}{\sqrt{1 - \left(\frac{f_c}{f}\right)^2}} > \lambda_0 \qquad \text{(Wavelength of guided wave)} \qquad (4.71)$$

The phase velocity v_{ph} is

$$v_{ph} = c_0 \frac{\lambda_W}{\lambda_0} = \frac{c_0}{\sqrt{1 - \left(\frac{f_c}{f}\right)^2}} > c_0 \qquad \text{(Phase velocity)} \qquad (4.72)$$

Like the rectangular waveguide, the circular waveguide is dispersive. The interested reader will find more information on circular waveguides in [1, 2].

Example 4.15 *Modes of a C104 Circular Waveguide*

Let us consider an air-filled C104 circular waveguide. The cut-off frequencies are listed in Table 4.7. The low-loss mode (TE_{01}*) can propagate for frequencies above* $f = 18.62\,\text{GHz}$. *At that frequency four other modes can propagate too:* TE_{11}, TM_{01}, TE_{21} *and* TM_{11}.

Table 4.7 Cut-off frequencies in a C104 waveguide ($R = 10.122\,\text{mm}$) (see Example 4.15)

Mode	TE_{11}	TM_{01}	TE_{21}	TM_{11}, TE_{01}	TE_{31}	TM_{21}
Cut-off freq.	8.68 GHz	11.34 GHz	14.40 GHz	18.62 GHz	19.80 GHz	24.21 GHz

Example 4.16 *Traffic Tunnels as Circular Waveguides*

The relations for circular waveguides can be used to estimate radio propagation in traffic tunnels [20], mining shafts and sewer shafts. Using typical radii we calculate $f_{c,TE_{11}} = 14.7\,\text{MHz}$ *for* $R = 6\,\text{m}$, $f_{c,TE_{11}} = 58.6\,\text{MHz}$ *for* $R = 1.5\,\text{m}$ *and* $f_{c,TE_{11}} = 293\,\text{MHz}$ *for* $R = 0.3\,\text{m}$.

We must be aware that in a real situation a traffic tunnel is crowded with vehicles that represent disturbances of the transmission line geometry. Therefore, reflections occur and the real behaviour may deviate significantly from the first approximation.

4.8 Two-Wire Line

A two-wire transmission line is a symmetrical (balanced) line with two conductors of equal size and shape (see Figure 4.25a). The cylindrical conductors have a diameter d and are separated by distance D (centre-to-centre). Such a transmission line supports a TEM-wave. Signal and return conductors carry opposite currents I^+ and I^- and are encircled by closed magnetic field lines $\vec{H}$ (see Figure 4.25b). Between the two conductors we have open electric field lines $\vec{E}$ due to different electric potentials on the conductors.

In a practical design the conductors are surrounded by dielectric isolating material in order to ensure a constant cross-section along the line. In this case the fundamental mode changes from TEM to quasi-TEM. We will neglect the influence of dielectric material in

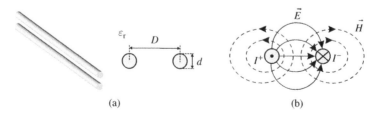

(a)   (b)

Figure 4.25 Geometry of a two-wire line and electric and magnetic field distribution of a TEM-wave.

our discussion. Ready-to-use formulas for the characteristic impedance of two-wire lines surrounded by dielectric material can be found in the literature [7]. Furthermore, EM simulation software can provide data for such transmission line geometries.

Modifications of the basic two-wire line concept are *twisted-pair* lines for high speed data networks. In order to minimize coupling to the electromagnetic environment these lines are *twisted* (with different lay lengths in different pairs) and *shielded*, that is surrounded by foil or braided conducting material. Furthermore, the line is driven by *differential signals*, that is the amplitudes of the electric potentials on the conductors will be of equal absolute value but opposite sign. We will look more deeply into differential and common mode signals in Section 4.9.

4.8.1 Characteristic Impedance

The characteristic impedance of a simple two-wire transmission line surrounded by a homogeneous material with a relative permittivity of ε_r is given by the following closed expression.

$$\boxed{Z_0 = \frac{Z_{F0}}{\pi\sqrt{\varepsilon_r}}\cosh^{-1}\left(\frac{D}{d}\right)}$$ (Characteristic impedance of two-wire line)   (4.73)

Figure 4.26 shows a graphical representation of the characteristic impedance as a function of the distance-to-diameter ratio D/d.

4.8.2 Areas of Application

As mentioned earlier, twisted-pair cables are modifications of the basic two-wire concept and are used in high speed data networks. Another variation of the two-wire concept

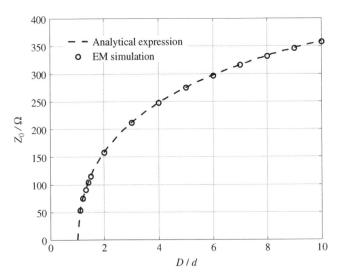

Figure 4.26 Characteristic impedance Z_0 of a two-wire line as a function of D/d (Distance-to-diameter-ratio) for $\varepsilon_r = 1$.

is the star-quad cable. In a star-quad configuration (see Figure 4.1f) two pairs are stranded to form a cable. Opposite conductors represent a two-wire line. The pairs are arranged orthogonally so as to minimize coupling. Such cables are often used in telecommunications.

Example 4.17 *Characteristic Impedance of a Two-wire Transmission Line*
A characteristic impedance of $Z_0 = 100\,\Omega$ is a commonly used value in data networks. According to Equation 4.73 the corresponding distance-to-diameter ratio is $D/d = 1.37$ for an air-filled line.

4.9 Three-Conductor Transmission Line

The transmission lines in the previous sections consist of *one* or *two* conductors. Rectangular and circular waveguides have one conductor and support TE modes as fundamental wave modes. Coaxial line, microstrip, strip line, coplanar and two-wire lines are two conductor lines that have a signal trace and a return trace and support TEM or quasi-TEM waves. Coaxial line, microstrip, strip line and coplanar line are unbalanced, where the signal and return trace are of different size or shape. The two-wire line is a balanced line, where the two conductors are of equal geometry.

We will now look at symmetrical *three*-conductor lines where we have two signal traces that are of equal geometry and a third conductor representing a common ground. As examples we look at three-conductor lines that are derived from a two-wire line and microstrip line.

4.9.1 Even and Odd Modes

A two-wire line is an open line that is susceptible to coupling from external electromagnetic fields. Furthermore, the electromagnetic fields from the waves on the transmission line reach out into the space around the two-wire line (see Figure 4.25b). We can improve the situation by surrounding the two-wire line by a cylindrical shielding enclosure (shielded two-wire line). Now the field of the wave on the line is limited to the interior of the cylindrical enclosure (see Figure 4.27a). Now we have a three-conductor line with two signal traces and a common ground.

Another way to get a symmetrical three-conductor line is to use *coupled microstrip lines* as shown in Figure 4.27b. The strips on the substrate (separated by a distance s) represent the signal traces and the metallic plate on the other side of the substrate is the common ground.

In a *two-conductor configuration* the two conductors 1 and 2 carry currents I_1 and I_2 of equal absolute value but opposite sign $I_1 = -I_2$. Between the conductors a voltage U_{12} exists.

In a *three-conductor configuration* (Figure 4.27c) there can be three different currents I_1, I_2 and I_3. Because of Kirchhoff's node rule the sum of the currents in a cross-section must be zero $I_1 + I_2 + I_3 = 0$. Hence, we have two independent currents (the third is then given by the node rule). Likewise we can define three voltages between the conductors. If we apply the voltage definition in Figure 4.27c Kirchhoff's loop rule requires $U_1 - U_2 - U_s = 0$. Hence, we end up with two independent voltages. If we

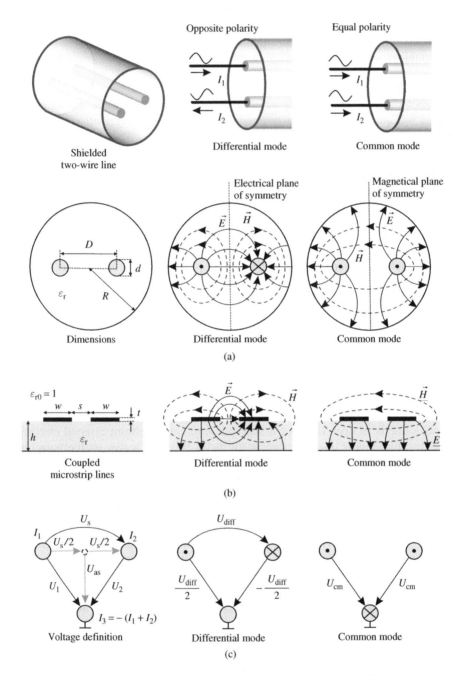

Figure 4.27 Three-conductor transmission line: (a) shielded transmission line and (b) coupled microstrip lines; (c) differential and common mode on a three-conductor line.

assume that the voltages U_1 and U_2 between the two signal traces and ground are the independent voltages, all voltages in Figure 4.27c are determined.

- The voltages U_1 and U_2 between signal lines and common ground are called *unsymmetrical* voltages.
- The voltage U_s between the signal lines is called *symmetrical* voltage.
- The voltage U_{as} is called *asymmetrical* voltage. It is the potential difference between the midpoint potential between the two lines and common ground.

We can calculate the symmetrical voltage U_s and the asymmetrical voltage U_s from the unsymmetrical voltages U_1 and U_2 as

$$\boxed{U_s = U_1 - U_2} \quad \text{and} \quad \boxed{U_{as} = \frac{U_1 + U_2}{2}} \tag{4.74}$$

In a three-conductor configuration any operating condition can be expressed as a *superposition* of two independent fundamental modes: *differential mode* and *common mode* (Figure 4.27).

Symmetrical (balanced) three-conductor lines are commonly operated in *differential mode*. The currents I_1 and I_2 on the signal lines are of equal amplitude but opposite sign. There is no net current on the common ground ($I_3 = 0$).

Likewise the voltages U_1 and U_2 are of equal amplitude but opposite sign. The symmetrical voltage between the signal line is called *differential* voltage. The asymmetrical voltage is zero.

The differential mode can be excited by differential circuits that provide opposite voltages and currents at the signal terminals.

$$\boxed{U_1 = -U_2} \quad \text{and} \quad \boxed{I_1 = -I_2}$$
$$\boxed{U_s = U_{\text{diff}}} \quad \text{and} \quad \boxed{U_{as} = 0} \qquad \text{(Differential mode)} \tag{4.75}$$

The second fundamental mode is the *common mode*. The currents I_1 and I_2 on the signal lines are equal. The common ground acts as a return conductor. Furthermore, the voltages U_1 and U_2 are equal. The differential voltage is zero ($U_s = 0$). In a cross-section the signal lines are of the same potential.

Common mode signals are usually unwanted.

$$\boxed{U_1 = U_2} \quad \text{and} \quad \boxed{I_1 = I_2}$$
$$\boxed{U_s = 0} \quad \text{and} \quad \boxed{U_{as} = U_{\text{cm}}} \qquad \text{(Common mode)} \tag{4.76}$$

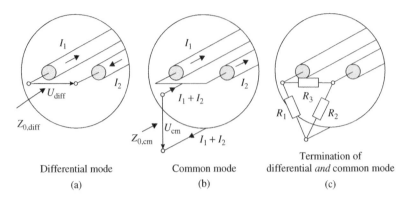

Differential mode   Common mode   Termination of
                                                differential *and* common mode

(a)   (b)   (c)

Figure 4.28 Characteristic line impedance of (a) differential mode and (b) common mode; (c) termination of both modes with a network of three resistors.

4.9.2 Characteristic Impedances and Propagation Constants

Differential and common mode generally have different characteristic impedances $Z_{0,\text{diff}}$ and $Z_{0,\text{cm}}$, respectively. The characteristic impedance $Z_{0,\text{diff}}$ of the differential mode equals the input impedance of an infinitely long line when we feed the line between signal conductor 1 and 2 (see Figure 4.28a). Correspondingly, the characteristic impedance $Z_{0,\text{cm}}$ of the common mode equals the input impedance of an infinitely long line when we combine the signal conductor 1 and 2 and feed the line between the combined signal conductor and common ground (see Figure 4.28b).

In the literature we often find related measures called *odd mode* and *even mode* characteristic impedances $Z_{0\text{o}}$ and $Z_{0\text{e}}$, respectively. These impedances are linked to differential and common mode characteristic impedances $Z_{0,\text{diff}}$ and $Z_{0,\text{cm}}$ by the following equations.

$$Z_{0,\text{diff}} = 2Z_{0\text{o}} \quad \text{and} \quad Z_{0,\text{cm}} = \frac{1}{2}Z_{0\text{e}} \tag{4.77}$$

The odd mode (even mode) characteristic impedance $Z_{0\text{o}}$ ($Z_{0\text{e}}$) is the input impedance of an infinitely long line fed between signal conductor 1 and common ground under the assumption that signal conductor 2 is excited by a signal of the same amplitude and opposite (equal) polarity.

Furthermore, if inhomogeneous dielectric materials are involved, differential and common mode have different propagation constants γ_{diff} and γ_{cm}, respectively. Therefore the modes generally differ in propagation speed and attenuation coefficient.

Perturbations of the cross-sectional area of a three-conductor line can cause *mode conversions*. For instance, if a pure differential signal is fed into the line and propagates through a deformed section of the line the common mode can be excited too. So at the end of the line we see a combination of differential and common mode signals. The superposition of differential mode and common mode can deteriorate the received signal. Therefore, in order to avoid mode conversion a constant cross-section has to be maintained in cable production and cable installation.

Example 4.18 *Even and Odd Mode Impedance of Coupled Microstrip Lines*

We investigate the characteristic impedances of a coupled microstrip line with a distance s between the traces. Each single ended (uncoupled) microstrip line has a characteristic impedance of $Z_0 = 50\,\Omega$. The parameters of the microstrip line are: substrate height $h = 635\,\mu m$, relative permittivity $\varepsilon_r = 9.8$, width of metallic strip $w = 600\,\mu m$, thickness of metallic strip $t = 10\,\mu m$, frequency of operation $f = 5\,GHz$.

Figure 4.29a shows differential and common mode characteristic impedances as a function of the ratio s/h. The even and odd mode impedances are displayed in Figure 4.29b. If we push the traces further away from each other (i.e. we increase the ratio s/h), odd and even mode impedances approach $Z_0 = 50\,\Omega$ because the coupling decreases. (The differential and common mode impedances approach $2Z_0 = 100\,\Omega$ and $Z_0/2 = 25\,\Omega$, respectively.) Bringing the lines together leads to increasing Z_{0e} and decreasing Z_{0o}.

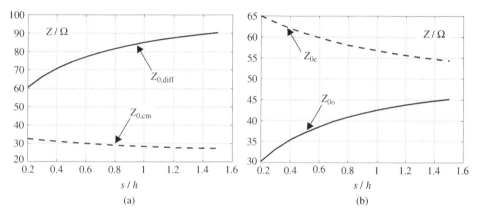

Figure 4.29   (a) Characteristic line impedances of differential and common mode; (b) characteristic line impedances of even and odd mode.

Coupled microstrip lines represent a balanced line that can be used to support differential signals where trace 1 is the signal conductor and trace 2 is the return conductor. The metallic plane on the back side of the substrate represents the common ground. Furthermore, coupled microstrip lines are used to build filters (see Section 6.5) and directional couplers (see Section 6.10). Each of which use coupled quarter wavelength sections to achieve a particular transmission performance. In circuit design an important parameter is the *coupling coefficient k* which represents the coupling for quarter wavelength sections [7]. The following set of equations relates the coupling coefficient k, single ended impedance Z_0 as well as odd and even mode impedance Z_{0o} and Z_{0e}.

$$Z_{0e} = Z_0\sqrt{\frac{1+k}{1-k}} \qquad ; \qquad Z_{0o} = Z_0\sqrt{\frac{1-k}{1+k}} \tag{4.78}$$

$$Z_0 = \sqrt{Z_{0e}Z_{0o}} \qquad ; \qquad k = \frac{Z_{0e} - Z_{0o}}{Z_{0e} + Z_{0o}} \tag{4.79}$$

Although coupling of lines can be used in circuit design to achieve a particular performance, coupling can cause problems in other applications. For example, in integrated electronic circuits coupling of lines may occur unintentionally due to their close proximity. Furthermore, long distance transmission lines may show coupling effects (crosstalk) due to non-ideal manufacturing processes.

4.9.3 Line Termination for Even and Odd Modes

Let us now consider how to terminate a three conductor transmission line. If the line is driven by differential signals (i.e. only the differential mode exists) it is sufficient to terminate the line between the two signal traces by a resistor with a value that equals the differential impedance $Z_A = Z_{0,\text{diff}}$.

In a non-ideal cable a part of the differential mode may be converted into the common mode (mode conversion). If the common mode reaches the end of the line it 'sees' an *open circuit* between the signal traces (which are on the same potential for the common mode) and common ground. Hence, the terminal impedance Z_A between signal trace 1 and 2 has no effect. The common mode is fully reflected back.

In order to avoid reflections of differential *and* common mode waves we can use the following three-resistor network [7] shown in Figure 4.28c.

$$R_1 = R_2 = 2Z_{0,\text{cm}} = Z_{0e} \tag{4.80}$$

$$R_3 = \frac{4Z_{0,\text{cm}}Z_{0,\text{diff}}}{4Z_{0,\text{cm}} - Z_{0,\text{diff}}} = \frac{2Z_{0e}Z_{0o}}{Z_{0e} - Z_{0o}} \tag{4.81}$$

4.10 Problems

4.1 Derive the relations in Equations 4.80 and 4.81 for the termination of differential and common mode. Calculate values of R_1, R_2 and R_3 for the coupled microstrip line from Example 4.18 and a ratio of $s/h = 1$. Estimate differential and common mode impedance from the graphs in Figure 4.29.

4.2 Calculate the cut-off frequencies of the first six modes in a R260 rectangular waveguide ($a = 8.636\,\text{mm}$, $b = 4.318\,\text{mm}$). Determine the characteristic line impedance of a TE_{10} wave at a frequency of $f = 1.5 f_{c,\text{TE}_{10}}$.

4.3 Consider two coaxial lines with characteristic impedances of $Z_{01} = 75\,\Omega$ and $Z_{02} = 125\,\Omega$. Both lines are filled with homogeneous dielectric material. The propagation speed of electromagnetic waves on the line is $0.81c_0$ (81% of the vacuum light speed). Both lines have the same outer radius $R_{o1} = R_{o2} = 2\,\text{mm}$. Determine

a) the relative permittivity ε_r of the dielectric material.
b) the inner radii R_{i1} and R_{i2} of the coaxial lines.

At a frequency of $f = 10\,\text{GHz}$ the two coaxial lines shall be matched by a coaxial quarter-wave transformer with an outer radius $R_{o3} = 2\,\text{mm}$ that equals the outer radii of the other two lines. The inner radius of the coaxial quarter-wave transformer equals the inner radius of line 2: $R_{i3} = R_{i2}$.

a) Calculate the characteristic impedance Z_{03} of the quarter-wave transformer.

b) What is the value of the relative permittivity ε_{r3} of the dielectric material inside the quarter-wave transformer coaxial line?

c) Determine the length ℓ_3 of the quarter-wave transformer.

4.4 A rectangular cavity is defined by the following dimensions: width $a = 7$ cm, height $b = 5$ cm and length $c = 9$ cm. Determine the first five resonance frequencies of the cavity. Use an EM simulation software to visualize the electric and magnetic field strength inside the cavity.

References

1. Pozar DM (2005) *Microwave Engineering*. John Wiley & Sons.
2. Kark K (2010) *Antennen und Strahlungsfelder*, Vieweg.
3. Meinke H, Gundlach FW (1992) *Taschenbuch der Hochfrequenztechnik*. Springer.
4. Dupont International (2010) http://www.dupont.com.
5. Jansen W (1992) *Streifenleiter und Hohlleiter*. Huethig.
6. Rogers Corporation (2010) http://www.rogerscorp.com.
7. Wadell BC (1991) *Transmission Line Design Handbook*. Artech House.
8. AWR (2003) *TX-Line Transmission Line Calculator*. AWR Corporation.
9. Heuermann H (2009) *Hochfrequenztechnik*. Vieweg.
10. Maas SA (1998) *The RF and Microwave Circuit Design Cookbook*. Artech House.
11. Zinke O, Brunswig H (2000) *Hochfrequenztechnik 1*. Springer.
12. Wheeler HA (1978 Transmission-Line Properties of a Strip Line Between Parallel Planes. *IEEE Trans. MTT*, Vol. 26, No. 11.
13. Simons RE (2001) *Coplanar Waveguide Circuits, Components, and Systems*. John Wiley & Sons.
14. Wolff I (2006) *Coplanar Microwave Integrated Circuits*. John Wiley & Sons.
15. Gustrau F, Manteuffel D (2006) EM *Modeling of Antennas and RF Components for Wireless Communication Systems*. Springer.
16. Williams DF, Alpert BK, Arz U, Walker DK, Grabinski H (2003) Causal Characteristic Impedance of Planar Transmission Lines. *IEEE Trans. Adv. Packaging*, Vol. 26, No. 2.
17. Ilyinsky AS, Slepyan GY, Slepyan AY (1993) *Propagation, Scattering and Dissipation of Electromagnetic Waves*. Peregrinus.
18. Golio M, Golio J (2008) *RF and Microwave Handbook*. CRC Press.
19. Bialkowski ME (1995) Analysis of a Coaxial-to-Waveguide Adaptor Including a Discended Probe and a Tuning Post. *IEEE Trans. MTT*, Vol. 43, No. 2.
20. Kraus JD, Fleisch DA (1999) *Electromagnetics with Applications*. McGraw-Hill.

5

Scattering Parameters

The electrical behaviour of linear multi-port networks is generally described by matrix equations. Different matrices like the impedance matrix $\mathbf{Z}$ or admittance matrix $\mathbf{Y}$ are commonly used. The voltages and currents of the different network terminals are then related by a linear system of equations.

In the RF and microwave frequency range we most often use the scattering matrix $\mathbf{S}$ that follows a slightly different approach. At higher frequencies network ports are generally connected to transmission lines that bridge distances between the multi-port network under consideration and other circuits. From our discussion in Chapter 3 we know that forward and backward propagating voltage and current waves can exist on such lines. By means of a suitable normalization of the voltage waves we finally arrive at power waves. Scattering parameters now relate these incoming and reflected waves at the ports of the terminal.

Scattering parameters bring practical benefits: in order to determine the elements of the S-matrix we connect the ports to transmission lines and measure incoming and reflected power waves. A vector network analyser (VNA) can perform this task up to very high frequencies. Direct measurement of voltages and currents at the terminals of a network becomes more and more difficult with increasing frequency. Furthermore, the voltage is not uniquely defined for non-TEM wave modes. By connecting the multi-port network to transmission lines the network is measured under real operating conditions. Measurement of Z- or Y-parameters requires open or short circuited ports. This may be a problem with active circuits.

5.1 Multi-Port Network Representations

In circuit theory we use matrix equations to describe the electrical input-output behaviour (black-box description) of n-port networks, where n is the number of ports. The phasors of voltages U_i and currents I_i at the ports are related by a system of linear equations. To keep things simple at the beginning of our discussion we limit ourself to two-port networks. The results can easily be transferred to more than two ports.

Figure 5.1 shows a two-port network with voltage and current definition. At each port the sum of the currents is zero $I_i = I_i'$. Most often in general circuit theory we use the

RF and Microwave Engineering: Fundamentals of Wireless Communications, First Edition. Frank Gustrau.
© 2012 John Wiley & Sons, Ltd. Published 2012 by John Wiley & Sons, Ltd.

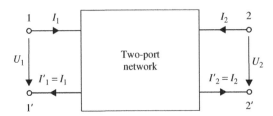

Figure 5.1   Definition of voltages and currents on a two-port network.

Z-parameter (impedance) matrix $\mathbf{Z}$, the Y-parameter (admittance) matrix $\mathbf{Y}$, the chain or ABCD-matrix $\mathbf{A}$ and the hybrid matrix $\mathbf{H}$ [1].

For the impedance matrix we write

$$
\begin{aligned}
U_1 &= Z_{11}I_1 + Z_{12}I_2 \\
U_2 &= Z_{21}I_1 + Z_{22}I_2
\end{aligned}
\quad \text{or} \quad
\underbrace{\begin{pmatrix} U_1 \\ U_2 \end{pmatrix}}_{\mathbf{U}} = \underbrace{\begin{pmatrix} Z_{11} Z_{12} \\ Z_{21} Z_{22} \end{pmatrix}}_{\mathbf{Z}} \underbrace{\begin{pmatrix} I_1 \\ I_2 \end{pmatrix}}_{\mathbf{I}}
\quad \text{or} \quad \mathbf{U} = \mathbf{ZI} \tag{5.1}
$$

The admittance matrix $\mathbf{Y}$ is given by

$$
\begin{aligned}
I_1 &= Y_{11}U_1 + Y_{12}U_2 \\
I_2 &= Y_{21}U_1 + Y_{22}U_2
\end{aligned}
\quad \text{or} \quad
\underbrace{\begin{pmatrix} I_1 \\ I_2 \end{pmatrix}}_{\mathbf{I}} = \underbrace{\begin{pmatrix} Y_{11} Y_{12} \\ Y_{21} Y_{22} \end{pmatrix}}_{\mathbf{Y}} \underbrace{\begin{pmatrix} U_1 \\ U_2 \end{pmatrix}}_{\mathbf{U}}
\quad \text{or} \quad \mathbf{I} = \mathbf{YU} \tag{5.2}
$$

The ABCD matrix is defined by

$$
\begin{aligned}
U_1 &= A_{11}U_2 + A_{12}(-I_2) \\
I_1 &= A_{21}U_2 + A_{22}(-I_2)
\end{aligned}
\quad \text{or} \quad
\begin{pmatrix} U_1 \\ I_1 \end{pmatrix} = \underbrace{\begin{pmatrix} A_{11} A_{12} \\ A_{21} A_{22} \end{pmatrix}}_{\mathbf{A}} \begin{pmatrix} U_2 \\ -I_2 \end{pmatrix} \tag{5.3}
$$

The hybrid matrix takes the following form

$$
\begin{aligned}
U_1 &= h_{11}I_1 + h_{12}U_2 \\
I_2 &= h_{21}I_1 + h_{22}U_2
\end{aligned}
\quad \text{or} \quad
\begin{pmatrix} U_1 \\ I_2 \end{pmatrix} = \underbrace{\begin{pmatrix} h_{11} h_{12} \\ h_{21} h_{22} \end{pmatrix}}_{\mathbf{H}} \begin{pmatrix} I_1 \\ U_2 \end{pmatrix} \tag{5.4}
$$

The impedance matrix is most suitable for series connection of two-port networks and the admittance matrix for parallel connections. In RF applications we often find cascading structures as shown in Appendix A.2.3. Figure A.2 shows a cascading structure of generator, mixer, amplifier, antenna, transmission line and receiver. The chain or ABCD matrix is advantageous for such cascading two-port networks. Transistors are often represented by a hybrid matrix.

In order to calculate or measure the individual matrix elements the ports are terminated by open or short circuits. As an example we will look at the matrix element Z_{11}. From Equation 5.1 we get the following relation

$$
Z_{11} = \left. \frac{U_1}{I_1} \right|_{I_2=0} \tag{5.5}
$$

To put the equation into words: the impedance parameter Z_{11} equals the input impedance of port 1 *under the constraint that* the current I_2 at port 2 is zero. We can satisfy the constraint $I_2 = 0$ easily by letting the terminals of port 2 be unconnected (open circuit).

If we look at active circuits or step into the microwave frequency range a number of problems with these voltage and current based network parameters may arise.

- Active circuits may not be stable if terminated by open or short circuits. It would be desirable to measure these components under real operating (load) conditions.
- Ideal open (and short) circuits are difficult to create in microwave circuits due to their high frequency. Open circuited ($Y = 0$) terminals show stray capacitance due to electric fields between the terminal connections. Hence, we see capacitive behaviour ($Y = j\omega C \neq 0$). Furthermore, unintentional radiation may occur. The terminals act as antennas. Short circuits ($Z = 0$) on the other hand carry currents accompanied by magnetic fields. Hence, there is an inductive component that cannot be neglected at high frequencies ($Z = j\omega L \neq 0$).
- Voltages and currents are not uniquely defined for non-TEM waves and no longer accurately characterize electromagnetic behaviour of circuits. Considering a rectangular waveguide, it is practically impossible to measure a voltage on the line.

In order to overcome the limitations of the voltage and current based matrix representations we advance to power based scattering parameters in the following sections.

5.2 Normalized Power Waves

Let us consider the two-port network in Figure 5.2. The network is connected to two transmission lines with characteristic impedances Z_{01} and Z_{02}.

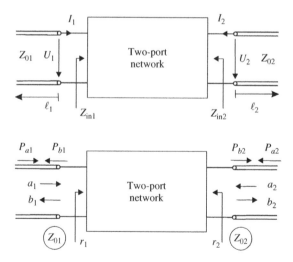

Figure 5.2 Definition of incident and outgoing power waves a_i and b_i at a two-port network attached to transmission lines with characteristic impedances Z_{01} and Z_{02}.

In Section 3.1 we discussed transmission lines and propagating voltage and current waves. We introduced a forward propagating voltage wave U_f and a reflected (backward travelling) voltage wave U_r. Based on these voltage waves we now derive *normalized power waves* by dividing the voltage waves by the square-root of the characteristic line impedances [1].

We define a forward propagating normalized power wave as

$$a_i = \frac{U_{fi}}{\sqrt{Z_{0i}}} \quad \text{(Forward propagating normalized power wave)} \quad (5.6)$$

where i is the port index ($i \in [2, 3]$). Furthermore, we define a backward propagating normalized power wave as

$$b_i = \frac{U_{ri}}{\sqrt{Z_{0i}}} \quad \text{(Backward propagating normalized power wave)} \quad (5.7)$$

For loss-less transmission lines the characteristic impedances are real-valued.[1] The impedances are also referred to as *port reference impedances*. Most often we use port reference impedances that equal the common[2] system impedance of $Z_0 = 50\,\Omega$. Throughout this book we assume port reference impedances of $50\,\Omega$, if not otherwise stated. In order to remind ourselves that we can use port reference impedances Z_{0i} with other values than $50\,\Omega$, we will write the port reference impedance with a circle around it near the port (see Figure 5.2).

The unit of a normalized power wave is $[a_i] = [b_i] = \sqrt{W}$. For a real port reference impedance there is a simple relation between power wave and active power associated with that wave. The wave a_i carries an active (time-averaged) power of

$$P_{ai} = \frac{1}{2}a_i a_i^* = \frac{U_{fi}U_{fi}^*}{2Z_{0i}} = \frac{1}{2}|a_i|^2 \quad \text{(Effective power impinging upon port } i) \quad (5.8)$$

Likewise, the wave b_i carries an active power of

$$P_{bi} = \frac{1}{2}b_i b_i^* = \frac{U_{ri}U_{ri}^*}{2Z_{0i}} = \frac{1}{2}|b_i|^2 \quad \text{(Effective power passing off port } i) \quad (5.9)$$

In Section 3.1 we derived formulas (Equations 3.46 and 3.47) that related the voltage U_0 and the current I_0 at the end of the line to the complex amplitudes of the propagating voltage waves U_f and U_r. For convenience we repeat these equations here.

$$U_f = \frac{1}{2}\left(U_0 + I_0 Z_0\right) \quad \text{and} \quad U_r = \frac{1}{2}\left(U_0 - I_0 Z_0\right) \quad (5.10)$$

If we interpret the voltage U_i and current I_i at port i as the voltage and current at the end of the line U_0 and I_0, we can relate voltage and current at that port with the power

---

[1] We will limit our discussion to real-valued port reference impedances since complex-valued port reference impedances are of minor practical importance. Practical low-loss transmission lines have characteristic impedances that are – with good approximation – real.

[2] Most likely the vector network analyser in your laboratory has $50\,\Omega$-ports.

waves a_i and b_i. The resulting formulas are

$$a_i = \frac{U_{fi}}{\sqrt{Z_{0i}}} = \frac{U_i + Z_{0i}I_i}{2\sqrt{Z_{0i}}} \qquad \text{and} \qquad b_i = \frac{U_{ri}}{\sqrt{Z_{0i}}} = \frac{U_i - Z_{0i}I_i}{2\sqrt{Z_{0i}}} \qquad (5.11)$$

By rearranging Equation 5.11 we get voltage and current at the port i as functions of the power waves a_i and b_i as

$$U_i = (a_i + b_i)\sqrt{Z_{0i}} \qquad \text{and} \qquad I_i = \frac{a_i - b_i}{\sqrt{Z_{0i}}} \qquad (5.12)$$

From Equation 5.12 we see that for a given port reference impedance we always can calculate power waves even if there is no transmission line connected to the ports. Power waves are a *general concept* that can be applied to lumped element port terminations as well, for instance a circuit that is terminated by a load or concentrated source with internal resistance. However, the idea of transmission lines is helpful for associating a physical meaning with the concept of power waves. In RF applications the concept of power waves comes naturally since transmission lines are very often connected to the port.

5.3 Scattering Parameters and Power

The power waves a_i and b_i of a two-port network are related by the following system of linear equations as

$$b_1 = s_{11}a_1 + s_{12}a_2 \qquad (5.13)$$

$$b_2 = s_{21}a_1 + s_{22}a_2 \qquad (5.14)$$

where the coefficients s_{ij} are the *scattering parameters* [4]. We can write the equations in matrix form as

$$\underbrace{\begin{pmatrix} b_1 \\ b_2 \end{pmatrix}}_{\mathbf{b}} = \underbrace{\begin{pmatrix} s_{11} s_{12} \\ s_{21} s_{22} \end{pmatrix}}_{\mathbf{S}} \underbrace{\begin{pmatrix} a_1 \\ a_2 \end{pmatrix}}_{\mathbf{a}} \qquad \text{or} \qquad \mathbf{b} = \mathbf{Sa} \qquad (5.15)$$

where $\mathbf{S}$ is the *scattering matrix*. Matrix elements s_{ii} with equal indices are *reflection coefficients* describing reflection of waves at port i. Matrix elements s_{ij} with different indices $i \neq j$ are *transmission coefficients*, describing the transmission of waves from port j to port i.

From Section 3.1 we already know reflection coefficients r. In order to understand that the scattering parameters with two identical indices s_{ii} are reflection coefficients too, we look at Equation 5.13, which is given by $b_1 = s_{11}a_1 + s_{12}a_2$. Let us assume that port 2 is terminated by an infinite transmission line with a characteristic impedance of Z_{02}. On this transmission line no incident wave a_2 exists. Hence, $a_2 = 0$. (Alternatively, we can terminate port 2 with a resistor that equals the characteristic line impedance of that port $R = Z_{02}$.) Due to Ohm's law the resistor enforces $U_2 = Z_{02}(-I_2)$. The minus sign stems from our definition of voltage and current at the terminals in Figure 5.2. From Equation 5.11 we get $a_2 = 0$ for $U_2 + Z_{02}I_2 = 0$.

Under the condition $a_2 = 0$ (i.e. port 2 terminated with a load equal to the port reference impedance Z_{02} (matched termination)) the scattering parameter s_{11} becomes

$$s_{11} = \left.\frac{b_1}{a_1}\right|_{a_2=0} = \frac{U_{r1}/\sqrt{Z_{01}}}{U_{f1}/\sqrt{Z_{01}}} = \frac{U_{r1}}{U_{f1}} = r_1 \tag{5.16}$$

where we use the definitions of a_1 and b_1 from Equations 5.6 and 5.7 and the definition of the reflection coefficient $r = U_r/U_f$ at the end of a line in Equation 3.130. So the scattering parameter s_{11} equals the reflection coefficient r_1 at the end of the feeding line at port 1.

Now we derive the following meaning for the scattering parameters

$$s_{11} = \left.\frac{b_1}{a_1}\right|_{a_2=0} = r_1 \qquad \text{(Input reflection coefficient for matched output)} \tag{5.17}$$

$$s_{21} = \left.\frac{b_2}{a_1}\right|_{a_2=0} \qquad \text{(Forward transmission coefficient for matched output)} \tag{5.18}$$

$$s_{12} = \left.\frac{b_1}{a_2}\right|_{a_1=0} \qquad \text{(Backward transmission coefficient for matched input)} \tag{5.19}$$

$$s_{22} = \left.\frac{b_2}{a_2}\right|_{a_1=0} = r_2 \qquad \text{(Output reflection coefficient for matched input)} \tag{5.20}$$

Figure 5.3 visualizes the signal flow graphically through arrows indicating reflection and transmission paths. The ports are represented as *single poles*. Single pole representation is often used in RF engineering in order to draw clearly structured circuit layouts and is especially suited for power wave display. A port in single pole display may represent two terminals (e.g. for connecting a coaxial line) or a waveguide port (for connecting a rectangular or circular waveguide).

The *single pole representation* of ports reminds us that in RF engineering ports are not fully characterized by two idealized terminals (as in the low-frequency range). At higher frequencies the port cross-sectional area and the electromagnetic wave mode most often have to be considered. Therefore, high-frequency simulation software tools provide different types of ports:

- *Lumped* or *concentrated ports* defined by two terminals that are suitable for lower frequencies.
- *Waveguide ports* that define a transmission line cross-section where a specified wave mode interacts with the simulation domain. This port-type generally provides greater accuracy due to the physical modelling aspects.

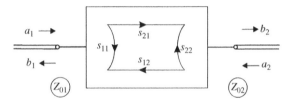

Figure 5.3   Scattering parameters (simplified representation of a port by a single pole).

Absolute values of scattering parameters are often given in logarithmic scale as

$$\boxed{\frac{s_{ij}^{\ell}}{\mathrm{dB}} = 20 \lg |s_{ij}|}$$ (Scattering parameters expressed in logarithmic scale)   (5.21)

The index ℓ is introduced formally to indicate a logarithmic value. However, in praxis the index is dropped, and there is no formal distinction, since the pseudo unit dB indicates the logarithmic scale and avoids possible confusion. Therefore, in this book we drop the index ℓ too.

The reciprocal values of the scattering parameters are referred to as losses, usually given in logarithmic scale. From the reflection coefficients we derive the *return loss* as

$$\frac{RL}{\mathrm{dB}} = 20 \lg \left| \frac{1}{s_{ii}} \right| = -20 \lg |s_{ii}| = -\frac{s_{ii}^{\ell}}{\mathrm{dB}}$$ (5.22)

The reciprocal value of a transmission coefficient is named *insertion loss*

$$\frac{IL}{\mathrm{dB}} = 20 \lg \left| \frac{1}{s_{ij}} \right| = -20 \lg |s_{ij}| = -\frac{s_{ij}^{\ell}}{\mathrm{dB}}$$ (5.23)

In logarithmic scale scattering parameters and losses have opposite signs and equal dB-values (see Example 5.1).

Equations 5.8 and 5.9 relate normalized power waves a_i and b_i to incident active power P_{ai} and reflected active power P_{bi}. Scattering parameters, on the other hand, relate normalized power waves a_i and b_i at different ports. Therefore, scattering parameters relate incident and reflected power at different ports.

The squared absolute value of s_{11} gives us the ratio of reflected to incident active power at port 1.

$$|s_{11}|^2 = \frac{|b_1|^2}{|a_1|^2} = \frac{P_{b1}}{P_{a1}}$$ (5.24)

The squared absolute value of s_{21} is the ratio of active power passing off port 2 to incident active power at port 1.

$$|s_{21}|^2 = \frac{|b_2|^2}{|a_1|^2} = \frac{P_{b2}}{P_{a1}}$$ (5.25)

The *squared absolute values* of the scattering parameters describe the reflected and transferred active power with reference to the incident active power. So from the scattering matrix we can easily derive power transfer relations between the different ports of a multi-port network.

Example 5.1 *Reflection Coefficient and Power*

Let us consider a network port with an incident active power of P_{a1} and a reflected power P_{b1}. The difference of incident and reflected power is accepted by the port

$$P_{1,acc} = P_{a1} - P_{b1} \qquad (5.26)$$

We put some numbers in here and assume a reflection coefficient of $|s_{11}| = 0.1$ and an incident active power of $P_{a1} = 1\,\mathrm{W}$. Then the reflected power is $P_{b1} = |s_{11}|^2 P_{a1} = 0.01\,P_{a1} = 10\,\mathrm{mW}$. In words 1% of the incident power is reflected. Hence, 99% of the incident power is accepted. The accepted power is given by

$$P_{1,acc} = P_{a1}\left(1 - |s_{11}|^2\right) \qquad \text{(Accepted power)} \qquad (5.27)$$

The reflection coefficient in logarithmic scale is $s_{11} = -20\,\mathrm{dB}$ (return loss $RL = +20\,\mathrm{dB}$). (Section A.2 gives an overview of logarithmic values.)

5.4 S-Parameter Representation of Network Properties

Networks may have special electric properties like reciprocity or loss-lessness. These properties can be directly identified by looking at the scattering parameters as we will show in the following sections.

5.4.1 Matching

A matched port has a zero-valued reflection coefficient. A network that is *matched at all ports* has only reflection coefficients equal to zero.

$$\boxed{s_{ii} = 0 \qquad \forall i} \qquad \text{(Matched circuit} \,\widehat{=}\, \text{all ports are matched)} \qquad (5.28)$$

According to Equation 5.17 a reflection coefficient that is zero means that there is no reflected wave b_i at that port. The input impedance $Z_{\mathrm{in}i}$ at port i equals the port reference impedance Z_{0i}. Matching (i.e. the absence of reflections) is generally desirable in RF circuits. Perfect matching is practically impossible, so we try to keep reflections under a certain limit depending on our application. Typical limits where we consider matching to be sufficient are $s_{11} \leq -20\,\mathrm{dB}$ (1% reflected power), $s_{11} \leq -10\,\mathrm{dB}$ (10% reflected power) or $s_{11} \leq -6\,\mathrm{dB}$ (25% reflected power). The frequency range where we find our requirements satisfied is called (impedance) bandwidth.

Figure 5.4 shows as an example a frequency dependent reflection coefficient of an antenna. In the frequency range from the lower frequency $f_{L,10\,\mathrm{dB}}$ to the upper frequency $f_{U,10\,\mathrm{dB}}$ the reflection coefficient is $-10\,\mathrm{dB}$ or less (the return loss is $+10\,\mathrm{dB}$ or more).

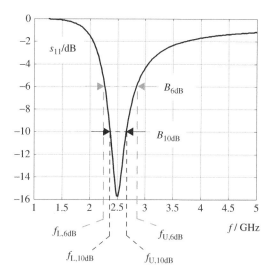

Figure 5.4 Definition of 6 dB and 10 dB bandwidth.

So the bandwidth is referred to as 10 dB bandwidth $B_{10\,\text{dB}}$. More generally we write

$$\boxed{B_{x\text{dB}} = f_{\text{U},x\text{dB}} - f_{\text{L},x\text{dB}}} \qquad \text{(Bandwidth)} \qquad (5.29)$$

where $f_{\text{U},x\text{dB}}$ and $f_{\text{L},x\text{dB}}$ are the maximum and minimum frequency, respectively. The *relative bandwidth* is given by

$$\boxed{B_{\text{rel}.x\text{dB}} = \frac{f_{\text{U},x\text{dB}} - f_{\text{L},x\text{dB}}}{f_0}} \qquad \text{(Relative bandwidth)} \qquad (5.30)$$

where f_0 is the centre frequency $f_0 = 1/2(f_{\text{U},x\text{dB}} + f_{\text{L},x\text{dB}})$.

We use the term *mismatched port* if the reflection coefficient happens to be higher than the specified limit. In this case we may apply special two-port networks to ensure sufficient matching conditions in a desired frequency range. In our discussion of the Smith chart (Section 3.1.11) we saw that transmission lines and reactive components can be used efficiently to transform an arbitrary impedance to a desired input impedance. Further *matching circuits* will be discussed in Section 6.3.

However, although matching is generally desirably selective *mismatch* can be used in filter design to reflect unwanted signal components. We will investigate such filters in Section 6.4.

5.4.2 Complex Conjugate Matching

In the previous section we considered real-valued port references since we assume loss-less lines to be connected to the ports. Loss-less lines have real-valued characteristic impedances. In this case reflection coefficients equal to zero ensure that the incident active power running down the line towards our port is fully accepted by the port. No reflected wave occurs.

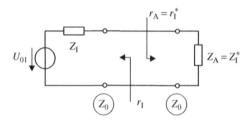

Figure 5.5   Reflection coefficients for conjugate complex matching (for maximum power transfer).

Now we consider a slightly different case. We connect an equivalent source (i.e. an ideal voltage source in series with a complex internal impedance Z_I) to a port with an input impedance of Z_A (see Figure 5.5). We further assume a real-valued port reference impedance Z_0. If we now want to deliver the maximum power from the source to the load we use complex-conjugate matching as

$$\boxed{r_A = r_I^*}\qquad \text{(Complex conjugate matching)}\tag{5.31}$$

We will investigate three different cases and compare impedances Z and reflection coefficients r.

Case 1 First we assume that the load impedance Z_A and the source impedance Z_i equal the port reference impedance Z_0.

$$Z_A = Z_I = Z_0 \in \mathbb{R}\tag{5.32}$$

In this case the reflection coefficient r_A at the load and the reflection coefficient at the source r_I are equal to zero.

$$r_A = \frac{Z_A - Z_0}{Z_A + Z_0} = 0 \quad \text{and} \quad r_I = \frac{Z_I - Z_0}{Z_I + Z_0} = 0\tag{5.33}$$

Hence, maximum power is transferred because Equation 5.31 is satisfied.

$$\underline{r_A = r_I^*}\tag{5.34}$$

Case 2 Next we assume that the load impedance Z_A and the source impedance Z_I are equal and real as in the first case, but we choose a different real-valued port reference impedance Z_0.

$$Z_A = Z_I \neq Z_0 \quad \text{and} \quad Z_A, Z_I, Z_0 \in \mathbb{R}\tag{5.35}$$

Now, the reflection coefficients r_A and r_I are non-zero, equal and real.

$$r_A = r_I \in \mathbb{R}\tag{5.36}$$

Therefore, maximum power is transferred. Compared to case 1 the circuit is identical, the only thing that has changed is the port reference impedance for the calculation of the scattering parameters. So, again Equation 5.31 is satisfied.

$$\underline{r_A = r_I^*}\tag{5.37}$$

Case 3 Now we choose the interesting case of complex conjugate impedances of load and source.

$$\boxed{Z_A = Z_I^*} \qquad \text{(Complex conjugate impedance)} \tag{5.38}$$

where Z_A and Z_I are complex impedances that can be written using resistances R and reactances X as

$$Z_A = R_A + jX_A \qquad \text{and} \qquad Z_I = R_I + jX_I = Z_A^* = R_A - jX_A \tag{5.39}$$

From circuit theory we know that Equation 5.38 represents the necessary condition for maximum power transfer. We choose a real-port reference impedance Z_0 and calculate the reflection coefficients as

$$r_A = \frac{Z_A - Z_0}{Z_A + Z_0} = \frac{R_A + jX_A - Z_0}{R_A + jX_A + Z_0} = \frac{R_A^2 - Z_0^2 + X_A^2 + j2X_AZ_0}{R_A^2 + Z_0^2 + X_A^2} \tag{5.40}$$

$$r_I = \frac{Z_I - Z_0}{Z_I + Z_0} = \frac{R_A - jX_A - Z_0}{R_A - jX_A + Z_0} = \frac{R_A^2 - Z_0^2 + X_A^2 - j2X_AZ_0}{R_A^2 + Z_0^2 + X_A^2} \tag{5.41}$$

If we compare the reflection coefficients in Equations 5.40 and 5.41 we recognize that their only difference is the opposite sign in the imaginary part. Hence, we can write

$$r_A = r_I^* \tag{5.42}$$

So we conclude that complex conjugate reflection coefficients correspond to complex conjugate input impedances.[3]

5.4.3 Reciprocity

A networks is said to be *reciprocal* if the transmission between two ports i and j is the same regardless of the direction. The transmission coefficient does not change if we swap the indices.

$$\boxed{s_{ij} = s_{ji} \quad \forall i, j \quad \text{with} \quad i \neq j} \qquad \text{(Reciprocity)} \tag{5.43}$$

Networks made of *passive* and *isotropic* materials are generally reciprocal. Passive filters, couplers and power dividers consisting of transmission lines, resistors, capacitors and so on are generally reciprocal circuits. Typical non-reciprocal networks are amplifiers (use of active components) and circulators (using anisotropic magnetic materials). Furthermore, satellite communication links may be non-reciprocal due to wave propagation through anisotropic plasma.

---

[3] Indeed we have already seen this when we discussed the Smith chart diagram in Section 3.1.11. If we mirror a point in the Smith chart at the real (horizontal) axis we get complex conjugate reflection coefficient and complex conjugate impedance.

5.4.4   Symmetry

A *symmetrical* network is reciprocal *and* has equal reflection coefficients at all ports.

$$\boxed{s_{ii} = s_{jj} \quad \forall i, j} \quad \text{and} \quad \boxed{s_{ij} = s_{ji} \quad \forall i, j \quad \text{with} \quad i \neq j} \qquad \text{(Symmetry)} \qquad (5.44)$$

Symmetry is most often a consequence of a symmetric circuit layout.

5.4.5   Passive and Loss-less Circuits

When a passive network is loss-less its scattering matrix S is a *unitary matrix* that satisfies the following equation

$$\boxed{\mathbf{S}^T \cdot \mathbf{S}^* = \mathbf{I}} \qquad \text{(Unitary matrix condition)} \qquad (5.45)$$

where $\mathbf{S}^T$ is the transpose of $\mathbf{S}$ and $\mathbf{S}^*$ is the complex conjugate of $\mathbf{S}$. The *identity matrix* $\mathbf{I}$ has ones in the main diagonal. All other matrix elements are zero. Equation 5.45 expresses mathematically a relation between all scattering parameters, so that all power that impinges upon a network is either reflected or passed through to other ports. No power is absorbed inside the network.

In order to better understand Equation 5.45 we derive some useful formulas for the absolute values of reflection and transmission coefficients at a two-port network. Let's look at the passive, loss-less two-port network in Figure 5.6 with an incoming active power P_{a1} at port 1, a reflected power P_{b1} at port 1 and an outgoing power P_{b2} at port 2.

Since the network is loss-less the incoming and outgoing power must be balanced. We write

$$P_{b1} + P_{b2} = P_{a1} \qquad (5.46)$$

or

$$\frac{P_{b1}}{P_{a1}} + \frac{P_{b2}}{P_{a1}} = 1 \qquad (5.47)$$

With Equations 5.24 and 5.25 we get the following relation for the absolute values of s_{11} and s_{21}

$$\boxed{|s_{11}|^2 + |s_{21}|^2 = 1} \qquad \text{(Loss-less two-port network)} \qquad (5.48)$$

Likewise we obtain for an incident power P_{a2} at port 2

$$\boxed{|s_{22}|^2 + |s_{12}|^2 = 1} \qquad \text{(Loss-less two-port network)} \qquad (5.49)$$

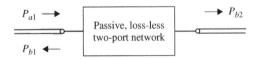

Figure 5.6   Incident, reflected and transmitted power on a passive, loss-less two-port network.

Equations 5.48 and 5.49 are derived from power considerations and are essential parts of Equation 5.45 as we can show by applying this equation to a two-port network.

$$\mathbf{S}^{\mathrm{T}} \cdot \mathbf{S}^* = \begin{pmatrix} s_{11} & s_{21} \\ s_{12} & s_{22} \end{pmatrix} \begin{pmatrix} s_{11}^* & s_{12}^* \\ s_{21}^* & s_{22}^* \end{pmatrix} = \begin{pmatrix} 1 & 0 \\ 0 & 1 \end{pmatrix} \tag{5.50}$$

If we expand the matrix equation we get the following set of four equations.

$$|s_{11}|^2 + |s_{21}|^2 = 1 \tag{5.51}$$

$$s_{11}s_{12}^* + s_{21}s_{22}^* = 0 \tag{5.52}$$

$$s_{12}s_{11}^* + s_{22}s_{21}^* = 0 \tag{5.53}$$

$$|s_{22}|^2 + |s_{12}|^2 = 1 \tag{5.54}$$

So Equation 5.51 corresponds to Equation 5.48 and Equation 5.54 corresponds to Equation 5.49.

Example 5.2 *Passive and Loss-less Two-Port Network*

Let us consider a passive, loss-less two-port network where we measure an absolute value of the reflection coefficient of $|s_{11}| = 0.3$ (return loss $RL = 10.46$ dB). According to Equation 5.48 we can calculate the absolute value of the transmission coefficient s_{21} as

$$|s_{21}| = \sqrt{1 - |s_{11}|^2} = 0.954 \tag{5.55}$$

which corresponds to an insertion loss of $IL = 0.41$ dB. Consequently, 9% $(= |s_{11}|^2)$ of the incident power is reflected and 91% $(= |s_{21}|^2)$ of the incident power is transmitted.

5.4.6 Unilateral Circuits

A two-port network is *unilateral* if it transmits signals only in one direction, for example from port 1 to port 2. A unilateral network has one transmission coefficient equal to zero and the other transmission coefficient is non-zero.

$$\boxed{s_{12} = 0} \quad \text{and} \quad \boxed{s_{21} \neq 0} \quad \text{(Unilateral two-port network)} \tag{5.56}$$

Unilateral networks are easier to analyse than non-unilateral networks, since no internal feedback has to be considered. Amplifiers and isolators are circuits that can be regarded as unilateral networks in a first approximation.

5.4.7 Specific Characteristics of Three-Port Networks

A three-port network cannot satisfy the following conditions simultaneously

- matching (at all ports),
- reciprocity and
- loss-lessness.

In Problem 5.8 we show that assuming that all conditions are met leads to contradictory equations. So, in a realizable three-port network one condition must be dropped. There are two particularly interesting applications: If we abstain from reciprocity we can construct a *circulator* that is matched at all ports (see Section 6.6). On the other hand, if we allow losses we can build a *power divider* that is matched at all ports (see Section 6.7).

5.5 Calculation of S-Parameters

The calculation of scattering parameters is similar to the calculation of impedance parameters and follows the common rules of circuit analysis (Ohm's Law, Kirchhoff's law, ...). It is important to terminate all ports by their port reference impedance Z_{0i} during calculation.

5.5.1 Reflection Coefficients

We have already calculated reflection coefficients in Section 3.1. So we can directly adopt the relations from our previous discussion.

The reflection coefficient s_{ii} at port i is given by

$$s_{ii} = \frac{Z_{\text{in}i} - Z_{0i}}{Z_{\text{in}i} + Z_{0i}} \qquad \text{(Reflection coefficient)} \qquad (5.57)$$

where $Z_{\text{in}i}$ is the input impedance of port i and Z_{01} is the port reference impedance. Figure 5.7 shows the circuit for the calculation of s_{11} on a two-port network. Port 2 is terminated by the port reference impedance Z_{02}.

5.5.2 Transmission Coefficients

In order to calculate the transmission coefficient from port i to port j we connect an equivalent voltage source (ideal voltage source U_{0i} and internal resistance Z_{0i}) to port i. All other ports are terminated by their port reference impedances. The transmission coefficient s_{ji} is then given by

$$s_{ji} = \frac{2U_j}{U_{0i}} \sqrt{\frac{Z_{0i}}{Z_{0j}}} \qquad \text{(Transmission coefficient)} \qquad (5.58)$$

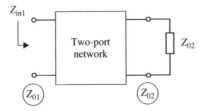

Figure 5.7 Circuit for the calculation of reflection coefficient s_{11}.

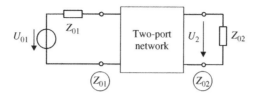

Figure 5.8   Circuit for the calculation of transmission coefficient s_{21}.

Figure 5.8 shows the corresponding circuit for a two-port network. We now look at two examples in order to demonstrate how to apply Equations 5.57 and 5.58 practically in circuit analysis. Further examples are given in Problem 5.3.

Example 5.3 *S-Parameters of a Two-Port Network with Shunt Impedance*
Let us consider a two-port network with a shunt impedance Z and port reference impedances Z_0 at both ports (Figure 5.9).
First we determine the input impedance Z_{in1} at port 1 as the parallel circuit of the shunt impedance Z and the port reference impedance Z_0

$$Z_{in1} = Z||Z_0 = \frac{Z \cdot Z_0}{Z + Z_0} \tag{5.59}$$

Next we apply Equation 5.57 to calculate s_{11} as

$$s_{11} = \frac{Z_{in1} - Z_0}{Z_{in1} + Z_0} = \frac{ZZ_0 - Z_0(Z + Z_0)}{ZZ_0 + Z_0(Z + Z_0)} = \frac{-Z_0}{2Z + Z_0} = s_{22} \tag{5.60}$$

Since the circuit has a symmetric layout, we conclude that $s_{11} = s_{22}$.
With Equation 5.58 we determine the transmission coefficient s_{21}. First we use the voltage divider rule to calculate the output-to-source-voltage-ratio U_2/U_{01}.

$$\frac{U_2}{U_{01}} = \frac{Z||Z_0}{Z_0 + Z||Z_0} = \frac{ZZ_0}{Z_0(Z + Z_0) + ZZ_0} = \frac{Z}{2Z + Z_0} \tag{5.61}$$

By using Equation 5.58 the transmission coefficient becomes

$$s_{21} = \frac{2U_2}{U_{01}} \sqrt{\frac{Z_0}{Z_0}} = \frac{2Z}{2Z + Z_0} = s_{12} \tag{5.62}$$

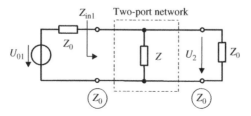

Figure 5.9   Two-port network with shunt impedance Z (see Example 5.3).

The circuit consists only of an impedance Z. Hence, the circuit is reciprocal ($s_{12} = s_{21}$).
We combine our results and write the scattering matrix **S** *as*

$$\boxed{\mathbf{S} = \frac{1}{2Z + Z_0} \begin{pmatrix} -Z_0 & 2Z \\ 2Z & -Z_0 \end{pmatrix}}$$ (S-matrix of a shunt impedance Z)   (5.63)

It is good practice to test the results against canonical cases where the solution is known.
First we consider an open circuit $Z \to \infty$. The scattering matrix then becomes

$$\mathbf{S} = \begin{pmatrix} 0 & 1 \\ 1 & 0 \end{pmatrix}$$ (5.64)

The two-port network is a simple through connection. As expected, no reflection occurs
($s_{11} = s_{22} = 0$) and the transmission coefficient is equal to one ($s_{12} = s_{21} = 1$), that is, all
power is transferred through the network.
    Next we consider a short circuit ($Z = 0$). The scattering parameter then is

$$\mathbf{S} = \begin{pmatrix} -1 & 0 \\ 0 & -1 \end{pmatrix}$$ (5.65)

For $Z = 0$ the input impedances at the ports are short circuits. As expected, the reflec-
tion coefficients are $s_{11} = s_{22} = -1$ (see Section 3.1), all power is reflected. No power
is transferred through the network. So the transmission coefficients are equal to zero
($s_{12} = s_{21} = 0$).

Example 5.4 *S-Parameters of a Matched Loss-less Transmission Line*
    Let us consider a loss-less line that is matched at both ends (see Figure 5.10). In order
to apply Equation 5.57 we need the input impedance at the ports. The input impedance
of a matched line is the characteristic line impedance $Z_{in1} = Z_0$ (see Section 3.1). So the
input reflection coefficient is

$$s_{11} = \frac{Z_{in1} - Z_0}{Z_{in1} + Z_0} = 0 = s_{22}$$ (5.66)

According to the symmetric structure the output reflection coefficient is also zero
($s_{22} = s_{11} = 0$).
    Equation 5.58 defines the transmission coefficient s_{21} based on the ratio of the out-
put voltage to the source voltage U_2/U_{01}. We can derive the voltage relation from our

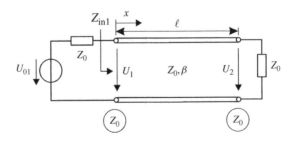

Figure 5.10 Two-port network with matched transmission line (see Example 5.4).

knowledge of transmission line theory (Section 3.1). On a matched transmission line a voltage wave propagates from the input to the output terminals. There is no reflected wave. We define x to be the coordinate along the line. At port 1 the coordinate is $x = 0$. Now we can write the following equation for the voltage wave

$$U(x) = U_1 e^{-j\beta x} \tag{5.67}$$

where U_1 is the voltage at port 1. If we substitute the line length ℓ for x we find the voltage at port 2 as

$$U_2 = U(\ell) = U_1 e^{-j\beta \ell} \tag{5.68}$$

Since the input impedance of the line is $Z_{in1} = Z_0$ the voltage divider rule can be used to find a relation between the source voltage U_{01} and the voltage U_1 at port 1.

$$U_1 = \frac{U_{01}}{2} \tag{5.69}$$

So the transmission coefficient s_{21} becomes

$$s_{21} = \frac{2U_2}{U_{01}} \sqrt{\frac{Z_0}{Z_0}} = e^{-j\beta \ell} = s_{12} \tag{5.70}$$

A transmission line is a reciprocal network, so $s_{12} = s_{21}$. The scattering matrix of the matched and loss-less transmission line summarizes our results.

$$\boxed{\mathbf{S} = \begin{pmatrix} 0 & e^{-j\beta \ell} \\ e^{-j\beta \ell} & 0 \end{pmatrix}} \qquad \text{(S-matrix of a matched and loss-less line)} \tag{5.71}$$

A matched and loss-less transmission line only changes the phases of the voltage waves on the line due to finite propagation speed. All power is transferred from one port to the other, since the absolute values of the transmission coefficients are equal to one.

5.5.3 Renormalization

Scattering matrices describe the operating characteristics of a network when the ports are terminated by impedances that equal the port reference impedances ($Z_i = Z_{0i}$). If the network is terminated by impedances ($Z_{i,\text{new}} \neq Z_{0i}$) other than the port reference impedances the networks show a different electrical behaviour. We can calculate the modified behaviour by renormalizing the scattering matrix to new port reference impedances ($Z_{0i,\text{new}} = Z_{i,\text{new}}$).

In practice most often port reference impedances are the same for all ports. Therefore, we look at a network with uniform port reference impedance Z_0. The network will be used in a different environment with a modified uniform port reference impedance $Z_{0,\text{new}}$. According to [5] the modified scattering matrix $\mathbf{S}_{\text{new}}$ is determined by

$$\mathbf{S}_{\text{new}} = (\mathbf{S} - r\mathbf{I})(\mathbf{I} - r\mathbf{S})^{-1} \tag{5.72}$$

where $\mathbf{I}$ is the identity matrix, the exponent -1 denotes the inverse of a matrix and the factor r is given as

$$r = \frac{Z_{0,\text{new}} - Z_0}{Z_{0,\text{new}} + Z_0} \tag{5.73}$$

For a two-port network we get the following relation

$$\mathbf{S}_{\text{new}} = \begin{pmatrix} s_{11} - r & s_{12} \\ s_{21} & s_{22} - r \end{pmatrix} \cdot \begin{pmatrix} 1 - rs_{11} & -rs_{12} \\ -rs_{21} & 1 - rs_{22} \end{pmatrix}^{-1} \tag{5.74}$$

The inverse of a matrix [6] is generally given by

$$\mathbf{A}^{-1} = \begin{pmatrix} a_{11} & a_{12} \\ a_{21} & a_{22} \end{pmatrix}^{-1} = \frac{1}{\det \mathbf{A}} \begin{pmatrix} A_{11} & A_{12} \\ A_{21} & A_{22} \end{pmatrix}^{\mathsf{T}} \tag{5.75}$$

where $\det \mathbf{A}$ is the determinant of the matrix $\mathbf{A}$. The term A_{ij} denotes the adjoint of element a_{ij} multiplied by a factor $(-1)^{i+j}$. The adjoint is the determinant of the matrix that results from deleting row i and column j of matrix $\mathbf{A}$. For a two-port network the renormalized matrix becomes

$$\mathbf{S}_{\text{new}} = \frac{1}{\det \mathbf{A}} \begin{pmatrix} (s_{11} - r)(1 - rs_{22}) + rs_{12}s_{21} & s_{12}(1 - r^2) \\ s_{21}(1 - r^2) & (s_{22} - r)(1 - rs_{11}) + rs_{12}s_{21} \end{pmatrix} \tag{5.76}$$

with the determinant

$$\det \mathbf{A} = (1 - rs_{11})(1 - rs_{22}) - r^2 s_{12}s_{21} \tag{5.77}$$

Renormalization of scattering parameters is conveniently done by using RF circuit simulation software. More general formulas for the renormalization of s-parameters can be found in [7].

Example 5.5 *S-Parameter Renormalization of a Two-Port Network*
Let us consider a two-port network with a shunt resistor of $Z = 75\,\Omega$ according to Figure 5.9. The port reference impedance for all ports shall be $Z_0 = 50\,\Omega$. With Equation 5.63 we calculate the scattering matrix as

$$\mathbf{S}_{50\Omega} = \frac{1}{2Z + Z_0} \begin{pmatrix} -Z_0 & 2Z \\ 2Z & -Z_0 \end{pmatrix} = \begin{pmatrix} -0.25 & 0.75 \\ 0.75 & -0.25 \end{pmatrix} = \begin{pmatrix} s_{11} & s_{12} \\ s_{21} & s_{22} \end{pmatrix} \tag{5.78}$$

Let us now renormalize the scattering matrix $\mathbf{S}_{50\Omega}$ to a different port reference impedance of $Z_{0,\text{new}} = 100\,\Omega$. We easily calculate the renormalized matrix $\mathbf{S}_{100\Omega}$ from Equation 5.76 using the determinant in Equation 5.77 and the factor r in Equation 5.73.

$$\mathbf{S}_{100\Omega} = \begin{pmatrix} -0.4 & 0.6 \\ 0.6 & -0.4 \end{pmatrix} \tag{5.79}$$

Alternatively, we may – for comparison purposes – directly calculate the scattering matrix of a shunt impedance Z for a port reference impedance of $Z_{0,\text{new}} = 100\,\Omega$ by substituting the numerical values in Equation 5.63. Of course the results are identical.

5.6   Signal Flow Method

The scattering matrix $\mathbf{S}$ interrelates the incoming and outgoing normalized power waves a_i and b_i at all ports i of a network. If we interconnect several networks to form a circuit we can write down a set of equations that interrelate all power waves in that circuit. Figure 5.11 shows a circuit with an equivalent source, a two-port network and a load. By setting up and solving the resulting system of equations we derive the electrical behaviour of the circuit.

As an alternative to the algebraic approach we can figure out the electrical behaviour by the *signal flow method*. This method represents a more direct and visual approach, especially when only simple feedback loops are involved. In the following we will discuss the basic rules and apply them to an illustrative example.

The signal flow method uses *signal flow graphs* to graphically represent scattering parameters and power waves. The scattering parameters are represented by arrows and the power waves are represented by nodes.

Signals that are passed along directed edges in the signal flow graph can be described by a set of rules given in Figure 5.12. The rules correlate signal flow graph and algebraic relation. The first three rules (directed, parallel and series connection) are intuitively clear. It is important to note that – for example – the equation $b = sa$ is an *assignment*. One is *not permitted* to rearrange the equation and write $a = (1/s)b$.

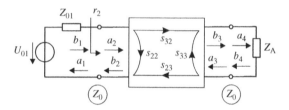

Figure 5.11   Calculation of RF circuits using power waves (two-port network with source and load).

Description	Signal flow graph	Equations
Directed connection	$a \quad\;\; s \quad\;\; b$	$b = sa$
Parallel connection	$a \quad\; s_y \quad\; b$ / s_x	$b = s_x a + s_y a = (s_x + s_y)a$
Serial connection	$a \quad c \quad b$ / $s_x \quad s_y$	$b = s_x s_y a$
Feedback	$a \quad s_y \quad b$ / s_x	$b = \dfrac{s_x}{1 - s_x s_y} a$

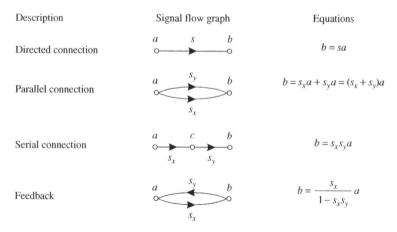

Figure 5.12   Basic rules for signal flow graphs.

The feedback rule is derived in Problem 5.6 and known from basic control technology for calculating gain in closed loop systems.

$$G_{\text{loop}} = \frac{G_f}{1 - G_{cl}} \tag{5.80}$$

where G_f is the *open loop forward gain* and G_{cl} is the *closed loop gain*.

The basic rules in Figure 5.12 are useful for analysing many practical networks quite efficiently. However, if networks become more complex and include feedback loops with nested loops the *Mason rule* is required. A description of the Mason rule is outside the scope of this book and can be found in the literature [2, 5, 8, 9]. In professional life as an RF engineer one commonly applies RF circuit simulation software to analyse and optimize more complex circuits. Nevertheless, we will look at an example to demonstrate that a signal flow graph is suitable for quick analysis.

Before we start with the example we will look at signal flow graph representations of common RF components. The basic sub-networks are either one-port networks (load, source), two-port networks (transmission line, filter, amplifier, matching circuit, line transition), three-port networks (power divider, circulator) or four-port networks (coupler).

5.6.1   One-Port Network/Load Termination

The signal flow graph of a simple one-port network and the corresponding algebraic equation are shown in Figure 5.13. The port number (4) is used in order to be consistent with the circuit in Figure 5.11. A one-port network has only one scattering parameter r_A. The reflection coefficient r_A describes the relation between incoming and outgoing wave a and b at that port.

5.6.2   Source

Figure 5.11 shows an ideal voltage source U_{01} with internal resistance Z_{01}. Such a source can be described by power waves and scattering parameters. (We will derive the relations in Problem 5.5.) The results are shown in Figure 5.14. The power wave b_{01} represents the active part of the source.

5.6.3   Two-Port Network

The signal flow graph of a two-port network and the corresponding algebraic equations are shown in Figure 5.15. The port numbers (2 and 3) are chosen in order to be consistent with the circuit in Figure 5.11. A two-port network has four scattering parameters. In Figure 5.15 it is quite obvious that the signal flow graph is a graphical representation of

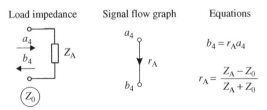

Load impedance   Signal flow graph   Equations

$$b_4 = r_A a_4$$

$$r_A = \frac{Z_A - Z_0}{Z_A + Z_0}$$

Figure 5.13   Signal flow graph of a load impedance.

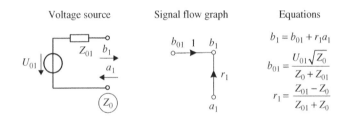

Figure 5.14 Signal flow graph of a voltage source.

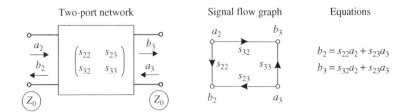

Figure 5.15 Signal flow graph and algebraic equations of a two-port network.

the algebraic equations. For example, the outgoing wave at node b_2 is the arrival point of incoming wave a_2 weighted by s_{22} and incoming wave a_3 weighted by s_{23}. Hence, the algebraic equation is $b_2 = s_{22}a_2 + s_{23}a_3$.

5.6.4 Three-Port Network

The signal flow graph of a three-port network and the corresponding algebraic equations are shown in Figure 5.16. A three-port network has nine scattering parameters.

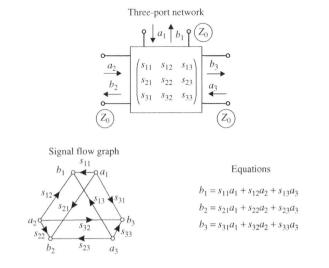

Figure 5.16 Signal flow graph and algebraic equations of a three-port network.

Four-port network

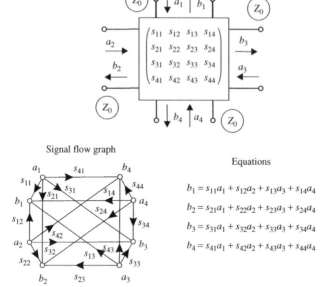

Figure 5.17   Signal flow graph and algebraic equations of a four-port network.

5.6.5   Four-Port Network

The signal flow graph of a four-port network and the corresponding algebraic equations are shown in Figure 5.17. A four-port network has sixteen scattering parameters.

Example 5.6 *Circuit with Matched Line and Shunt Impedance*

We will investigate the cascade connection of two two-port networks and a load termination as shown in Figure 5.18a. Two-port network 1 contains a shunt impedance Z and two-port network 2 contains a matched transmission line. We have already determined the corresponding scattering matrices in previous examples. The matrices are given by

$$\mathbf{S}_1 = \begin{pmatrix} s_{22} & s_{23} \\ s_{32} & s_{33} \end{pmatrix} = \frac{1}{2Z + Z_0} \begin{pmatrix} -Z_0 & 2Z \\ 2Z & -Z_0 \end{pmatrix}$$

$$\mathbf{S}_2 = \begin{pmatrix} s_{44} & s_{45} \\ s_{54} & s_{55} \end{pmatrix} = \begin{pmatrix} 0 & e^{-j\beta\ell} \\ e^{-j\beta\ell} & 0 \end{pmatrix} \tag{5.81}$$

We will calculate the reflection coefficient r_2 at port 2 using two different methods. First, we compile the set of equations and solve for the reflection coefficient r_2. Second, we use the signal flow method to demonstrate how efficient this graphical method is.

Let's start with the equations. Beginning at the load we find the following set of equations.

$$b_6 = r_A a_6 \tag{5.82}$$

$$a_5 = b_6 \quad \text{and} \quad a_6 = b_5 \tag{5.83}$$

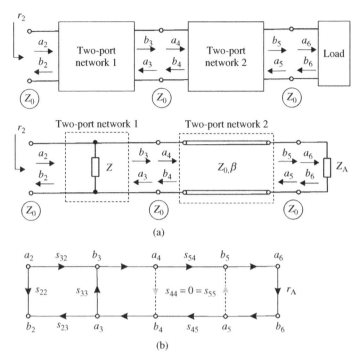

Figure 5.18   (a) Circuit with shunt impedance, matched line and load (see Example 5.6), (b) signal flow graph.

$$b_5 = a_4 s_{54} \quad \text{and} \quad b_4 = a_5 s_{45} \tag{5.84}$$

$$a_3 = b_4 \quad \text{and} \quad a_4 = b_3 \tag{5.85}$$

$$b_2 = s_{33} a_2 + s_{23} a_3 \tag{5.86}$$

$$b_3 = s_{32} a_2 + s_{33} a_3 \tag{5.87}$$

We solve the set of equations for the ratio of $r_2 = b_2/a_2$ and get, after some algebraic conversions,

$$r_2 = s_{22} + s_{32} s_{23} \frac{r_A s_{45} s_{54}}{1 - s_{33} r_A s_{45} s_{54}} \tag{5.88}$$

Using the signal flow method we can write down the resulting formula for r_2 directly by inspecting the signal flow graph in Figure 5.18b. If we follow the arrows in the signal flow graph we recognize that there are two signal paths for the incoming signal a_2 to become the outgoing signal b_2:

1. *First, there is a direct path following the reflection coefficient s_{22}.*
2. *Furthermore, there is an additional path (following s_{32} and returning over s_{23}) that includes a loop. The loop is defined by the following scattering parameters: s_{33}, s_{54}, r_A*

and s_{45}. (Arrows without scattering parameters have unity value, e.g. $a_4 = b_3$.) Due to the matching conditions of the transmission line (network 2) the reflection coefficients of two-port network 2 are zero $s_{44} = s_{55} = 0$. The loop gain G_{loop} can be determined by using Equation 5.80 where the forward open loop gain is $G_f = s_{45}r_A s_{54}$ and the closed loop gain is $G_{cl} = s_{33}s_{45}r_A s_{54}$.

Putting the direct path and the path that includes the loop in one formula we can immediately write down the solution for r_2 as

$$r_2 = s_{22} + s_{32}s_{23} \frac{r_A s_{45}s_{54}}{1 - s_{33}r_A s_{45}s_{54}} \qquad (5.89)$$

Of course the result equals our previous relation in Equation 5.88.

In order to demonstrate a possible area of application of the circuit we put some numbers in here.

- *Shunt impedance: $Z = 1/(j\omega C)$ where $C = 5\,\text{pF}$*
- *Transmission line: length $\ell = 0.363\lambda$, characteristic impedance equal to port reference impedance $Z_0 = 50\,\Omega$*
- *Load impedance $Z_A = 100\,\Omega + j\omega L$ where $L = 15.92\,\text{nH}$*
- *Frequency of operation: $f = 1\,\text{GHz}$*

The absolute value of the reflection coefficient r_2 is plotted in Figure 5.19 using Matlab [10] from Mathworks for a frequency range from DC to 2 GHz.

The circuit is a matching circuit for an operating frequency of $f = 1\,\text{GHz}$. The load impedance is matched via a transmission line and a shunt reactance to the port reference impedance of 50 Ω. Hence, $r_2(1\,\text{GHz}) = 0$. We look at matching circuits in more detail in Section 6.3.

In Figure 5.19 we see that for very low frequencies $f \to 0$ the reflection coefficient approaches a value of $1/3$. At DC the capacitance in two-port network 1 represents an open circuit. Hence, network 1 becomes a simple through connection. The transmission line

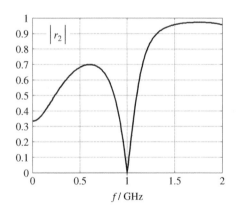

Figure 5.19   Reflection coefficient as a function of frequency (see Example 5.6).

has an electrical length of $\ell = 0.363\lambda$ at a frequency of $f = 1\,\text{GHz}$. For low frequencies the transmission line becomes an electrically short line. Hence, network 2 also becomes a simple through connection. So at DC the input impedance of our circuit is $Z_{\text{in}}(f \to 0) = 100\,\Omega$. Hence, the corresponding reflection coefficient is

$$r_2(f \to 0) = \frac{Z_{\text{in}}(f \to 0) - Z_0}{Z_{\text{in}}(f \to 0) + Z_0} = \frac{100 - 50}{100 + 50} = \frac{1}{3} \tag{5.90}$$

5.7 S-Parameter Measurement

Scattering parameters of RF components and circuits are commonly measured by a *vector network analyser* (VNA) [11, 12]. The term '*vector*' underlines that the amplitude *and* phase of reflection and transmission coefficients are determined. Figure 5.20 shows a vector network analyser with high-precision coaxial measurement lines and a test-fixture holding the device under test (DUT), in this case a microstrip filter on an alumina substrate.

A vector network analyser usually has two coaxial measurement ports with a port reference impedance of $Z_0 = 50\,\Omega$. A one-port network or a two-port network can be directly connected to the measurement ports using transmission lines. If the network has three or more ports all unconnected ports have to be terminated by the port reference impedance. (In Figure 5.21 the tapered symbol indicates a reflectionless termination $(r = 0)$ of port 3 to N.) By systematically connecting all network ports (arranged in pairs) to the measurement ports of the VNA all scattering parameters of a multi-port network can be measured successively with a two-port network analyser.

A modern vector network analyser features an easy to handle user interface. The entire measurement procedure and data-processing is computer-controlled. Figure 5.21 shows

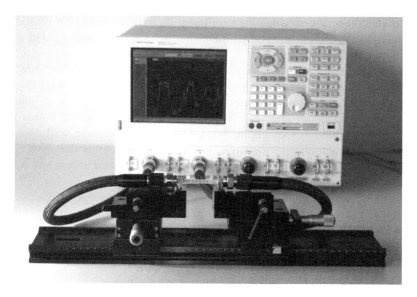

Figure 5.20 Vector network analyser Agilent N5230A (for measurements in the frequency range 300 kHz–20 GHz) and test fixture with microstrip filter.

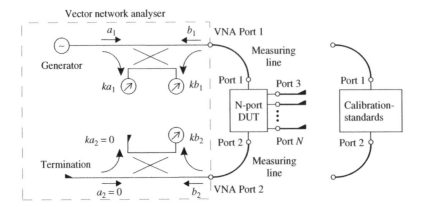

Figure 5.21 Vector network analyser Agilent N5230A (for measurements in the frequency range 300 kHz–20 GHz) and test fixture with microstrip filter.

the fundamental components and the basic structural design of a vector network analyser. In order to measure scattering parameters in a predefined frequency range a sine-wave signal generator steps through that frequency range. At each frequency point the power wave a_1 is routed to the device under test. A directional coupler[4] separates a fraction of wave a_1 as a reference signal (ka_1). At the device under test a reflected wave b_1 and a transmitted wave b_2 occur. The directional coupler splits off the signal kb_1 from the reflected wave b_1. So, the directional coupler allows the determination of the incoming and reflected wave. After measuring ka_1 and kb_1 by a detector we can determine the reflection coefficient s_{11}.

At port 2 we measure kb_2. From the transmitted signal kb_2 and the reference signal ka_1 we can determine the transmission coefficient s_{21}. So one reflection coefficient and one transmission coefficient are known. Likewise we can determine s_{22} and s_{12} if we swap generator and termination, that is port 2 is active and port 1 is passive. This is automatically done by the software, so all four scattering parameters of a two-port network are determined.

In order to perform accurate measurements of amplitude and phase it is important to *calibrate* a network analyser prior to a measurement. There are many different methods that apply precise standards (see Figure 5.21). Measurement of these well-defined standards allows the compensation of losses and reflections from transmission lines, non-ideal connectors and line transitions (e.g. coax-to-microstrip transition). Furthermore calibration allows the definition of a reference plane for phase measurements.

A common calibration method (SOLT) applies short-circuits, open-circuits, matched load terminations and through connections. Real network analysers are very complex equipment including mixers, filters and so on. Further information on network analyser design and calibration methods can be found in [12].

---

[4] Directional couplers are four-port networks that split off small amounts of waves travelling in different directions on a line to separated output ports. An important parameter of a direction coupler is the coupling coefficient k. We will discuss directional couplers in more detail in Section 6.10.

Modern vector analysers can export s-parameter measurement results in standardized data formats. So measurement data can be imported into RF circuit simulation software in order to design additional circuits (e.g. a matching circuit for a measured antenna). A commonly used data format is the *Touchstone SnP-format* where n indicates the number of ports.

Figure 5.22 shows a *.s2p-file of a two-port network. The first four lines of text are comment lines beginning with an exclamation mark (!) and contain information on equipment, date, time, calibration and measured s-parameter set. The fifth line starts with a sharp sign (#) and declares the format for the data that follows in the subsequent lines. The first column is the frequency in Hertz (Hz). The scattering parameters are given in logarithmic scale (dB)[5] and the phase is in degree. The port reference impedance is $Z_0 = 50 \, \Omega$. All subsequent lines of text comprise nine values indicating frequency and all four s-parameters s_{11}, s_{21}, s_{12} and s_{22} with amplitude in dB and phase in degree.

In communication engineering we note the increased use of components that use symmetric three-conductor lines (balanced lines) like coupled microstrip lines. As shown in Section 4.9 such transmission lines support two fundamental wave modes: differential mode and common mode. Commercial vector network analysers have unbalanced coaxial measurement ports.

By using a *balun* network (balun = *bal*anced-*un*balanced) we can measure the differential behaviour of a balanced network (see Figure 5.23a). Baluns are discussed in Section 6.11. Unfortunately, we get no information on the common mode behaviour. By using a four-port network analyser we can measure common and differential behaviour of the symmetrical circuit (mixed-mode parameter).

```
!Agilent Technologies,N5230A,MY12345678,A.00.00

!Date: Saturday, August 18, 2012 13:25:40

!Correction: S11(Full 2 Port(1,2)) S21(Full 2 Port(1,2)) S12(Full 2 Port(1,2)) S22(Full 2 Port(1,2))

!S2P File: Measurements: S11, S21, S12, S22:

# Hz S  dB  R 50

8000000000 -7.068530e-001 -1.088347e+002 -5.033566e+001 -9.525459e+001

                          -5.119051e+001 -9.444051e+001 -6.366544e-001 -1.155149e+002

8022500000 -7.183312e-001 -1.117752e+002 -5.052730e+001 -1.052149e+002

                          -5.108871e+001 -9.723354e+001 -6.364053e-001 -1.187093e+002

...

12477500000 -6.654899e-001 4.224392e+001 -3.169779e+001 1.142805e+002

                          -3.153713e+001 1.160675e+002 -8.575924e-001 3.742066e+001

12500000000 -6.905157e-001 3.936354e+001 -3.129497e+001 1.095876e+002

                          -3.115040e+001 1.089006e+002 -8.625705e-001 3.447514e+001
```

Figure 5.22 Example of a Touchstone SnP-data format (two-port network, *.s2p-file).

---

[5] Instead of 'dB' the term could be 'RI' (for s-parameters given by real and imaginary parts) or 'MA' (for s-parameters in linear scale with phase in degree).

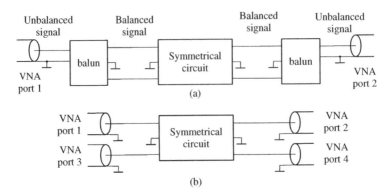

Figure 5.23 S-parameter measurement of a symmetrical circuit (a) by using baluns and a two-port network analyser, (b) by using a four-port network analyser.

5.8 Problems

5.1 The scattering matrix $\mathbf{S}$ of a two-port network is given as

$$\mathbf{S}_{50\,\Omega} = \begin{pmatrix} 5/13 & j12/13 \\ j12/13 & 5/13 \end{pmatrix} \tag{5.91}$$

The port reference impedance is $Z_0 = 50\,\Omega$.

a) Determine the properties of the network: is it matched? Is it reciprocal? Is it symmetric? Is is loss-less? Give reasons for your answers.

b) Calculate the reflection loss (RL) and the insertion loss (IL).

c) Renormalize the scattering matrix to a new port reference impedance of $Z_{0,\text{new}} = 100\,\Omega$.

5.2 With a vector network analyser we measure a reflection coefficient of $r_A = 0.4\,e^{-j20°}$ at the input terminals of an antenna. Port reference impedance is $Z_0 = 50\,\Omega$.

a) Determine the input impedance Z_A of the antenna.

We connect the antenna to a transmission line with a characteristic impedance of $Z_{0L} = 75\,\Omega$. The incident active power on the transmission line is set to $P_{a1} = 1\,\text{W}$.

b) Calculate the reflected power P_{b1} at the antenna terminals and the power P_{acc} that is accepted by the antenna.

5.3 Determine the scattering matrices of the two-port networks given in Figure 5.24. The port reference impedance is Z_0 for all ports.

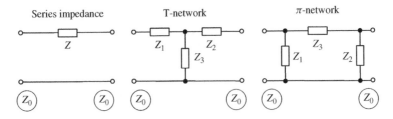

Figure 5.24 Series impedance, T- and π-network (Problem 5.3).

5.4 Figure 5.25 shows a cascade connection of three two-port networks. An unknown two-port network $\mathbf{S}_x$ is embedded between two matched lines. The port reference impedance is Z_0 for all ports.

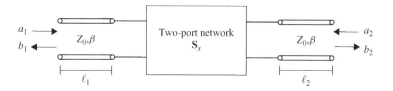

Figure 5.25 Two-port network embedded between two matched lines (Problem 5.4).

The scattering matrices are given by

$$\mathbf{S}_1 = \begin{pmatrix} 0 & e^{-j\beta\ell_1} \\ e^{-j\beta\ell_1} & 0 \end{pmatrix} \quad , \quad \mathbf{S}_2 = \begin{pmatrix} 0 & e^{-j\beta\ell_2} \\ e^{-j\beta\ell_2} & 0 \end{pmatrix} \tag{5.92}$$

and

$$\mathbf{S}_x = \begin{pmatrix} s_{x,11} & s_{x,12} \\ s_{x,21} & s_{x,22} \end{pmatrix} \tag{5.93}$$

a) Determine the scattering matrix $\mathbf{S}$ of the resulting network.

Let us assume that we have measured the matrix $\mathbf{S}$. Furthermore we know the lengths of the lines ℓ_1 and ℓ_2.

b) Is it possible to determine the matrix $\mathbf{S}_x$ of the embedded two-port network?

5.5 Derive the equation for the wave source b_{01} in Figure 5.14.

5.6 Deduce the feedback rule for signal flow graphs in Figure 5.12.

5.7 Figure 5.26 shows a circuit with disconnected terminals in order to calculate input impedances and reflection coefficients. We already studied this circuit in Example 5.6. By using the electrical parameters from that example and a frequency of $f = 1\,\text{GHz}$ determine

a) the input impedances Z_{in1} and Z_{in2} and
b) the reflection coefficients r_1 and r_2.

How could you interpret the results?

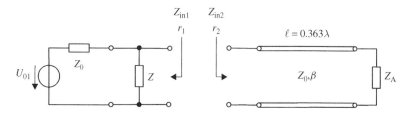

Figure 5.26 Calculation of reflection coefficients and impedances (Problem 5.7).

5.8 Demonstrate that a three-port network cannot satisfy the following conditions simultaneously

- matching (at all ports),
- reciprocity, and
- loss-lessness.

References

1. Ludwig R, Bogdanov G (2009) RF *Circuit Design, Theory and Applications*. Prentice Hall.
2. Ahn HR (2006) *Asymmetric Passive Components in Microwave Integrated Circuits*. John Wiley & Sons.
3. Bowick C (2008) *Circuit Design*. Newnes.
4. Pozar DM (2005) *Microwave Engineering*. John Wiley & Sons.
5. Michel H-J (1981) *Zweitor-Analyse mit Leistungswellen*. Teubner.
6. Bronstein IN, Semendjajew KA, Musiol G, Muehlig H (2008) *Taschenbuch der Mathematik*. Harri Deutsch.
7. Zinke O, Brunswig H (2000) *Hochfrequenztechnik 1*. Springer.
8. Gronau G (2001) *Hoechstfrequenztechnik*. Springer.
9. Mason SJ (1956) Feedback Theory: Further Properties of Signal Flow Graphs. *IRE Proc.*, Vol. 44.
10. Mathworks Corporation (2010) *Matlab*. Mathworks Corporation.
11. Schiek B (1999) *Grundlagen der Hochfrequenzmesstechnik*. Springer.
12. Thumm M, Wiesbeck W, Kern S (1998) *Hochfrequenzmesstechnik - Verfahren und Messsysteme*. Vieweg/Teubner.

Further Reading

Detlefsen J, Siart U (2009) *Grundlagen der Hochfrequenztechnik*. Oldenbourg.
Hagen, JB (2009) *Radio-Frequency Electronics: Circuits and Applications*. Cambridge University Press.
Meinke H, Gundlach FW (1992) *Taschenbuch der Hochfrequenztechnik*. Springer.

6

RF Components and Circuits

In this chapter we will start with concentrated electrical components and explain equivalent circuits that describe element behaviour at higher frequencies. The main part of this chapter is about standard RF and microwave circuits that are composed of concentrated circuit elements and transmission lines. Throughout the following sections we will make extensive use of circuit simulation software to illustrate practical examples.

6.1 Equivalent Circuits of Concentrated Passive Components

At low frequencies the electrical behaviour of real concentrated electrical components like resistors, capacitors and inductors is approximated by idealized nominal values. With increasing frequency *parasitic effects* become more pronounced and have to be considered for realistic circuit design. Parasitic effects result from losses in conductors and dielectric or magnetic material, inductances of wire leads and capacitances of contact terminals.

We will derive simple and physically reasonable *equivalent circuits* of the most important concentrated elements: resistor, capacitor and inductor. Each of these equivalent circuits consists of three ideal elements and is quite useful and sufficiently accurate for many practical applications.

We can further improve the accuracy, extend the frequency range and expand the scope of applications by considering more complex equivalent circuits [1] including more than three elements. However, it is more difficult to determine the elements of the complex equivalent circuits, for example from measurements. Furthermore, with increasing frequency the physical mounting condition effects the component behaviour. Coupling effects between the electrical component and the circuit environment have to be considered.

6.1.1 Resistor

Resistors come in different shapes and package types. High-frequency circuits commonly use *surface mounted device* (SMD) components. These components are rectangular in shape and exhibit low parasitics due to small size (typical length ~mm) and direct mounting (no wire leads) onto the surface of the printed circuit board (PCB).

RF and Microwave Engineering: Fundamentals of Wireless Communications, First Edition. Frank Gustrau.
© 2012 John Wiley & Sons, Ltd. Published 2012 by John Wiley & Sons, Ltd.

Figure 6.1a shows the basic construction of a SMD resistor: an isolating material (backing material) supports a thin resistive layer with a typical thickness of 50 μm. At both ends metallic contacts represent the terminals that can be soldered to the surface and easily combined with microstrip and coplanar technology. Table 6.1 lists standardized dimensions of SMD components. The first two digits of the four-digit identifier specify the length (in $1/100$ inch ≈ 0.254 mm) and the last two digits specify the width. The dimensions may have some tolerances that are not listed in the table.

In order to understand the equivalent circuit in Figure 6.1b we need to look at the construction of a SMD resistor in Figure 6.1a. The resistive layer represents the resistor R that denotes the nominal resistance of the component. The resistance can be assumed to be constant in a certain frequency range if the thickness of the resistive layer is small compared to the skin depth in the frequency range of usage. The amplitude of the current density is then uniformly distributed in the resistive material.

Due to Ampere's law the current density that flows through the resistive layer is encircled by magnetic field lines. Therefore, we include a series inductance L in the equivalent circuit. Across the resistive layer we see a drop in voltage U. The voltage is caused by an electric field in the resistive layer. At the ends of the resistive layer we find two metallic caps. The arrangement reminds us of a capacitor. Therefore, we include a parallel

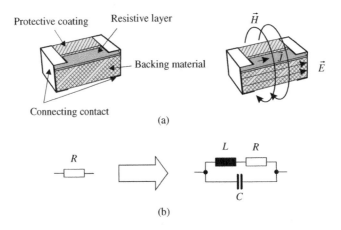

(a)

(b)

Figure 6.1   (a) Geometry of a SMD-resistor and (b) equivalent circuit.

Table 6.1   Dimensions of SMD components (height: typical values)

SMD size	Length L	Width W	Height H
0201	0.5 mm	0.25 mm	0.23 mm
0402	1.0 mm	0.5 mm	0.35 mm
0603	1.6 mm	0.8 mm	0.45 mm
0805	2.0 mm	1.25 mm	0.5 mm
1206	3.1 mm	1.55 mm	0.55 mm
1506	3.8 mm	1.55 mm	0.55 mm

capacitance C in the equivalent circuit. The equivalent circuit represents a practical resistor and the impedance is

$$Z = (R + j\omega L) \| \frac{1}{j\omega C} = \frac{R + j\omega L}{j\omega C(R + j\omega L) + 1} \tag{6.1}$$

Let us put some numbers in here and look at an example.

Example 6.1 *Frequency Dependent Impedance of a SMD Resistor*
We consider three SMD resistors with nominal values of $R = 10\,\Omega$, $100\,\Omega$ and $1\,k\Omega$. For the parasitic capacitance and inductance we assume $C = 0.04\,pF$ and $L = 0.7\,nH$. The resulting frequency dependent impedances are shown in Figure 6.2 in the frequency range from $10\,MHz$ to $10\,GHz$. With increasing frequency absolute value and phase deviate from their nominal values and the effects of the parasitic elements become visible.

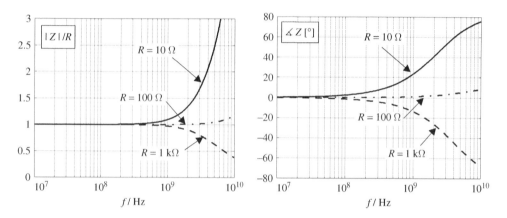

Figure 6.2 Frequency dependent impedance of a SMD resistor (see Example 6.1).

6.1.2 Capacitor

Capacitors, like resistors, come in different shapes and sizes. At high frequencies chip capacitors are commonly used with sizes equal to SMD resistors (Table 6.1). Figure 6.3 shows two possible realizations. The electrodes of the capacitor may lie in the same plane (Figure 6.3a). Alternatively, the electrodes may be arranged in different planes of a multilayer substrate (Figure 6.3b). (In order to better visualize the different planes the illustration is stretched in the vertical direction.)

The main element of the equivalent circuit in Figure 6.3c is the nominal capacitance C. Conduction and displacement current density in the metallic and dielectric parts of the capacitor lead to a surrounding magnetic field. Hence, a series inductor is included in the equivalent circuit. Dielectric and ohmic losses are summed up in a series resistance (*equivalent series resistance*, ESR). The equivalent circuit is a series-resonant circuit with an impedance of

$$Z = R_{\mathrm{ESR}} + j\omega L + \frac{1}{j\omega C} = R_{\mathrm{ESR}} + j\left(\omega L - \frac{1}{\omega C}\right) \tag{6.2}$$

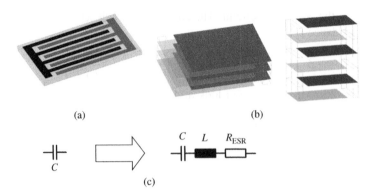

(a)   (b)

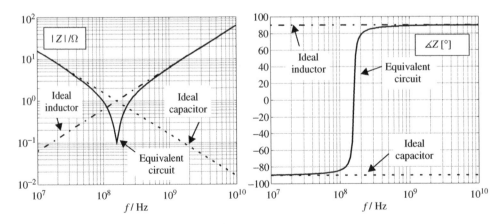

(c)

Figure 6.3 Construction of a (a) planar (interdigital) and (b) multilayer SMD capacitor and (c) equivalent circuit.

where f_0 is the resonance frequency

$$f_0 = \frac{1}{2\pi} \cdot \frac{1}{\sqrt{LC}} \qquad \text{(Resonance frequency)} \qquad (6.3)$$

At resonance ($f = f_0$) the imaginary part of the impedance is zero: $\omega L - 1/(\omega C) = 0$. The quality factor Q_C is a common measure for the losses of a capacitor. The reciprocal value is the loss factor $\tan \delta_C$.

$$Q_C = \frac{1}{\tan \delta_C} = \frac{1}{R_{ESR}\omega C} \qquad (6.4)$$

Example 6.2 *Frequency Dependent Impedance of a Chip Capacitor*
   Let us consider a chip capacitor with a nominal capacitance of $C = 1$ nF. The parasitic elements are given as $R = 0.1\,\Omega$ and $L = 1$ nH. Figure 6.4 shows the resulting frequency

Figure 6.4 Frequency dependent impedance of a SMD capacitor (see Example 6.2).

dependent impedance. For low frequencies the real component approaches the ideal capac-
itive behaviour. For comparison the impedance of an ideal capacitor is given as a dotted
line. At a resonance frequency $f_0 \approx 159$ MHz the impedance is ohmic and equals R_{ESR}.
Above resonance the impedance is inductive and approaches ideal inductive behaviour
(shown as a dash-dotted line) with increasing frequency.

6.1.3 Inductor

Configurations that store magnetic energy show inductive behaviour. A common way to
build inductors is by winding wires into coils. The inductance can be increased by using a
magnetic material with high relative permeability μ_r. Inductors with the small dimensions
of SMD components can be realized by miniaturizing this general concept. Figure 6.5a
shows a planar inductor on a substrate and Figure 6.5b shows a coil in a multilayer
substrate. (The figure is stretched in z-direction to demonstrate the arrangement of the
winding in different layers.)

The equivalent circuit shows the nominal inductance L. Losses in the conductive as
well as the magnetic material are summarized in a series resistance R_S. Between the
winding a capacitance exists that provides a current-bypass at higher frequencies. The
capacitance is therefore parallel to the nominal inductance and resistance. The simple
equivalent circuit is given in Figure 6.5c with an impedance of

$$Z = (R_S + j\omega L)||\frac{1}{j\omega C} = \frac{R_S + j\omega L}{j\omega C(R_S + j\omega L) + 1} \tag{6.5}$$

The serial loss resistance R_S is small compared to the impedance of the inductance.
Hence, the resonance frequency f_0 is approximately given by a parallel resonance
circuit as

$$f_0 = \frac{1}{2\pi} \cdot \frac{1}{\sqrt{LC}} \tag{6.6}$$

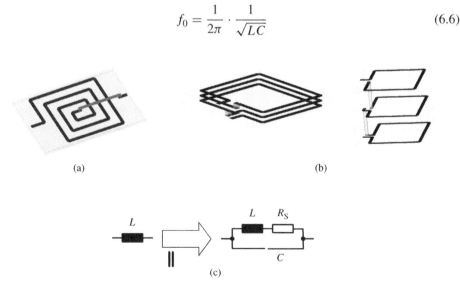

(a)   (b)

(c)

Figure 6.5   Construction of a planar (a) and multilayer (b) SMD-inductor and (c) equivalent circuit.

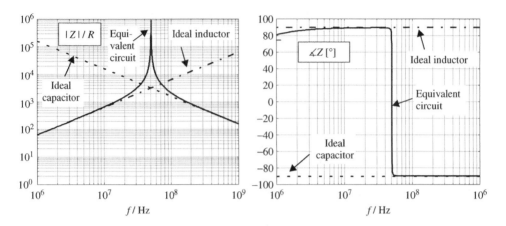

Figure 6.6 Typical frequency dependent impedance of a SMD inductor.

The quality factor Q_L and the loss factor $\tan \delta_L$ of an inductor are

$$Q_L = \frac{1}{\tan \delta_L} = \frac{\omega L}{R_S} \tag{6.7}$$

Figure 6.6 shows the typical frequency behaviour of an RF inductor. At low frequencies the impedance increases linearly with frequency ($Z \approx j\omega L$) until the parasitic capacitance leads to a parallel resonance (maximum impedance). Above resonance the behaviour is capacitive.

6.2 Transmission Line Resonator

By using two concentrated elements (capacitor C and inductor L) we can design parallel and series resonance circuits. Figure 6.7a shows as an example a simple parallel-resonant circuit. The resonance frequency is given as

$$f_0 = \frac{1}{2\pi} \cdot \frac{1}{\sqrt{LC}} \tag{6.8}$$

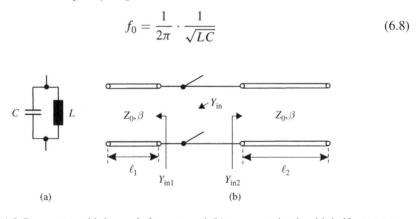

Figure 6.7 (a) LC resonator with lumped elements and (b) resonant circuit with half-wave transmission line ($\ell = \ell_1 + \ell_2 = \lambda/2$).

At higher frequencies it is more difficult to use concentrated elements due to the parasitic effects discussed in the previous section. In RF engineering transmission lines become an interesting alternative. Therefore, we will look at the transmission line resonators that are important for the filters that will be discussed in later sections.

Further resonating structures for filter design include the cavity resonators already discussed in Section 4.6.6. Oscillators (active circuits) employ quartzes and dielectric resonators. Further informations may be found in literature, for example [1–3].

6.2.1 Half-Wave Resonator

In Section 3.1 we have seen that loss-less lines that are terminated by an open-circuit or short-circuit exhibit a reactive input impedance. Let us start with an open-circuited transmission line (line length ℓ). According to Equation 3.110 the input impedance is

$$Z_{\text{in}} = -jZ_0 \cot(\beta\ell) = -j\frac{Z_0}{\tan(\beta\ell)} = jX_{\text{in}} = \begin{cases} \text{capacitive, for } \ell < \dfrac{\lambda}{4} \\ \text{inductive, for } \dfrac{\lambda}{4} < \ell < \dfrac{\lambda}{2} \end{cases} \tag{6.9}$$

In order to design a parallel-resonant circuit we connect two lines in parallel (see Figure 6.7b). The first line is shorter than a quarter wavelength (line length $\ell_1 < \lambda/4$) with a capacitive input impedance. The second line is longer than a quarter wavelength (line length $\ell_2 > \lambda/4$) with an inductive input impedance. The total line length is $\ell_1 + \ell_2$.

At the input terminals the admittance (see Figure 6.7b) is

$$Y_{\text{in}} = Y_{\text{in1}} + Y_{\text{in2}} = -\frac{1}{jZ_0}\left(\tan(\beta\ell_1) + \tan(\beta\ell_2)\right) \tag{6.10}$$

At (parallel) resonance the imaginary part of the input admittance is zero. By using the following relation

$$\tan x \pm \tan y = \frac{\sin(x \pm y)}{\cos x \cos y} \tag{6.11}$$

the input impedance at resonance is

$$\text{Im}\left\{Y_{\text{in}}\right\} = \frac{1}{Z_0} \cdot \frac{\sin\left(\beta(\ell_1 + \ell_2)\right)}{\cos(\beta\ell_1)\cos(\beta\ell_2)} \overset{!}{=} 0 \qquad \text{(at resonance)} \tag{6.12}$$

The imaginary part is zero if the numerator is zero. The zeros of the sine function are given by integer numbers of π

$$\beta(\ell_1 + \ell_2) = n\pi \qquad \text{where} \quad n \in \mathbb{Z} \tag{6.13}$$

Hence, the total line length ℓ at resonance becomes

$$\boxed{\ell = \ell_1 + \ell_2 = \frac{n\pi}{\beta} = \frac{n\pi\lambda}{2\pi} = n\frac{\lambda}{2}} \tag{6.14}$$

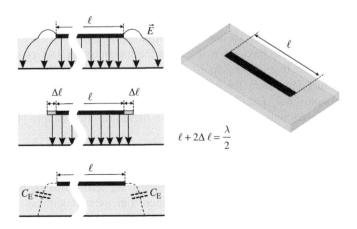

Figure 6.8 Microstrip $\lambda/2$-resonator: open end effect.

A transmission line with open-circuited terminations at both ends shows parallel resonance for a minimum line length equal to half a wavelength.

All the considerations made above are based on ideal lines where the electric and magnetic field is restricted to the region between the signal line and return conductor. On a practical transmission line the open ends of the line show parasitic capacitive behaviour (see Figure 6.8). This *open end effect* can be modelled by an end-capacitance C_E. As a consequence the physical line length ℓ must be slightly shorter than half a wavelength ($\ell + 2\Delta\ell = \lambda/2$). The factor for length reduction depends on the actual geometry and cannot be determined generally.

Example 6.3 *Half-Wave Resonator Using Microstrip Transmission Lines*

Figure 6.9a shows a half-wave microstrip resonator on an alumina substrate ($\varepsilon_r = 9.8$, $h = 635\,\mu m$). The resonator is capacitively coupled to two $50\,\Omega$-microstrip lines (line width $w = 625\,\mu m$, gap width $s = 0.2\,mm$). The geometric length is $\ell = 6\,mm$. Based on an effective relative permittivity of $\varepsilon_{r,eff} = 6.85$ the resonance frequency should be around 9.5 GHz. Due to the capacitive open end effect and additional coupling effects we get a resonance frequency of approximately 9 GHz (see Figure 6.9b). In Section 6.4 we will design filters based on transmission line resonators.

6.2.2 Quarter-Wave Resonator

In order to reduce the physical size of a half-wave resonator we can connect an open-ended line and a short-ended line in parallel (see Figure 6.10b). According to Equation 3.101 the input impedance of a short-ended transmission line is

$$Z_{in2} = jZ_0\tan(\beta\ell) = jX_{in2} = \begin{cases} \text{inductive, for } \ell < \dfrac{\lambda}{4} \\[2mm] \text{capacitive, for } \dfrac{\lambda}{4} < \ell < \dfrac{\lambda}{2} \end{cases} \qquad (6.15)$$

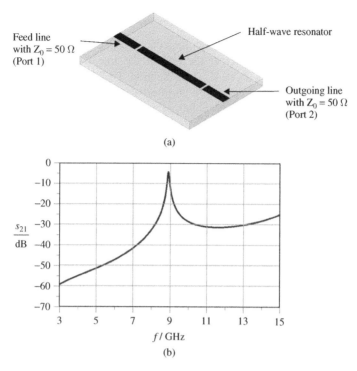

(a)

(b)

Figure 6.9   Microstrip $\lambda/2$−resonator ($\ell = 6$ mm, $s = 0.2$ mm): (a) three-dimensional view of geometry and (b) transmission coefficient (see Example 6.3).

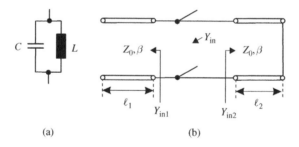

(a)   (b)

Figure 6.10   (a) LC resonator with lumped elements and (b) resonant circuit with quarter-wave transmission line ($\ell = \ell_1 + \ell_2 = \lambda/4$).

In Problem 6.1 we demonstrate that the shortest length for resonance is now a quarter of a wavelength $\ell = \ell_1 + \ell_2 = \lambda/4$.

An open-ended transmission line and a short-ended transmission line represent a resonant circuit. The shortest physical line length for resonance is a quarter wavelength $\ell = \lambda/4$. The quarter-wave resonator has half of the length of a half-wave resonator.

6.3   Impedance Matching

A transmission line with a characteristic impedance Z_0 that is terminated by a load impedance Z_A shows *no reflections* at the end of the line if characteristic impedance and load impedance are equal $Z_0 = Z_A$ (see Section 3.1.8). Furthermore, in order to obtain *maximum power* from a source with an internal impedance Z_I the load impedance Z_A should be the conjugate-complex value of the internal impedance $Z_A = Z_I^*$ (see Section 5.4.2).

If the load impedance is mismatched, special two-port networks (*matching circuits*) can be designed to transform the actual load impedance to the required input impedance Z_{in} (see Figure 6.11).

Low-loss matching networks can be constructed by using reactive components (capacitors and inductors) or transmission lines. Since these components exhibit frequency dependent behaviour the desired impedance transformation is limited to a certain frequency range.

We will start by looking at resistive load impedances ($Z_A = R_A$) and resistive target impedances ($Z_{in} = R_I$). Later on we will expand our view and include universal complex load impedances. Even today the classical Smith chart is a valuable tool when designing matching circuits. Therefore, we will use the Smith chart to illustrate the impedance transformation of different matching networks. Computerized Smith chart tools (e.g. [4]) support the graphical design of matching networks.

6.3.1   LC-Networks

A resistive load R_A can be transformed to a resistive input impedance R_I by a two-port network that consists of two reactive components, one of which is a capacitor C and the

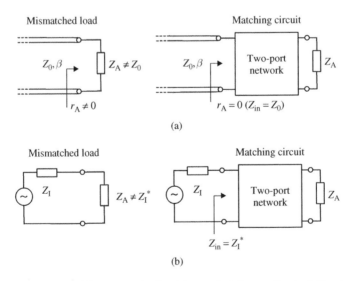

Figure 6.11   Two-port matching networks for (a) matched (non-reflective) line termination and (b) maximum power transfer.

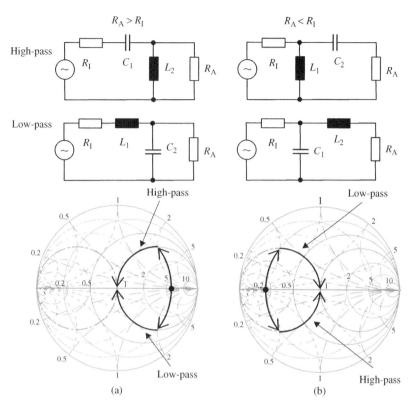

Figure 6.12   LC-matching networks for (a) $R_A > R_I$ and (b) $R_A < R_I$.

other one is an inductor L. One component is arranged as a serial element and the other component is arranged as a parallel element (see Figure 6.12). The circuit layout looks like a letter 'L' (or Greek symbol 'Γ'). Hence, they are often referred to as L-network or Γ-transformer [2, 5].

In the design process we distinguish between two cases: either the load resistor is greater than the input impedance ($R_A > R_I$) or the load resistor is smaller than the input impedance ($R_A < R_I$). In the first case we start seen from the load with a parallel component then we add a series element. In the second case we start with a serial component then we add a parallel element. For both cases two different circuit layouts are possible: high-pass and low-pass structures. Figure 6.12 shows the four different layouts as well as the corresponding transformation paths in the Smith chart diagram. (The reference impedance of the Smith chart is $Z_0 = R_I$. Hence, the matching point is the centre point of the diagram.) The components can be calculated by the following set of equations.

For $R_A > R_I$ the absolute values of the reactances $|X_1|$ and $|X_2|$ as well as the quality factor Q are

$$|X_1| = R_I \cdot Q \quad \text{and} \quad |X_2| = R_A/Q \quad \text{where} \quad Q = \sqrt{\frac{R_A}{R_I} - 1} \qquad (6.16)$$

For $R_A < R_I$ the equations are

$$|X_1| = R_I/Q \quad \text{and} \quad |X_2| = R_A \cdot Q \quad \text{where} \quad Q = \sqrt{\frac{R_I}{R_A} - 1} \qquad (6.17)$$

The reactances X_1 and X_2 are either inductive ($X = \omega L$) or capacitive ($X = -1/(\omega C)$). In a circuit the two elements are of different type (see Figure 6.12). For a given frequency f_0 we can derive the capacitance C and inductance L from the following relations.

$$L_{1,2} = \frac{|X_{1,2}|}{2\pi f_0} \quad \text{and} \quad C_{1,2} = \frac{1}{|X_{1,2}| 2\pi f_0} \qquad (6.18)$$

Example 6.4 *Matching Network with LC Elements*
At a frequency of $f_0 = 400\,\text{MHz}$ a load resistance of $R_A = 150\,\Omega$ shall be matched to a $50\,\Omega$-transmission line. Hence, the target (input) resistance is $R_I = 50\,\Omega$. So, for the circuit design we must consider the case $R_A > R_I$. According to Equation 6.16 we find

$$|X_1| = R_I \cdot \sqrt{2} \approx 70.71\,\Omega \quad ; \quad |X_2| = R_A/\sqrt{2} \approx 106.1\,\Omega \quad ; \quad Q = \sqrt{2} \qquad (6.19)$$

In accordance with Figure 6.12 there are two possible layouts: either a high-pass structure or a low-pass structure. With Equation 6.18 we find the element values L and C for both layouts. For the high-pass structure we get $C_1 = 5.63\,\text{pF}$ and $L_2 = 42.2\,\text{nH}$. For the low-pass structure we get $L_1 = 28.1\,\text{nH}$ and $C_2 = 3.75\,\text{pF}$. The reflection coefficients in the frequency range from DC to 800 MHz are given in Figure 6.13a ($R_A > R_I$).

For very low frequencies ($f \to 0\,\text{Hz}$) the low-pass structure approaches $-6\,\text{dB}$ since the capacitor represents an open-circuit and the inductor represents a short-circuit. Hence, the load resistor is directly connected to the transmission line and the reflection coefficient is calculated as

$$r_A(f=0) = s_{11}(f=0) = \frac{R_A - R_I}{R_A + R_I} = \frac{1}{2} \quad \to \quad \frac{s_{11}}{\text{dB}} = 20\lg|s_{11}| = -6 \qquad (6.20)$$

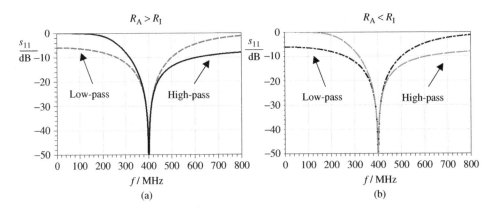

Figure 6.13   Reflexion coefficients (see Example 6.4).

As expected the high-pass structure approaches 0 dB *as the frequency converges to* 0 Hz.

In order to end our discussion on matching networks we swap the values of load and target impedance ($R_A = 50\,\Omega$, $R_I = 150\,\Omega$), *so that now we have* $R_A < R_I$. *The equation to be considered now is Equation 6.17.*

$$|X_1| = R_I/\sqrt{2} \approx 106.1\,\Omega \quad ; \quad |X_2| = R_A \cdot \sqrt{2} \approx 70.71\,\Omega \quad ; \quad Q = \sqrt{2} \qquad (6.21)$$

For the low-pass structure we get $C_1 = 5.63\,\text{pF}$ *and* $L_2 = 42.2\,\text{nH}$. *For the high-pass structure we get* $L_1 = 28.1\,\text{nH}$ *and* $C_2 = 3.75\,\text{pF}$. *The reflection coefficients in the frequency range from DC to* 800 MHz *are given in Figure 6.13b* ($R_A < R_I$).

Finally, we notice that according to Equation 6.20 the reflection coefficient without *matching circuit* would be equal to -6 dB *in the whole frequency band.*

Figure 6.13 shows that matching (e.g. specified by $|r| < -20\,\text{dB}$) is achieved only in a certain frequency band. The greater the difference between R_A and R_I, the smaller the matching bandwidth is. If the bandwidth of a simple LC circuit is not sufficient we can increase the bandwidth by *cascading* two or more LC networks.

Problem 6.3 gives an example of a network with increased bandwidth. In a two-stage LC network the load resistance R_A is transformed to an intermediate resistance R_m in the first instance. The intermediate resistance equals the geometric mean of load and target resistance

$$R_m = \sqrt{R_I R_A} \qquad (6.22)$$

The second LC network then transforms the intermediate resistance R_m to the target resistance R_I.

6.3.2 Matching Using Distributed Elements

6.3.2.1 Quarter-Wave Transformer

We have already discussed, in Section 3.1.9, quarter wavelength transmission lines for impedance transformation. Figure 6.14a shows a simple power divider in microstrip design with 50 Ω transmission lines at all ports. The power P delivered to port 1 is divided equally between port 2 and port 3. In order to provide a matched input at port 1 for a frequency of $f = 5\,\text{GHz}$ a quarter-wave transformer is incorporated in the circuit. At the end of the quarter-wave transformer the impedance is 25 Ω, since there are two 50 Ω transmission lines in parallel (see Figure 6.14b). The quarter wavelength line transforms this load resistance (25 Ω) to a target impedance of 50 Ω. According to Equation 3.120 the characteristic impedance of the quarter wavelength line must be the geometric mean of 25 Ω and 50 Ω, which is $Z_{0Q} = \sqrt{25\,\Omega\,50\,\Omega} \approx 35.4\,\Omega$. (Problem 6.4 shows the detailed design of the power divider.)

Figure 6.14c shows the reflection coefficient at port 1, 2 and 3 in the frequency range from 1 GHz to 10 GHz. As expected port 1 is matched at a frequency of 5 GHz. A drawback of the simple power divider concept is the mismatch at port 2 and 3. At a frequency of 5 GHz the output reflection coefficients are $s_{22} = s_{33} = -6\,\text{dB}$. Viewed from the output side the impedance of the quarter-wave transformer is 25 Ω. Therefore, the load

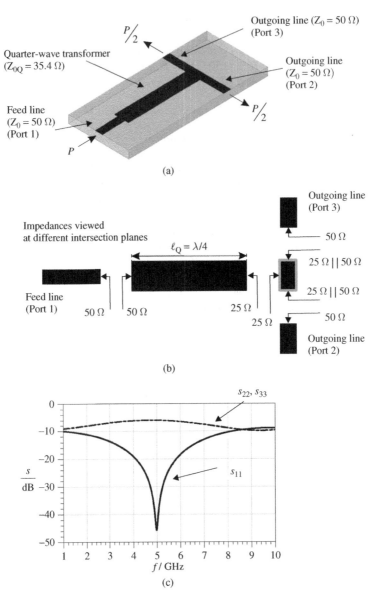

Figure 6.14 (a) Simple power divider with quarter-wave transformer for impedance matching, (b) view of impedances at different intersection planes and (c) reflection coefficients.

impedance of the outgoing lines is $25\,\Omega$ in parallel with $50\,\Omega$. The resulting reflection coefficient is

$$s_{22} = s_{33} = \frac{(25\,\Omega\|50\,\Omega) - 50\,\Omega}{(25\,\Omega\|50\,\Omega) + 50\,\Omega} = -0.5 \quad \rightarrow \quad |s_{22}| = |s_{33}| = -6\,\text{dB} \qquad (6.23)$$

As in the case of LC networks the bandwidth where port 1 is matched is limited. The bandwidth can be increased by cascading quarter-wave transformers, as we will show in Problem 6.4.

6.3.2.2   Stub Line and Butterfly Stubs

In practical designs we often find matching circuits that use a serial line and an open-ended parallel line (stub line) as shown in Figure 6.15a. The circuit is applicable to arbitrary complex load impedances Z_A. Figure 6.15b shows a Smith chart diagram to visualize the transformation path from the load impedance to the matching point.

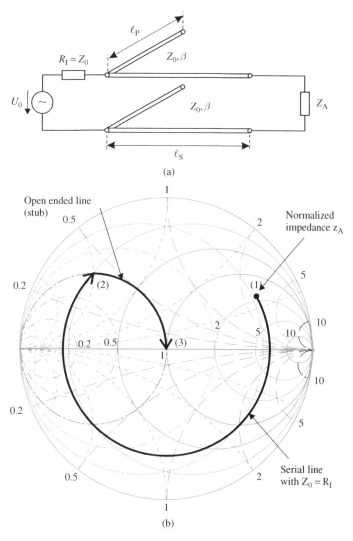

Figure 6.15   (a) Matching circuit with serial line and open ended parallel line (stub) and (b) Smith chart with transformation path.

Point (1) is given by the normalized load impedance $z_A = Z_A/R_I$ where R_I is the port reference impedance. First we use a serial line with a characteristic impedance that equals the port reference impedance. With increasing length we follow a circle around the centre point. We choose a length of the serial line ℓ_S that brings us to the circle of unity normalized conductance in the admittance plane. A parallel capacitance would now bring us to the centre point (3). Instead of using a concentrated capacitor we apply a parallel open-ended transmission line (stub line). Such a transmission line has a capacitive input impedance if the line length ℓ_P is shorter than a quarter wavelength. By choosing an appropriate length we achieve the desired capacitance. The Smith chart provides an intuitive and visual way to determine the appropriate line lengths ℓ_S and ℓ_P. Using a computerized Smith chart gives us a reliable solution within seconds. An example of dimensioning such a circuit is given in Problem 6.5. Figure 6.16a depicts a matching circuit in microstrip design.

A *radial line stub* represents an alternative to the straight stub line. Radial line double-stubs are commonly referred to as *butterfly stubs* (see Figure 6.16b). Radial line stubs have several advantages over straight stub lines, for example wider bandwidth and smaller parasitic reactance at the junction with the main line [6]. In Figure 6.16c the reflection coefficients for the straight stub line design and the butterfly stub design are displayed.

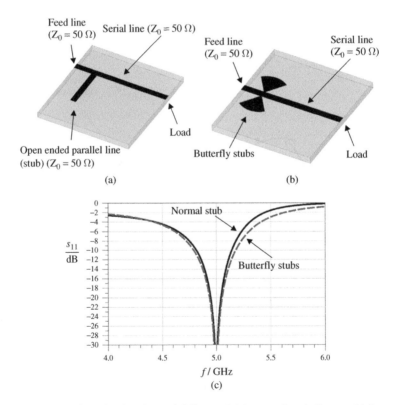

Figure 6.16 (a) Matching circuit with serial line and (a) normal stub line or (b) butterfly stub; (c) reflection coefficients for both circuits.

6.4 Filter

Filters are two-port networks with a frequency selective behaviour. A filter shall reject (or attenuate) signals in the stop band while signals in the pass band are unaffected. Figure 6.17 shows four basic filter types with distinctive pass and stop band schemes: *low-pass* filter, *high-pass* filter, *band-pass* filter, and *band-stop* filter [3, 7].

The desired frequency response (maximum attenuation in the pass band $A_{dB,p}^{max}$ and minimum attenuation in the stop band $A_{dB,s}^{min}$) is graphically displayed in *tolerance schemes*. Pass and stop band frequency limits are given by cut-off frequency f_c and stop band frequency f_S. The filter transfer function is represented by the transmission coefficient s_{21} or the insertion loss IL (equal to attenuation A).

In order to ensure low distortions in the pass band the phase of the transfer function should be as linear as possible. The group delay as the derivative of the phase should be constant [8].

$$t_{gr} = -\frac{d(\angle s_{21}(\omega))}{d\omega} \qquad \text{(Group delay)} \qquad (6.24)$$

6.4.1 Classical LC-Filter Design

Classical lumped element filter theory starts with a standard low-pass prototype filter where lumped elements (capacitance C and inductance L) are arranged in a ladder network structure (Figure 6.18a) [5, 9, 10]. The number of reactive elements (or resonant circuits) determines the order n of the filter.

The desired frequency response is achieved by choosing an appropriate filter order and selecting corresponding capacitance and inductance values. There are four common filter characteristics: *Butterworth*, *Chebyshev*, *Bessel* and *Cauer*. Figure 6.18b and Figure 6.18c show typical transfer functions and phase responses for filters with an order of $n = 5$.

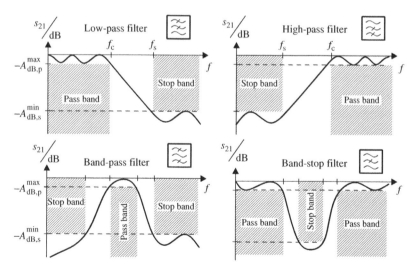

Figure 6.17 Transfer functions (transmission coefficients) of different types of filters: low-pass, high-pass, band pass and band stop filter.

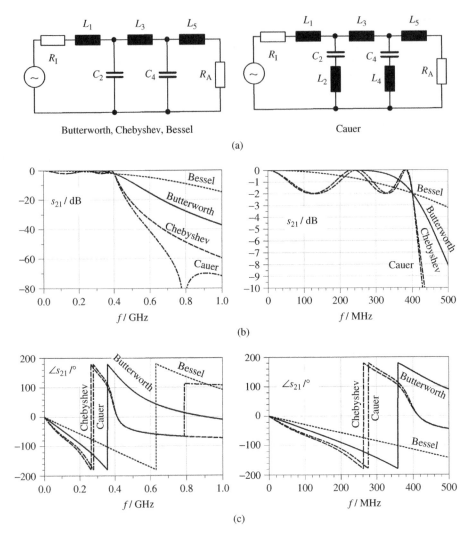

Figure 6.18 Fifth order low-pass filters: (a) topology of Butterworth, Chebyshev, Cauer and Bessel design, (b) absolute value of transfer function $s_{21}/$dB, (c) phase of transfer function $\angle s_{21}/°$.

The filter characteristics show the following properties

- *Butterworth filters* exhibit a monotonic attenuation response. Since there is no ripple in the pass band it is said to be *maximally flat*. The slope of the attenuation response is moderately steep in the transition zone from pass band to stop band. Distortions due to non-linear phase are small. Therefore, the Butterworth characteristic often represents an acceptable compromise between the requirements of high frequency selectivity (steep slope) and low distortions (linear phase).
- *Chebyshev filters* have ripples in the pass band. For example Figure 6.18b shows a Chebyshev response with ripples of 2 dB. The ripples can be controlled by selecting

appropriate capacitance and inductance values. Increasing ripples allow steeper slopes in the transition zone from pass band to stop band. For practically acceptable ripples Chebyshev filters exhibit much faster roll off than Butterworth filters thus producing high selectivity. However, the phase is less linear than that of a Butterworth filter. (Chebyshev filters of *even order* ($n = 2, 4, 6, \ldots$) require non-equal load and source impedances $R_A \neq R_I$ [5].)

- *Cauer filters* have ripples in the pass and stop band. Cauer filters are also known as *elliptical filters*. Zeros of transfer function in the stop band can be used to efficiently reject narrow band interference signals (see Figure 6.18b). Like Chebyshev filters, Cauer filters have a better rate of attenuation than Butterworth filters. The phase response is similar to Chebyshev filters for the same amplitude of ripples.
- *Bessel filters* have linear phase (constant group delay) and are optimal if distortions are to be minimized. The attenuation is the least steep compared to the previous filters, resulting in minimum selectivity.

Classical LC filter design is based on low-pass filter prototypes with normalized cut-off frequency f_c. Tables exist with coefficients for the calculation of capacitances and inductances for given cut-off frequencies as well as source and load impedances [5]. The design steps are described in detail in reference books such as [10]. On the following pages we will consider the design of Butterworth filters to illustrate the design process.

6.4.2 Butterworth Filter

6.4.2.1 Low-pass Filter

The design of Butterworth low-pass filters is particularly easy if the following conditions are met:

- The source resistance equals the load resistance $R_I = R_A = R$.
- The maximum attenuation in the pass band is $A_{dB,p}^{max} = 3\,dB$.

Then the attenuation response of a Butterworth filter of order n is given by

$$A_{dB} = -\frac{|s_{21}|}{dB} = 10\lg\left(1 + \left(\frac{f}{f_c}\right)^{2n}\right) \qquad (6.25)$$

From Equation 6.25 we calculate the order n of the filter to attain a sufficient attenuation at a given stop band frequency f_S. Table 6.2 shows examples of attenuation and

Table 6.2 Attenuation of the Butterworth filter of order n for stop band frequencies f_S that equal 1.2-, 1.5- and two-fold 3 dB cut-off frequencies f_c

Attenuation A \ Order n	$n = 2$	$n = 3$	$n = 4$	$n = 5$	$n = 6$
$A_{dB}\left(f_S = 1.2f_c\right)$	4.88 dB	6.01 dB	7.24 dB	8.57 dB	9.96 dB
$A_{dB}\left(f_S = 1.5f_c\right)$	7.83 dB	10.9 dB	14.3 dB	17.7 dB	21.2 dB
$A_{dB}\left(f_S = 2f_c\right)$	12.3 dB	18.1 dB	24.1 dB	30.1 dB	36.1 dB

filter order for stop band frequencies that equal 1.2-, 1.5- and two-fold 3 dB cut-off frequencies. After selecting an appropriate filter order n we determine the filter coefficients a_i as

$$a_i = 2 \sin\left(\frac{(2i-1)\pi}{2n}\right) \quad \text{for} \quad i = 1, 2, \ldots, n \qquad \text{(Butterworth filter coefficients)}$$

$$(6.26)$$

The capacitances C_i and inductances L_i are then given as

$$C_i = \frac{a_i}{\omega_c R} \quad ; \quad L_i = \frac{a_i R}{\omega_c} \quad ; \quad \omega_c = 2\pi f_c \qquad \text{(Low-pass filter)} \qquad (6.27)$$

where w_c is the angular cut-off frequency. There are two ways to build the ladder network: starting with a series inductance – seen from the source – as shown in Figure 6.18, or starting with a shunt capacitance. Let's look at an example.

Example 6.5 *Design of a Butterworth Low-pass Filter*
We will design a Butterworth low-pass filter with the following parameters

- $R_I = R_A = R = 50\,\Omega$ *(for the calculation of scattering parameters we use $Z_0 = R = 50\,\Omega$ as port reference impedance)*
- *3- dB cut-off frequency $f_c = 400\,\text{MHz}$*
- *minimum attenuation in the stop band $A_{\text{dB,s}}^{\min}(f \geq f_S) \geq 33\,\text{dB}$ with a stop band frequency of $f_S = 2f_c$*

Table 6.2 gives us a minimum filter order of $n = 6$ to fulfil the required attenuation of 33 dB at the stop band frequency f_S. Our filter consists of six reactive elements. According to Equation 6.26 the filter coefficients are $a_1 = a_6 = 0.5176$, $a_2 = a_5 = 1.414$ and $a_3 = a_4 = 1.932$. Following Equation 6.27 we calculate the elements for a filter topology that starts with a series inductance as

$$L_1 = 10.3\,\text{nH}, \quad C_2 = 11.3\,\text{pF}, \quad L_3 = 38.4\,\text{nH},$$
$$C_4 = 15.4\,\text{pF}, \quad L_5 = 28.1\,\text{nH}, \quad C_6 = 4.12\,\text{pF}$$

Alternatively, we can calculate the elements for a filter topology that starts with a shunt capacitance as

$$C_1 = 4.12\,\text{pF}, \quad L_2 = 28.1\,\text{nH}, \quad C_3 = 15.4\,\text{pF},$$
$$L_4 = 38.4\,\text{nH}, \quad C_5 = 11.3\,\text{pF}, \quad L_6 = 10.3\,\text{nH}$$

Figure 6.19 shows schematic circuit diagrams of both circuits using an RF simulator (ADS from Agilent [11]).
   Our design in Figure 6.19 fulfils the specifications (Transfer function $s_{21} = -3\text{dB}$ at cut-off frequency $f_c = 400\,\text{MHz}$ and sufficient attenuation $s_{21} \leq -33\,\text{dB}$ in the stop band). The phase response shows a comparatively linear behaviour in the pass band.

Butterworth low-pass filter (Order $n = 6$)

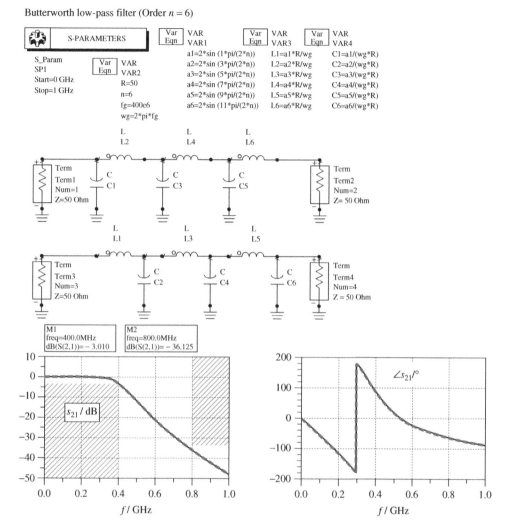

Figure 6.19 Butterworth sixth order low-pass filter (see Example 6.5).

In a circuit simulator we can usually generate user-specific templates and apply symbolic equations to calculate filter coefficients and element values. Since filter design is highly standardized, circuit simulators (like ADS) commonly contain *design guides* for automatic design and optimization of filters.

Standard filter design may result in *unrealistic* component values. Therefore, we may substitute the *idealized* capacitance and inductances by models of *real* capacitors and inductors from component manufacturers. Real components put extra constraints on the design process (available nominal values and undesired parasitics) and may introduce unwanted insertion loss in the path band due to losses. In subsequent optimizations the initial design can be adjusted to meet the specifications under more realistic conditions.

6.4.2.2 High-pass Filter

A high-pass Butterworth filter can be directly derived from the low-pass prototype described on the previous pages. Again we assume a maximum attenuation of 3 dB in the pass band and equal source and load resistances ($R_1 = R_A = R$).

The following set of rules leads us directly to the high-pass filter

1. The attenuation response of a Butterworth high-pass with 3 dB cut-off frequency f_c is given by

$$A_{dB,HP} = -\frac{|s_{21}|}{dB} = 10\lg\left(1 + \left(\frac{f_c}{f}\right)^{2n}\right) \tag{6.28}$$

With this relation we can determine the order n of the filter to obtain sufficient attenuation in the stop band.

2. In the circuit layout capacitances and inductances swap their places: the circuit consists of series capacitances and shunt inductances.

3. The modified formulas for the element values are given as

$$\boxed{C_i = \frac{1}{a_i \omega_c R} \quad ; \quad L_i = \frac{R}{a_i \omega_c} \quad ; \quad \omega_c = 2\pi f_c} \quad \text{(High-pass filter)} \tag{6.29}$$

Example 6.6 *Design of a Butterworth High-pass Filter*
We will design a Butterworth high-pass filter with the following parameters

- $R_1 = R_A = R = 50\,\Omega$ *(for the calculation of scattering parameters we use $Z_0 = R = 50\,\Omega$ as port reference impedance)*
- *3-dB cut-off frequency $f_c = 400\,\text{MHz}$*
- *minimum attenuation in the stop band $A_{dB,s}^{min}(f \le f_S) \ge 33\,\text{dB}$ with a stop band frequency of $f_S = 0.5 f_c$*

If we solve Equation 6.28 for the filter order n we get $n = 6$. The filter coefficients a_i equal the results in Example 6.5. Hence, $a_1 = a_6 = 0.5176$, $a_2 = a_5 = 1.414$, $a_3 = a_4 = 1.932$. By using Equation 6.29 we calculate the capacitances and inductances of the elements. For a filter topology that starts with a series capacitance we get

$$C_1 = 15.4\,\text{pF}, \quad L_2 = 14.1\,\text{nH}, \quad C_3 = 4.12\,\text{pF},$$
$$L_4 = 10.3\,\text{nH}, \quad C_5 = 5.63\,\text{pF}, \quad L_6 = 38.4\,\text{nH}$$

Alternatively, we calculate the elements for a filter topology that starts with a shunt inductance as

$$L_1 = 38.4\,\text{nH}, \quad C_2 = 5.63\,\text{pF}, \quad L_3 = 10.3\,\text{nH},$$
$$C_4 = 4.12\,\text{pF}, \quad L_5 = 14.1\,\text{nH}, \quad C_6 = 15.4\,\text{pF}$$

Figure 6.20 shows ADS schematic circuit diagrams of both circuits [11]. Again we use symbolic equations in the software to calculate the element values. As expected our design fulfils the specifications (Transfer function $s_{21} = -3$ dB at cut-off frequency $f_c = 400$ MHz and sufficient attenuation $s_{21} \leq -33$ dB in the stop band).

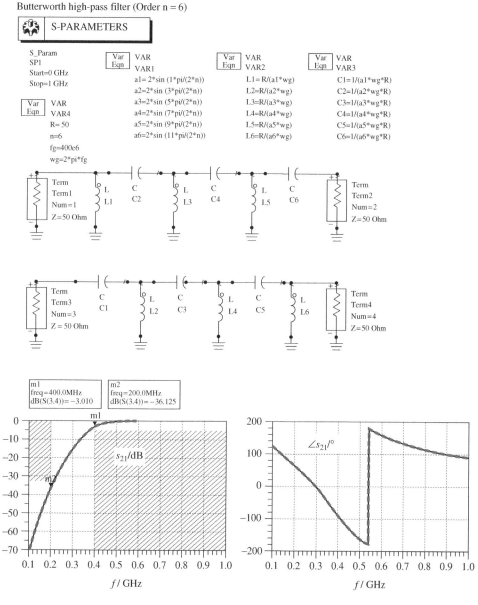

Figure 6.20 Butterworth sixth order high-pass filter (see Example 6.6).

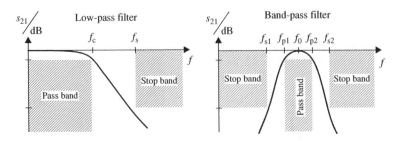

Figure 6.21   Band-pass and low-pass prototype frequency scaling.

6.4.2.3   Band-pass Filter

Band-pass Butterworth filters – like high-pass filters – can be directly derived from the low-pass prototypes described on the previous pages. Again we assume a maximum attenuation of 3 dB in the pass band and equal source and load resistances ($R_I = R_A = R$).
    The following set of rules leads us directly to the high-pass filter

1. Shunt elements are replaced by parallel-resonant circuits.
2. Series elements are replaced by series-resonant circuits.
3. The following relations establish a link between low-pass parameters (cut-off frequency f_c and stop band frequency f_s) and band-pass parameters (centre frequency f_0 and bandwidth B). Figure 6.21 illustrates the transfer functions of the low-pass prototype and band-pass filter.

$$f_0 = \sqrt{f_{p1} f_{p2}} \quad \text{and} \quad \omega_0 = 2\pi f_0 \quad \text{(Centre frequency)} \quad (6.30)$$

$$B = f_{p2} - f_{p1} \quad \text{and} \quad BW = 2\pi(f_{p2} - f_{p1}) \quad \text{(Bandwidth)} \quad (6.31)$$

$$B_{rel} = \frac{f_{p2} - f_{p1}}{f_0} \quad \text{(Relative bandwidth)} \quad (6.32)$$

The following equation converts low-pass frequency f_{LP} to band-pass frequency f_{BP} in order to determine the attenuation at a specific frequency.

$$\frac{f_{LP}}{f_c} = 2\frac{f_{BP} - f_0}{f_{p2} - f_{p1}} \quad \text{(Frequency conversion)} \quad (6.33)$$

With the frequency conversion formula in Equation 6.33 and the low-pass attenuation response of a Butterworth filter in Equation 6.25 we can calculate the required order n of the band-pass filter.
    The elements of the series-resonant circuits are given by the following equations

$$\boxed{L_{si} = \frac{a_i R}{BW} \quad ; \quad C_{si} = \frac{1}{a_i R} \cdot \frac{BW}{\omega_0^2}} \quad \text{(Band-pass filter; series elements)} \quad (6.34)$$

where $\omega_0 = 2\pi f_0$ is the angular centre frequency and $BW = 2\pi B$ is the angular bandwidth. Finally, the elements of the parallel-resonant circuits are given as

$$L_{pi} = \frac{R}{a_i} \cdot \frac{BW}{\omega_0^2} \quad ; \quad C_{pi} = \frac{a_i}{BW \cdot R} \qquad \text{(Band-pass filter; shunt elements)} \qquad (6.35)$$

Example 6.7 *Design of a Butterworth Band-pass Filter*
Design a filter with following specifications:

- *Pass band ($A_{dB,p}^{max} \leq 3$ dB) from $f_{p1} = 1.45$ GHz to $f_{p2} = 1.55$ GHz*
- *Stop band ($A_{dB,s}^{min} \geq 15$ dB) for $f < f_{s1} = 1.425$ GHz and $f > f_{s2} = 1.575$ GHz*

Equation 6.33 gives us a normalized low-pass stop frequency of $f_{s2,LP}/f_c \approx 1.5$. With Table 6.2 we conclude that the minimum filter order for $A \geq 15$ dB is $n = 5$. Figure 6.22 shows an ADS schematic with symbolic equations that apply to the previously discussed relations. The s-parameter plots indicate that the specifications are met.

6.5 Transmission Line Filter

LC filters are used in a wide frequency band. With increasing frequency the parasitic effects of the reactive components become more pronounced and the quality demands continue to rise. Hence, in the GHz frequency range transmission line filters become an increasingly attractive alternative to classical LC filters. Due to reduced wavelength at higher frequencies transmission line filters offer compact physical dimensions.

From transmission line theory (see Section 3.1.8) we know that open-ended and short-ended loss-less transmission lines show reactive behaviour and can therefore be used to mimic reactive elements. An open-ended transmission line that is shorter than a quarter of a wavelength ($\ell < \lambda/4$) has a capacitive input impedance. The same line with a length in the range of $\lambda/2 > \ell > \lambda/4$ is inductive. For $3\lambda/4 > \ell > \lambda/2$ the line is capacitive again, and so forth. So, a transmission line shows periodic behaviour.

There is a significant difference between LC and transmission line filters. LC filters have a non-periodic transfer function, that is a band-pass filter exhibits a single pass-band. The stop band that follows extends to infinity (Figure 6.23).

In contrast, the transfer functions of transmission line filters show a *cyclical pattern*: pass-bands and stop-bands occur *at regular intervals*. Figure 6.23 shows a typical transmission line filter response with an intended pass-band around 2 GHz and parasitic (spurious) pass-bands at higher frequencies (4 GHz, 6 GHz, ...).

The classical LC filter design methods can be developed further and result in a variety of filter topologies. In the following section we will – as an example – demonstrate the design of planar microstrip edge-coupled line filters. Planar transmission line types (microstrip, stripline) are particularly suitable for the design of integrated filters. Furthermore, we will look at microstrip hair-pin filters, stepped impedance filters and waveguide filters.

Butterworth band-pass filter (Order $n = 5$)

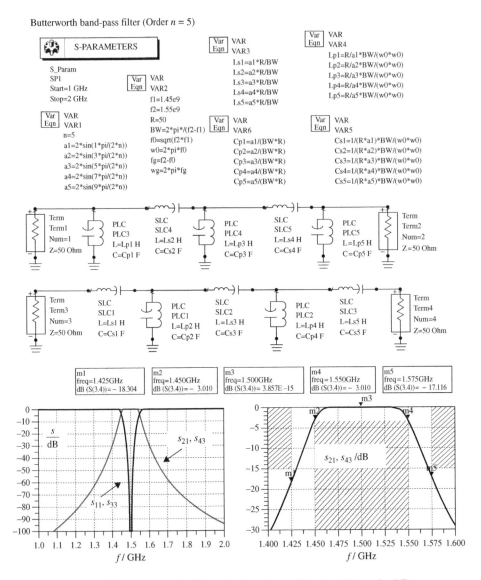

Figure 6.22 Butterworth fifth order band-pass filter (see Example 6.7).

6.5.1 Edge-Coupled Line Filter

Edge-coupled line filters are widely used band-pass transmission line structures. Figure 6.24 shows the geometry of such a filter realized with microstrip lines. The filter consists of open-ended $\lambda/2$-transmission line segments each of which overlaps by a quarter of a wavelength with adjacent elements. Unlike LC-filters (with ideal lumped components) transmission line filters show a cyclic pattern of passbands (Figure 6.23) due to the periodicity of transmission line resonances (Section 6.2).

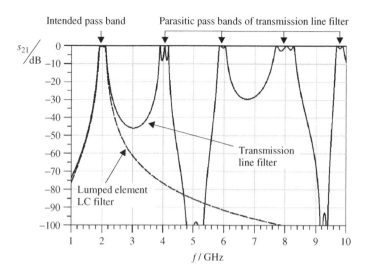

Figure 6.23 Transfer functions of band-pass filters: LC filter (single pass-band, dashed line) and transmission line filter (cyclic pattern with parasitic pass-bands, solid line).

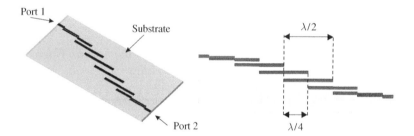

Figure 6.24 Microstrip edge-coupled line filter.

As usual we first calculate the required filter order n and the filter coefficients a_i for $i = 1, 2, \ldots n$. In the case of a Butterworth characteristic we use Equation 6.26 otherwise (e.g. Chebyshev response) we extract tabulated values from literature (e.g. [10]). In addition we define $a_0 = a_{n+1} = 1$. Let us further assume equal load impedance, source impedance and port reference impedance ($R = R_\mathrm{I} = R_\mathrm{A} = Z_0$).

Next we determine the following auxiliary quantities

$$J_{01} = \frac{1}{R}\sqrt{\frac{\pi\,BW}{2\omega_0}} \cdot \frac{1}{\sqrt{a_0 a_1}} \tag{6.36}$$

$$J_{i,i+1} = \frac{1}{R} \cdot \frac{\pi\,BW}{2\omega_0} \cdot \frac{1}{\sqrt{a_i a_{i+1}}} \qquad \text{for} \qquad i = 1, \ldots, n-1 \tag{6.37}$$

$$J_{n,n+1} = \frac{1}{R}\sqrt{\frac{\pi\,BW}{2\omega_0}} \cdot \frac{1}{\sqrt{a_n a_{n+1}}} \tag{6.38}$$

where ω_0 is the angular centre frequency and BW is the angular bandwidth given by

$$f_0 = \frac{f_{p1} + f_{p2}}{2} \quad \text{and} \quad \omega_0 = 2\pi f_0 \quad \text{(Center frequency)} \tag{6.39}$$

$$B = f_{p2} - f_{p1} \quad \text{and} \quad BW = 2\pi(f_{p2} - f_{p1}) \quad \text{(Bandwidth)} \tag{6.40}$$

With the auxiliary quantities J we calculate the even and odd mode characteristic impedances as

$$Z_{0e,i} = R\left(1 + RJ_{i,i+1} + \left(RJ_{i,i+1}\right)^2\right) \tag{6.41}$$

$$Z_{0o,i} = R\left(1 - RJ_{i,i+1} + \left(RJ_{i,i+1}\right)^2\right) \tag{6.42}$$

From these characteristic impedances we derive

• the trace width w_i of the microstrip (or stripline) transmission lines and
• the spacing s_i between coupled line segments.

Transmission line tools like [12] (see Section 4.3) and RF simulators like ADS [11] can provide conversions for specified substrate parameters. Example 6.8 illustrates the design of a microstrip edge-coupled line band-pass filter with Butterworth response on alumina substrate.

Example 6.8 *Design of an Edge-Coupled Line Filter with Butterworth Characteristic The filter shall be defined by the following characteristics:*

• *Load impedance and source impedance equal to port reference impedance $R = R_I = R_A = Z_0 = 50\,\Omega$*
• *Pass-band ($A_{dB,p}^{max} \leq 3\,dB$) from $f_{p1} = 4.8\,GHz$ to $f_{p2} = 5.2\,GHz$*
• *Stop-band ($A_{dB,s}^{min} \geq 15\,dB$) for $f < f_{s1} = 4.7\,GHz$ and $f > f_{s2} = 5.3\,GHz$ From the frequency conversion formula Equation 6.33 we get*

$$\frac{f_{s,LP}}{f_c} = 2\frac{f_{s2,BP} - f_0}{f_{p2} - f_{p1}} = 2\frac{5.3 - 5}{5.2 - 4.8} = 1.5 \tag{6.43}$$

With the attenuation response of a Butterworth low-pass filter prototype in Equation 6.25 (or alternatively by looking into Table 6.2) we derive a filter order of $n = 5$. By using Equation 6.26 we calculate the filter coefficients a_i. The first row in Table 6.3 shows the numerical values. Equations 6.36 to 6.42 provide the even and odd mode characteristic impedances of the coupled line sections, $Z_{0e,i}$ and $Z_{0o,i}$, respectively. The second and third row in Table 6.3 show the numerical values of the characteristic impedances.

In a first simulation with an RF circuit simulator we use mathematical models of ideal transmission lines that are coupled over the length of a quarter wavelength (electrical length $E = 90° = \lambda/4$). Figure 6.25 shows the schematic representation of the circuit and s-parameter results. The pass-band and stop-band specifications of the filter in the frequency range around $5\,GHz$ are met. As expected from our previous discussions transmission line filters show parasitic pass-bands. Our edge-coupled line filter with ideal

Table 6.3 Filter coefficients a_i, even and odd mode characteristic impedances. Design parameters of coupled microstrip lines: width w_i and spacing s_i (see Example 6.8)

a_i	$Z_{0e,i}/\Omega$	$Z_{0o,i}/\Omega$	$w_i/\mu\mathrm{m}$	$s_i/\mu\mathrm{m}$
$a_0 = 1$				
$a_1 = 0.618$	$Z_{0e,1} = 82.71$	$Z_{0o,1} = 37.62$	$w_1 = 359.9$	$s_1 = 162.9$
$a_2 = 1.618$	$Z_{0e,2} = 57.07$	$Z_{0o,2} = 44.51$	$w_2 = 572.6$	$s_2 = 739.9$
$a_3 = 2$	$Z_{0e,3} = 53.74$	$Z_{0o,3} = 46.75$	$w_3 = 587.8$	$s_3 = 1195.7$
$a_4 = 1.618$	$Z_{0e,4} = 53.74$	$Z_{0o,4} = 46.75$	$w_4 = 587.8$	$s_4 = 1195.7$
$a_5 = 0.618$	$Z_{0e,5} = 57.07$	$Z_{0o,5} = 44.51$	$w_5 = 572.6$	$s_5 = 739.9$
$a_6 = 1$	$Z_{0e,6} = 82.71$	$Z_{0o,6} = 37.62$	$w_6 = 359.9$	$s_6 = 162.9$

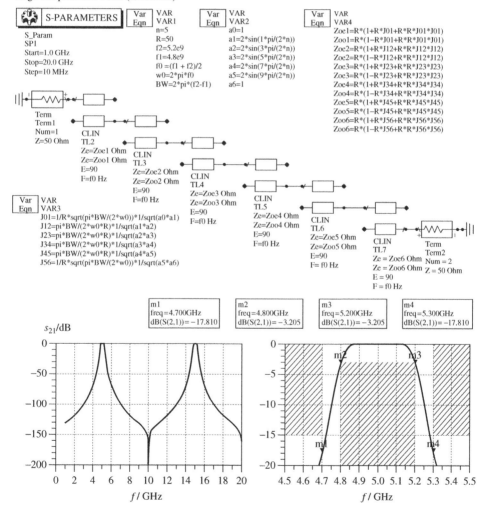

Figure 6.25 Edge-coupled line filter (with coupled ideal lines) (see Example 6.8).

lines shows extra pass-bands at odd multiples of the centre frequency $(3f_0, 5f_0, \ldots)$. Furthermore, we see zeros in the transfer function $s_{21} \to -\infty$ (i.e. attenuation pole $A \to \infty$) at even multiples of the centre frequency $(2f_0, 4f_0, \ldots)$.

Next, we will replace the ideal lines of the initial simulation by microstrip lines. We chose an alumina substrate with a relative permittivity of $\varepsilon_r = 9.8$ and a height of $h = 635\,\mu m$. We derive the trace widths w_i and the spacings s_i (i.e. distances between coupled line elements) from the even and odd mode characteristic impedances by using a transmission line tool (e.g. [12]). The last two rows in Table 6.3 show the numerical values of w_i and s_i.

The lengths ℓ_i of the quarter-wavelength line sections are around $\ell_i \approx 6\,mm$. In order to determine accurate values of the line length we must consider the following aspects of real microstrip lines:

- *Even mode and odd mode waves propagate with different speeds on coupled microstrip lines. The field patterns of even and odd modes in a cross-sectional area (see Figure 6.26) indicate that the fields distribute differently in substrate and air. Hence, the effective relative permittivities are different $\varepsilon_{r,eff}^{odd} \neq \varepsilon_{r,eff}^{even}$. Obviously, the coupled line segment has a unique physical length and cannot satisfy the quarter wavelength condition for both modes at the same time.*
- *On a microstrip line the open ends show parasitic capacitive behaviour (open end effect) due to fringing electric fields (see Figure 6.8). As a consequence the physical line length ℓ must be slightly shorter than a quarter of a wavelength.*
- *There is some field coupling between all $\lambda/4$ filter segments that is not considered in a circuit simulation. In a circuit simulator the different models are connected by idealized terminals. EM simulation software includes field coupling between all elements of a structure and is therefore best suited for final validation and optimization.*

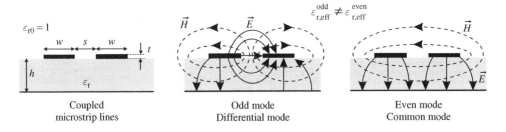

Figure 6.26 Coupled microstrip lines: field patterns of odd and even mode.

The result of the final microstrip design is shown in Figure 6.27 and compared to the initial design with ideal quarter-wave lines. There are some interesting points to note:

- *The microstrip version of the edge-coupled line filter exhibits* additional pass-bands *at even multiples of the centre frequency $(2f_0, 4f_0, \ldots)$. These pass-bands result from an unavoidable mismatch between line segments [10].*
- *The parasitic pass-band we would expect at – for example – $3f_0$ occurs at a slightly lower frequency due to dispersion.*
- *Altogether, the filter response is close to the intended characteristic. Optimization is still possible for further improvement of the filter.*

Edge-Coupled-Line-Filter (Microstrip)

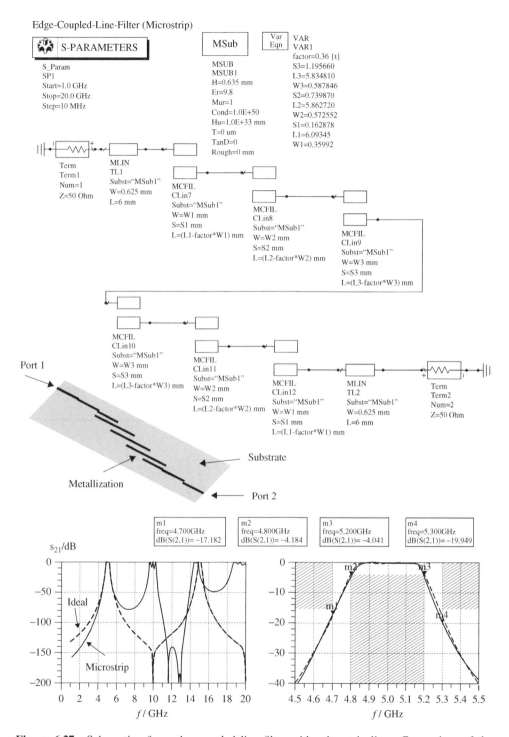

Figure 6.27 Schematic of an edge-coupled line filter with microstrip lines. Comparison of the transfer function of a microstrip line filter (solid line) and an ideal line filter (dashed line) (see Example 6.8).

6.5.2 Hairpin Filter

A disadvantage of the previously discussed edge-coupled line filters is their extended length resulting from the $\lambda/4$ coupling sections that are linked together. By rearranging the coupling sections we obtain a more compact filter known as a *hairpin* filter. Figure 6.28 shows an example to illustrate the filter topology.

6.5.3 Stepped Impedance Filter

The previously discussed edge-coupled line filter and hairpin filter show band-pass behaviour. Furthermore, these filters block DC currents due to disconnections of the transmission lines. *Stepped-impedance filters* on the other hand show low-pass-behaviour and provide a DC path from input port to output port in order to power active components behind the filter.

A stepped impedance filter consists of series transmission lines with different characteristic impedances. Figure 6.29a shows a microstrip stepped impedance filter with alternating sections of wide (low impedance) and narrow (high impedance) transmission lines. If these transmission line sections are short compared to the wavelength (i.e. at lower frequencies) the wide line segments behave like shunt capacitances since the electric field is concentrated between line and ground (like the electric field in a parallel-plate capacitor). Furthermore, the narrow line segments behave like series inductances since the narrow path of the current leads to high values of the magnetic field around the line. Figure 6.29b shows an equivalent circuit for the low-pass frequency range. At higher frequencies the behaviour becomes more complex and typically spurious pass-bands occur (see Figure 6.29c).

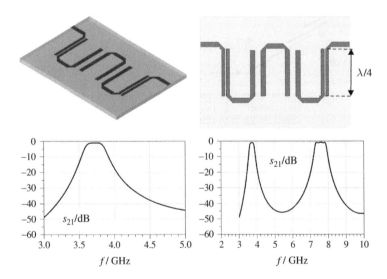

Figure 6.28   Hairpin filter.

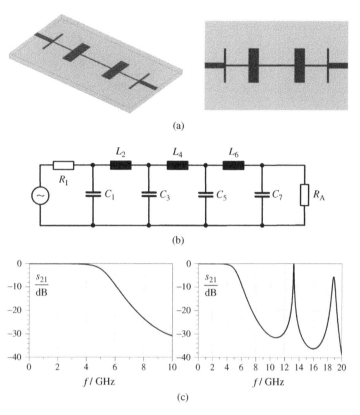

Figure 6.29 Stepped impedance filter: (a) microstrip design, (b) equivalent circuit for the low-pass frequency range and (c) transmission coefficients.

6.5.4   Parasitic Box Resonance

Planar filters and other circuits are often shielded by rectangular metallic enclosures in order to avoid coupling to surrounding structures and circuitry. These housings represent rectangular cavity resonators (see Section 4.6.6) with unwanted resonances.

Figure 6.30 shows as an example an edge-coupled line filter with a pass-band around 8.7 GHz. After integrating the filter in a metallic enclosure a parasitic pass-band occurs at a frequency of 14 GHz. In Figure 6.30 the arrows of the electric field strength at $f = 14$ GHz are plotted in a horizontal plane above the filter. The distribution shows the typical TE_{101} resonance pattern with vertical arrows and sinusoidal variation in both horizontal directions (see Figure 4.21a for comparison). With the following equation (also see Equation 4.68)

$$f_{R,101} = \frac{c_0}{2}\sqrt{\left(\frac{1}{a}\right)^2 + \left(\frac{1}{c}\right)^2} \tag{6.44}$$

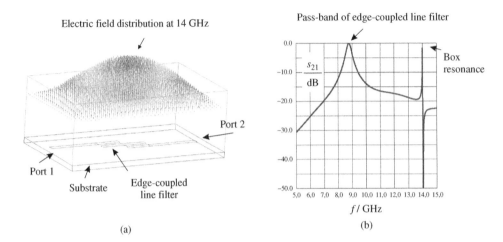

(a)

(b)

Figure 6.30 Edge-coupled line filter in a metallic enclosure: (a) distribution of the electric field strength in a plane above the substrate showing TE$_{101}$ cavity resonance, (b) transmission coefficient showing regular pass-band of edge-coupled line filter as well as narrow-band pass-band due to parasitic box resonance.

and the known length $a = 12$ mm and width $c = 19.6$ mm we estimate a resonance frequency of the (empty) cavity resonator of $f_{R,101} \approx 14.6$ GHz. The actual resonance frequency (14 GHz) deviates from the calculated value since the filter and substrate fill out a part of the cavity.

With increasing frequency further pass-bands occur due to higher order resonances (see Section 4.6.6). If the parasitic pass-bands lead to unacceptable behaviour, for example a minimum attenuation at resonance is required, the following constructive steps may be taken:

- By changing the dimensions of the enclosure the resonance can be shifted into a less critical frequency range. Larger dimensions shift the resonance to lower frequencies. Smaller dimensions shift the resonance to higher frequencies.
- By bringing lossy material into the enclosure the resonance effect can be reduced. However, the lossy material shall not influence the regular behaviour of the circuit. Hence, it would be advisable to attach an absorbing material underneath the cover (top side of the enclosure).

The example of a filter in an enclosure demonstrates that coupling to the surrounding structures can significantly modify the circuit behaviour. Fortunately, EM simulation software may assist in the systematic investigation of these effects.

6.5.5 Waveguide Filter

Edge-coupled line filters use resonant $\lambda/2$-lines with field-coupling between adjacent elements. This concept can be transferred to band-pass structures consisting of rectangular waveguides resulting in *iris-coupled cavity filters*. In order to build resonant structures, pairs of vertical metallic blocks are placed into the waveguide (width a, height b) that

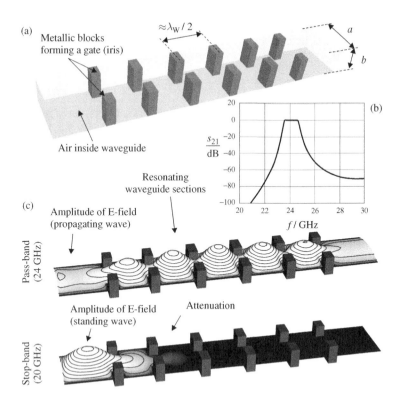

Figure 6.31 Rectangular waveguide with band-pass iris filter: (a) geometry, (b) transmission coefficient and (c) absolute values of the electric field at 24 GHz (pass-band) and 20 GHz (stop-band).

form cavities with a length of half a waveguide wavelength $\lambda_W/2$. Figure 6.31a shows an example with five coupled cavities. The size of a cavity is therefore $a \times b \times \lambda_W/2$. Adjacent cavities are coupled by the lateral opening (iris) between two bars. By varying the width of the opening we can control the coupling and accomplish desired band-pass response.

Figure 6.31b illustrates the transfer function of the filter example with a pass-band around 24 GHz. In Figure 6.31c we see the magnitude of the electric field strength E in the pass-band (24 GHz) and stop-band (20 GHz) in a horizontal cut-plane through the waveguide.

Let's look at the pass-band first. On the left side (ahead of the filter) we see a constant magnitude along the line. From Section 3.1 we know that this indicates a pure forward travelling wave. In the cavities of the filter we see resonant behaviour with increased magnitude. Behind the filter there is again a pure forward travelling wave that equals the incoming wave in magnitude. In the stop-band (20 GHz) we see a standing wave pattern with stationary maxima and minima ahead of the filter. This pattern indicates the superposition of a forward travelling and reflected wave of the same amplitude. The cavities are off-resonance and lead to the reflection of waves. Behind the filter there is virtually no outgoing wave.

Waveguide filters show low insertion loss and allow high slopes of the transfer function. Furthermore, waveguide structures tolerate comparatively high power.

6.6   Circulator

The previously discussed passive circuits (matching networks, filters) have *two* ports. We will now look at circuits with *three* ports. As discussed in Section 5.4.7 a three-port network cannot satisfy the following conditions simultaneously

- matching (at all ports),
- reciprocity and
- loss-lessness.

In order to construct a practical three-port network at least one condition must be dropped. If we allow losses we can build a *power divider* that is matched at all ports (see Section 6.7). Furthermore, an interesting non-reciprocal ($s_{ij} \neq s_{ji}$) network that is matched at all ports is a *circulator*.

An *ideal circulator* transfers signals from port 1 to port 2, from port 2 to port 3 and from port 3 to port 1 as shown in Figure 6.32a. The signal flow graph shows that the *forward* transmission factors are $s_{21} = s_{32} = s_{13} = 1$. All other scattering parameters are zero (Figure 6.32b). If we consider propagation delay, a more realistic term for the transmission coefficients includes a phase term $e^{-j\varphi}$. The corresponding scattering matrices are

$$\mathbf{S}_{\text{C,ideal}} = \begin{pmatrix} 0 & 0 & 1 \\ 1 & 0 & 0 \\ 0 & 1 & 0 \end{pmatrix} \qquad \mathbf{S}_{\text{C,ideal},\varphi} = \begin{pmatrix} 0 & 0 & e^{-j\varphi} \\ e^{-j\varphi} & 0 & 0 \\ 0 & e^{-j\varphi} & 0 \end{pmatrix} \qquad (6.45)$$

A non-ideal circulator is usually described by non-zero reflection coefficients ρ, non-zero *backward* transmission coefficients σ and *forward* transmission coefficients τ with a magnitude of less than one. The corresponding signal flow diagram is shown in Figure 6.32c.

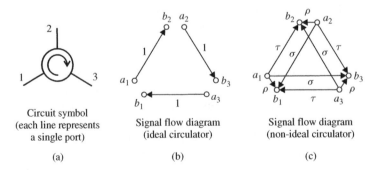

Circuit symbol
(each line represents
a single port)

(a)

Signal flow diagram
(ideal circulator)

(b)

Signal flow diagram
(non-ideal circulator)

(c)

Figure 6.32   (a) Graphical circuit symbol and signal flow diagram of (b) ideal and (c) non-ideal circulator.

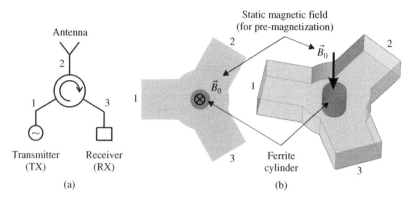

Figure 6.33 (a) Application of a circulator as a transmit-receive switch (duplexer) and (b) pre-magnetized ferrite cylinder in a rectangular waveguide configuration.

The scattering matrix is given as

$$S_C = \begin{pmatrix} \rho & \sigma & \tau \\ \tau & \rho & \sigma \\ \sigma & \tau & \rho \end{pmatrix} \tag{6.46}$$

A typical application of a circulator is the separation of transmitting and receiving signals at an antenna as shown in Figure 6.33a. The signal from the transmitter at port 1 is transferred to the antenna at port 2 and radiated into space. A signal that is received by the antenna is transferred to the receiver at port 3. Since the signals from the transmitter are usually stronger than the weak signals that are received by the antenna a good isolation from port 1 to port 3 is crucial.

Circulators are commonly based on microstrip, stripline and waveguide technology [13, 14]. The core element of such a circulator is a magnetized ferrite. Figure 6.33b shows the basic construction of a circulator with three rectangular waveguides connected to a cylindrical cavity with a central cylindrical ferrite. The ferrite is magnetized by a static magnetic field $\vec{B}_0$ usually generated by an external permanent magnet. An incoming TE_{10}-wave at port 1 excites a TE_{110}-resonant mode in the cylindrical cavity. The resonant mode in the cavity can be described by two counter rotating fields. The non-reciprocal properties of the magnetized ferrite lead to different propagation characteristics of the rotating field components. By choosing the right dimensions and magnetic field strength $\vec{B}_0$ we can guide the incoming signal to port 2 and decouple port 3.

6.7 Power Divider

6.7.1 Wilkinson Power Divider

In the previous section we considered a *non-reciprocal*, loss-less three-port network that was matched at all ports. Now we introduce a three-port network that is matched at all ports, reciprocal but *lossy*: a power divider.

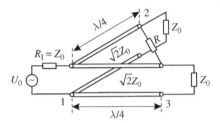

Figure 6.34 Basic design of a Wilkinson power divider.

In Section 6.3.2 we looked at a simple power divider using two outgoing lines and a quarter-wave transformer for impedance matching at the input. The simple power divider was *not matched* at the output ports. With two quarter-wave transformers and a resistor $R = 2Z_0$ we can design a power divider that is matched at all ports. Figure 6.34 shows the basic construction of a *Wilkinson* power divider.

The power that is delivered to port 1 is split into equal parts and transferred to port 2 and port 3. Due to the symmetrical construction there is no voltage drop across the resistor R if the circuit is fed at port 1. Hence, there is no power loss in the resistor R. The resistor is only necessary to ensure the matching of port 2 and port 3.

The quarter-wave lines have characteristic impedances of $Z_{0Q} = \sqrt{2}Z_0$ and transform the output impedances Z_0 to input impedances of $2Z_0$. The parallel circuit of the two quarter-wave transformers at the input terminal leads to $Z_{in,1} = (2Z_0 || 2Z_0) = Z_0$. Hence, port 1 is matched.

Figure 6.35a depicts a Wilkinson power divider with microstrip lines and a SMD resistor. The scattering parameters in Figure 6.35b demonstrate the equal power split between port 2 and port 3 ($s_{21} = s_{31} = -3\,\text{dB}$). Furthermore, we see that matching (typical requirement $s_{11} < -20\,\text{dB}$) is achieved in a certain frequency band (bandwidth). The bandwidth can be increased by a multistage design (see [2]).

The scattering matrix of an *ideal* Wilkinson power divider at the centre frequency f_0 is given as

$$\mathbf{S}_{\text{Wilkinson}} = \frac{-j}{\sqrt{2}} \begin{pmatrix} 0 & 1 & 1 \\ 1 & 0 & 0 \\ 1 & 0 & 0 \end{pmatrix} \tag{6.47}$$

The scattering parameters reveal reciprocity ($s_{ij} = s_{ji}$) and matching at all ports ($s_{ii} = 0$). It is apparent that the Wilkinson power divider is lossy due to the incorporation of an ohmic resistor R. (However, losses do not occur if we feed the network at port 1 and ideally terminate the output ports by Z_0.) Of course we can demonstrate lossy behaviour by showing that the scattering matrix is not a unitary matrix (see Problem 6.6).

6.7.2 Unequal Split Power Divider

The Wilkinson power divider concept can be modified in order to split the power *unequally* among the output ports. As an example we will use design formulas from [15]. According

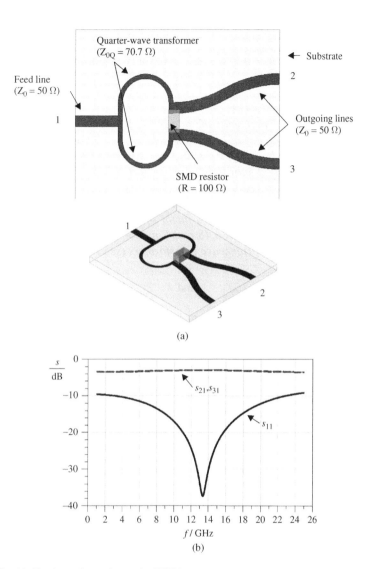

Figure 6.35 (a) Design of a microstrip Wilkinson power divider with SMD resistor and (b) scattering parameters s_{11}, s_{21} and s_{31}.

to Figure 6.36a we choose the characteristic impedances of the quarter-wave transformers as

$$Z_{0b} = Z_0 \sqrt{1 + \left(\frac{k}{1-k} \right)^2} \tag{6.48}$$

$$Z_{0a} = \frac{1-k}{k} Z_{0b} \tag{6.49}$$

$$R = 2Z_0 \tag{6.50}$$

where $k = P_{b2}/P_{a1}$ is the fraction of the input power (fed to port 1) that is delivered to port 2.

For an equal power split ($k = 0.5$) the design formulas result in the classical Wilkinson design with $Z_{0a} = Z_{0b} = \sqrt{2}Z_0$.

Example 6.9 *Power Divider (2:1)*

We design a power divider that splits the power fed to port 1 between port 2 and port 3 with a ratio of 2:1. The frequency of operation shall be $f = 1\,\mathrm{GHz}$.

The design parameter is $k = P_{b2}/P_{a1} = 2/3 = |s_{21}|^2$ (see Equation 5.25). Hence, the ideal value of the transmission factor from port 1 to port 2 should be $s_{21} = -1.76\,\mathrm{dB}$. Furthermore, we have $P_{b3}/P_{a1} = 1/3 = |s_{31}|^2$. So, the ideal value of the transmission factor from port 1 to port 3 should be $s_{31} = -4.77\,\mathrm{dB} = s_{21} - 3.01\,\mathrm{dB}$. The difference between s_{21} and s_{31} equals $3\,\mathrm{dB}$ corresponding to a factor of 2 in power (2:1 power divider).

With the design formulas Equation 6.48 and Equation 6.49 we can calculate the characteristic impedances of the quarter-wave transformers (see Figure 6.36a) as

$$Z_{0b} = \sqrt{5}Z_0 \quad \text{and} \quad Z_{0a} = \frac{\sqrt{5}}{2}Z_0 \tag{6.51}$$

The length of the quarter-wave lines have to be $\lambda/4$ at $f = 1\,\mathrm{GHz}$.

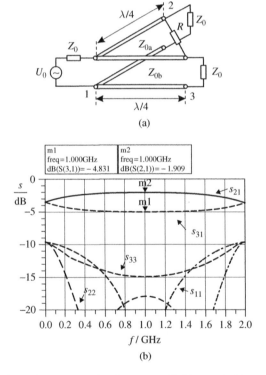

(a)

(b)

Figure 6.36 (a) Design of an unequal-split power divider and (b) s-parameter results (see Example 6.9).

With a circuit simulator we can calculate scattering parameters shown in Figure 6.36b. The results of the transmission coefficients are close to the theoretical values. Matching at port 1 (s_{11}) is nearly ideal at $f = 1\,\text{GHz}$. The output ports (s_{22}, s_{33}) are sufficiently matched for most practical applications.

6.8 Branchline Coupler

6.8.1 Conventional 3 dB Coupler

A *branchline* coupler is a four-port network that can be used either as a power divider or as a signal combiner. Figure 6.37a shows the *power divider* operation: the signal fed into port 1 is split into equal shares between port 2 and port 3. Port 4 is isolated. The signals at the output ports 2 and 3 are $90°$ out of phase. Figure 6.37b shows the *combiner* operation: Signals with equal magnitude but a phase difference of $\Delta\varphi = 90°$ at port 1 and 4 are summed up at port 2. Port 3 is isolated.

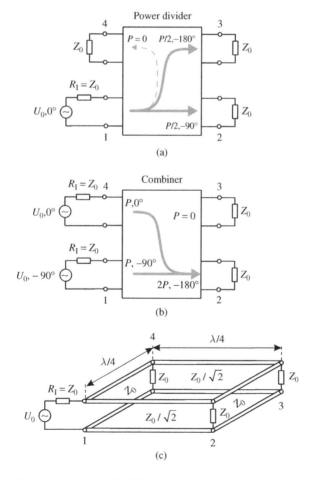

(a)

(b)

(c)

Figure 6.37 Branchline coupler: (a) 3 dB/$90°$ power divider, (b) combiner, and (c) basic design.

The branchline coupler consists of two series lines (characteristic impedance $Z_0/\sqrt{2}$) and two shunt lines (characteristic impedance Z_0), each of which is a quarter wavelength long (see Figure 6.37c).

The scattering matrix of an *ideal* branchline coupler (at the centre frequency f_0) is given as

$$\mathbf{S}_{\text{Branchline}} = \frac{-1}{\sqrt{2}} \begin{pmatrix} 0 & j & 1 & 0 \\ j & 0 & 0 & 1 \\ 1 & 0 & 0 & j \\ 0 & 1 & j & 0 \end{pmatrix} \tag{6.52}$$

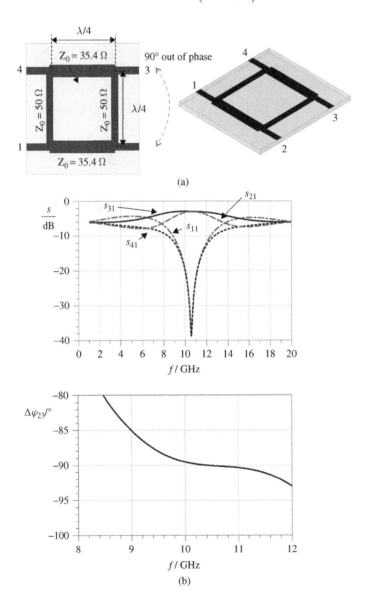

Figure 6.38 Microstrip branchline coupler: (a) geometry and (b) s-parameter results.

Figure 6.38 shows a branchline coupler with microstrip lines. At the centre frequency the scattering parameters s_{11} and s_{41} are low. The transmission coefficients are $s_{21} = s_{31} = -3\,\text{dB}$ indicating an equal power split. The phase difference between the output signals is $90°$. These properties apply approximately in a frequency band around f_0. The bandwidth can be increased using three shunt lines (double box branchline coupler) [16].

6.8.2 Unequal Split Branchline Coupler

If we change the characteristic impedances of series and shunt quarter-wave lines we can establish an unequal power split. At the same time we can match different impedances at input and output ports. The output ports 2 and 3 have a characteristic impedance Z_{02} and the input port 1 as well as the isolated port 4 have a characteristic impedance of Z_{01} (see Figure 6.39a).

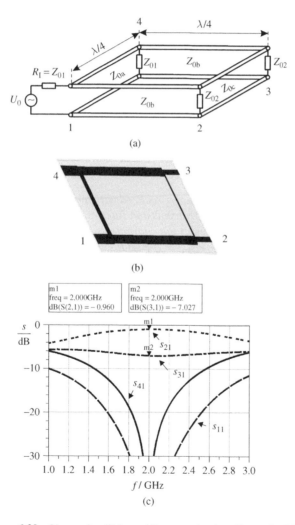

Figure 6.39 Unequal split branchline coupler (see Example 6.10).

According to [15] the design formulas are

$$Z_{0a} = \frac{Z_{01}}{k} \qquad \text{(Input shunt quarter-wave line)} \qquad (6.53)$$

$$Z_{0b} = \sqrt{\frac{Z_{01} Z_{02}}{1 + k^2}} \qquad \text{(Series quarter-wave lines)} \qquad (6.54)$$

$$Z_{0c} = \frac{Z_{02}}{k} \qquad \text{(Output shunt quarter-wave line)} \qquad (6.55)$$

where

$$k = \left| \frac{s_{31}}{s_{21}} \right| = \sqrt{\frac{P_{b3}}{P_{b2}}} \qquad (6.56)$$

defines the square-root of the power ratio delivered to the output ports ($P_{b3} = k^2 P_{b2}$).

Example 6.10 *Unequal Split Branchline Coupler with Impedance Transformation*
A branchline coupler shall be match to $Z_{01} = 50\,\Omega$ at port 1 and 4 and $Z_{02} = 75\,\Omega$ at port 2 and 3. The power ratio between port 2 and 3 shall be 4:1 ($P_{b2} = 4P_{b3}$), that is 80% of the power is delivered to port 2 and 20% of the power is delivered to port 3. Hence, the design parameter becomes $k = \sqrt{P_{b3}/P_{b2}} = 0.5(= -6\,dB)$. The frequency of operation shall be $f = 2\,GHz$.
With Equation 6.53 to 6.55 the characteristic line impedances are

$$Z_{0a} = 100\,\Omega \quad ; \quad Z_{0b} = 54.8\,\Omega \quad ; \quad Z_{0c} = 150\,\Omega \qquad (6.57)$$

Figure 6.39b shows the branchline coupler realized with microstrip lines (substrate parameter: $h = 1.524\,mm$, $\varepsilon_r = 2.2$). The transmission lines at port 1 and 4 have characteristic impedances of $50\,\Omega$ and at port 2 and 3 the characteristic impedances are $75\,\Omega$. The theoretical transmission coefficients are $s_{21} = \sqrt{P_{b2}/P_{a1}} = \sqrt{0.8} = -0.969\,dB \approx -1\,dB$ and $s_{31} = \sqrt{P_{b3}/P_{a1}} = \sqrt{0.2} = -6.990\,dB \approx -7\,dB$. The simulation results in Figure 6.39c are in good agreement with the theoretical values.

With decreasing k-value (i.e. increased power transfer to port 2 and decreased power transfer to port 3) the characteristic impedances of the shunt lines increase leading to smaller microstrip lines. Therefore, inevitable production tolerances limit the design of circuits with extreme power ratios. In this case directional couplers (see Section 6.10) provide an attractive alternative.

Interestingly, the design formulas Equation 6.53 to 6.55 cover a particular case when k approaches zero ($k \to 0$). In this case the characteristic impedances of the shunt lines become infinite and the corresponding line width is zero. Hence, there are no existing shunt lines. Furthermore, the series quarter-wave transformer between port 1 and port 2 is $Z_{0b} = \sqrt{Z_{01} Z_{02}}$. From Section 3.1.9 we know that this relation ensures impedance matching between port 1 and port 2.

6.9   Rat Race Coupler

Just like the branchline coupler, the *rat race coupler* is a four-port network with equal power-split between the two output ports. Figure 6.40 shows – as an example – a rat race coupler in microstrip technology with corresponding line lengths and characteristic impedances. The annular arrangement of $\lambda/4$- and $3\lambda/4$- lines has characteristic impedances of $Z_{0Q} = \sqrt{2}Z_0$. The power fed into port 1 is equally split between port 2 and port 3. Port 4 is isolated. In contrast to the branchline coupler, the rat race coupler provides output signals with a phase difference of $180°$.

The principal behaviour can be derived by investigating the signal paths on the annular arrangement of lines. An incident wave from port 1 separates into two waves each of which travels either in a clockwise or counter-clockwise direction along the ring. At port 4 both waves cancel out due to a $\lambda/2$ phase shift. At port 2 and port 3 the waves add up constructively. Like the branchline coupler the rat race coupler can be used as a combiner.

The scattering matrix of an ideal rat race coupler at the centre frequency f_0 is given as

$$\mathbf{S}_{\text{Rat-race}} = \frac{j}{\sqrt{2}} \begin{pmatrix} 0 & -1 & 1 & 0 \\ -1 & 0 & 0 & -1 \\ 1 & 0 & 0 & -1 \\ 0 & -1 & -1 & 0 \end{pmatrix} \tag{6.58}$$

Rat race couplers can be designed with unequal power split (see [15]).

6.10   Directional Coupler

If the amount of power that is separated is rather small or input and output require galvanic isolation, *directional couplers* can be used. A directional coupler basically consists of two coupled parallel lines with a length of $\ell = \lambda/4$ separated by a distance s (see Figure 6.41).

This kind of coupler is a *backward coupler*, since the fraction of power $k^2 P$ that is coupled in the adjacent line is delivered to port 4. Port 3 at the other end of the line is isolated. The main part $(1 - k^2)P$ of the incident power is transferred onto the direct path to port 2.

The coupling coefficient is commonly expressed in logarithmic scale as

$$\frac{k}{\text{dB}} = |20 \lg(k)| \tag{6.59}$$

A 20 dB directional coupler ($k = 0.1$) decouples 1% of the incident power to port 4. 99% of the incident power propagates through the main line to port 2. Directional couplers are used in measurement, for example network analysers use directional couplers to evaluate forward travelling and backward travelling waves on transmission lines (see Section 5.7).

According to Section 4.9 coupled lines can be described by their even and odd mode characteristic impedances, Z_{0e} and Z_{0o}, respectively. These impedances are linked to the

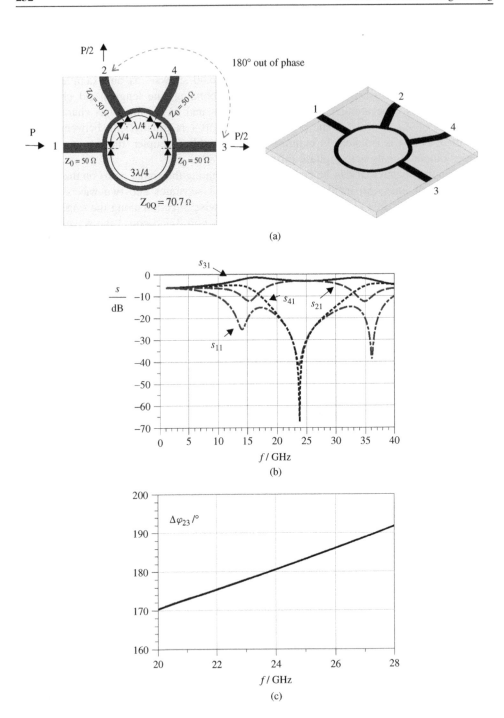

Figure 6.40   Microstrip rat race coupler.

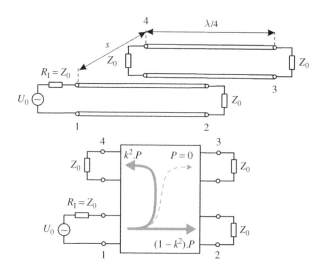

Figure 6.41 Directional coupler consisting of two parallel lines ($\ell = \lambda/4$).

coupling coefficient k and the single-ended characteristic impedance Z_0 by the following relations:

$$Z_{0e} = Z_0\sqrt{\frac{1+k}{1-k}} \tag{6.60}$$

$$Z_{0o} = Z_0\sqrt{\frac{1-k}{1+k}} \tag{6.61}$$

Solving these equations for k and Z_0 yields

$$Z_0 = \sqrt{Z_{0e}Z_{0o}} \tag{6.62}$$

$$k = \frac{Z_{0e} - Z_{0o}}{Z_{0e} + Z_{0o}} \tag{6.63}$$

The scattering matrix of an ideal directional coupler (at centre frequency f_0) is given as

$$\mathbf{S}_{\text{Directional}} = \begin{pmatrix} 0 & -j\sqrt{1-k^2} & 0 & k \\ -j\sqrt{1-k^2} & 0 & k & 0 \\ 0 & k & 0 & -j\sqrt{1-k^2} \\ k & 0 & -j\sqrt{1-k^2} & 0 \end{pmatrix} \tag{6.64}$$

As an example we will design a directional coupler based on striplines.

Example 6.11 *Stripline 20 dB Directional Coupler*

A 20 dB-*directional coupler with striplines* ($Z_0 = 50\,\Omega$) *shall be designed for a frequency of* $f = 6\,\text{GHz}$. *The substrate parameters are: height* $h = 1.6\,\text{mm}$, *relative permittivity* $\varepsilon_r = 2.5$ *and thickness of metallization* $t = 10\,\mu\text{m}$.

A coupling coefficient of 20 dB corresponds to $k = 0.1$. With Equations 6.60 and 6.61 we calculate the even mode characteristic impedance $Z_{0e} = 55.28\,\Omega$ and the odd mode characteristic impedance $Z_{0o} = 45.23\,\Omega$. With a transmission line tool (e.g. TX-Line [12] or LineCalc [11]) we get line length $\ell = 7.9\,mm$, width $w = 1.148\,mm$ and spacing $s = 0.535\,mm$. The geometry and scattering parameters are shown in Figure 6.42a and 6.42b. The results are in good agreement with the specification.

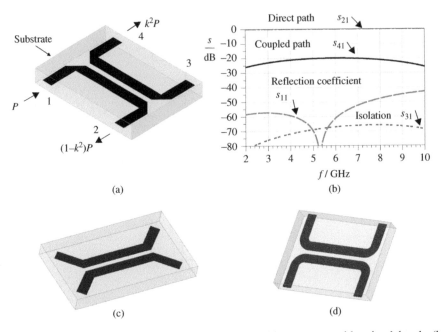

Figure 6.42 20 dB directional coupler with striplines: (a) geometry with mitred bends (b) s-parameter results; alternative geometry with (c) 45° bends or (d) curved elements.

In order to minimize reflections mitred 90°-bends are used to feed the signals to the coupled transmission lines (Figure 6.42a). Alternatively, 45°-bends (Figure 6.42c) or curved transmission line elements (Figure 6.42d) can be used.

6.11 Balanced-to-Unbalanced Circuits

In Section 4.9 we discussed balanced and unbalanced transmission lines. Microstrip lines, striplines and coaxial lines are *unbalanced* transmission lines. Symmetrical two-wire lines and coupled microstrip lines represent *balanced* transmission lines. A balanced line is commonly driven with symmetrical signals where the voltages at the terminals have equal magnitude but opposite signs with respect to ground $U_1 = -U_2$ (see Equation 4.75).

If we connect a balanced transmission line or component to an unbalanced transmission line or component *baluns* (balanced-to-unbalanced transformers) are used. For low and medium frequencies a *transformer with a secondary centre tap* provides a broadband conversion between unbalanced and balanced systems (Figure 6.43a). Such a transformer

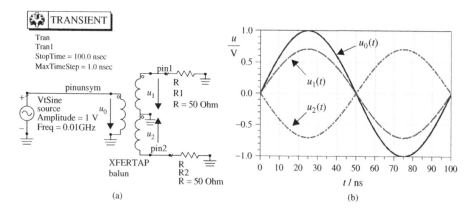

Figure 6.43   Balanced-to-unbalanced circuit (balun) with transformer.

can additionally be used for impedance matching. Figure 6.43b shows an unbalanced input signal $u_0(t)$ and balanced output signals $u_1(t)$ and $u_2(t)$ where $u_1(t) = -u_2(t)$.

At higher frequencies transformers become unattractive due to increased losses. Therefore, transmission line structures are commonly used to provide *narrowband* conversion. Figure 6.44a shows a balun with two $\lambda/4$-lines to provide a $180°$ phase

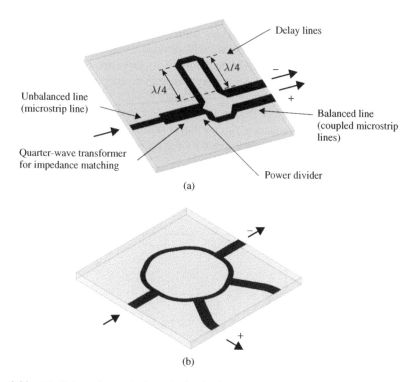

Figure 6.44   (a) Balanced-to-unbalanced circuit (balun) with two $\lambda/4$-delay lines to provide a $180°$ phase shift; (b) rat race $180°$-coupler.

shift. The unbalanced signal on a microstrip line is matched to a power divider and split into two signals of equal magnitude and phase. Both signals run over lines with different length. The difference in line length is $\lambda/4 + \lambda/4 = \lambda/2$. At the end of the lines the signals are $180°$ out of phase and excite the differential mode on a coupled microstrip line (balanced line).

Furthermore, a rat race coupler (already discussed in Section 6.9 can be used to provide two signals of equal magnitude but with opposite polarity (see Figure 6.44b). Therefore, a rat race coupler may be used as a balanced-to-unbalanced circuit.

6.12 Electronic Circuits

This book focuses on linear, passive structures described by Maxwell's equations or circuit theory. If we combine these elements with active, non-linear components (diodes, transistors) we can design active circuits like oscillators, amplifiers and mixers.

As intensively discussed in the previous sections linear, passive circuits are understood on the basis of electromagnetic fields, propagating waves and linear circuit theory. The design is based on EM simulation or (linear) circuit simulation.

Active and non-linear RF circuits on the other hand are based on general (non-linear) circuit theory. RF electronics require knowledge of basic design concepts for amplifiers, transistor parameters, stability considerations, non-linear behaviour (1 dB compression point, 3rd-order intercept point), noise figures, and so on. RF circuit simulators can usually handle these aspects. However, it is beyond the scope of this book to go into the details of RF electronics. The following list provides sources for further reading: [1, 17–21].

Nevertheless, in the following sections we will provide a short overview of important, active key elements in RF transmission technology to give the reader a first orientation. Figure 6.45 shows a simplified elementary transmitter/receiver unit (transceiver) for wireless communications.

In radio systems the original (low-frequency) signal (e.g. speech, data) with a frequency f_{TX} is transferred to a higher frequency $f_{RF, TX}$ that is suited for wireless transmission. The frequency range for transmission is determined by the radio standard used. In the *transmit*

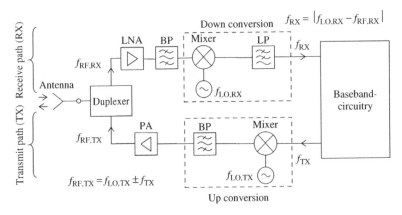

Figure 6.45 Basic design of an RF front-end for a simple wireless communication system.

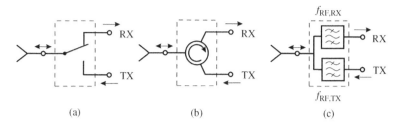

Figure 6.46 Duplexer with (a) RF switch, (b) circulator and (c) frequency-separating filters (diplexer).

path the *mixer* converts the low-frequency (baseband) signal into the RF frequency range supported by a *local oscillator* with a frequency of $f_{LO, TX}$. A *band-pass filter* BP selects the desired frequency component $f_{RF, TX}$ that is then increased in magnitude by a *power amplifier* PA. Finally, a *duplexer* routes the signal to the antenna where it is radiated into space.

There are different concepts in order to realize a *duplexer* (see Figure 6.46). If the transmit and receive signals are in the same frequency range an *RF switch* (e.g. using PIN-diodes) can be used for discontinuous communication. Depending on the position of the switch either transmission or reception is possible (see Figure 6.46a). Alternatively, a *circulator* (see Section 6.6) can be used to separate transmit and receive signals at the antenna. A circulator allows transmission and reception at the same time (see Figure 6.46b). If transmit and receive signals are in separated frequency bands a *diplexer* may be used. A diplexer basically consists of two frequency selective band-pass filters with a common terminal where the antenna is connected (see Figure 6.46c).

In the *receive path* the weak signal captured by the antenna is intensified by a low-noise amplifier (LNA) and carried to a mixer. The mixer converts the RF signal ($f_{RF, RX}$) to a

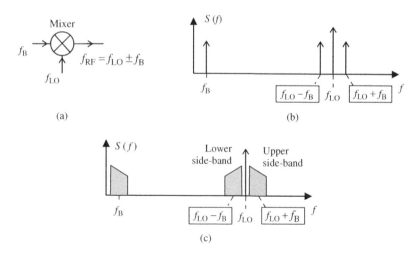

Figure 6.47 Ideal mixer multiplying two signals.

base-band signal (f_{RX}). The local oscillator provides a monofrequent signal ($f_{LO, RX}$) for frequency conversion. The low-pass filter (LP) selects the base-band signal and suppresses unwanted higher frequency content.

Electronic circuits can be miniaturized in different ways [18]. *Monolithic Microwave Integrated Circuits* (MMIC) use semiconductor substrate material like Gallium Arsenide (GaAs) in order to integrate passive structures (transmission lines, filters, couplers) and active non-linear structures (transistors, diodes) on a single chip. This approach enables small integrated RF modules for radio equipment. *Hybrid circuits* combine passive planar structures on dielectric substrates (e.g. alumina, LTCC (Low-Temperature Cofired Ceramics)) with concentrated elements mounted to the surface of the substrate (e.g. transistors, ICs, SMD-components).

6.12.1 Mixers

The ideal mixer is basically a multiplier. Let us consider a harmonic local oscillator signal $u_{LO}(t) = U_{LO} \cos(\omega_{LO}t)$ and a harmonic base-band input signal $u_B(t) = U_B \cos(\omega_B t + \varphi_B)$. By using the following mathematical relation

$$\cos(\alpha)\cos(\beta) = \frac{1}{2}\left[\cos(\alpha - \beta) + \cos(\alpha + \beta)\right] \tag{6.65}$$

we get an output signal of an ideal mixer equal to the product of the input signal $u_B(t)$ and the local oscillator signal $u_{LO}(t)$ as

$$u_{LO}(t) \cdot u_B(t) = U_{LO} \cos(\omega_{LO}t) \cdot U_B \cos(\omega_B t + \varphi_B)$$
$$= \frac{1}{2}U_{LO}U_B\left[\cos\left((\omega_{LO} - \omega_B)t - \varphi_B\right) + \cos\left((\omega_{LO} + \omega_B)t + \varphi_B\right)\right] \tag{6.66}$$

The output signal of the mixer is proportional to the amplitude U_B of the input signal. Furthermore, the output signal contains the phase φ_B of the input signal. The frequency f_B of the input signal is shifted by the local oscillator frequency f_{LO}: we can find the sum and difference of the frequencies (see Figure 6.47a and 6.47b).

Next we consider an input signal that covers a certain frequency band around f_B as shown in Figure 6.47c. After mixing the base-band signal appears above and below the local oscillator. These frequency bands are denoted as upper and lower side-bands. The upper side-band is in the normal position and the lower side-band is in the inverted position, that is the maximum frequency component in the original base-band signal is now at the minimum frequency in the lower side-band.

Both side-bands contain the full information content of the original base-band signal, so only one side-band is selected to maintain spectral efficiency. If we select the upper side-band we speak of *up-conversion* in the normal position and if we use the lower side-band we speak of up-conversion in the reverse frequency position. Transferring an RF (RX) signal into the base-band by a mixer is denoted as *down-conversion*. If the RF signal has upper and lower side-bands around the carrier frequency both side-bands

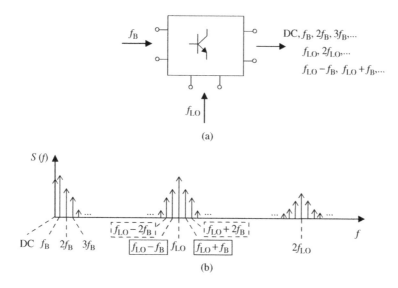

Figure 6.48   Frequency spectrum of a non-ideal mixer.

Table 6.4   Frequency components of non-ideal mixer output signal (where f_B and f_{LO} are frequencies of input signals)

0 Hz (DC)					
f_B	f_{LO}				
$2f_B$	$2f_{LO}$	$f_{LO} \pm f_B$			
$3f_B$	$3f_{LO}$	$f_{LO} \pm 2f_B$	$2f_{LO} \pm f_B$		
$4f_B$	$4f_{LO}$	$f_{LO} \pm 3f_B$	$2f_{LO} \pm 2f_B$	$3f_{LO} \pm f_B$	
$\cdots$	$\cdots$	$\cdots$	$\cdots$	$\cdots$	$\cdots$

superimpose at low frequencies. Therefore, we usually preselect one of the side-bands by a filter.

The ideal mixer described above cannot be realized. A real mixer feeds input and local oscillator signals to a non-linear element like a transistor or diode. Mixing with a non-linear element creates not only the wanted sum and difference frequencies discussed above but a great number of undesired frequencies. Figure 6.48 illustrates the frequencies on a frequency axis and Table 6.4 lists the frequencies systematically.

In Figure 6.48b the wanted frequencies are presented with a solid-line frame. The most unwanted frequency components (third-order products) are shown with a dashed-line frame. The latter are closest to the useful signal components so they are most likely to cause interference.

The most commonly used parameters to characterize non-linear behaviour are the *1 dB compression point* and the *third-order intercept point* (IP$_3$). Their definitions are illustrated

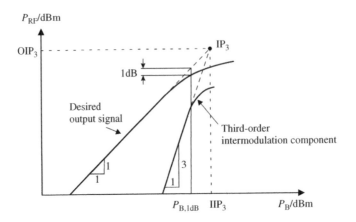

Figure 6.49 Illustration of the 1 dB compression point and third-order intercept point (IP_3).

in Figure 6.49 where we find the power of the desired output signal and the power of the third-order component as a function of the input-power P_B in double-logarithmic scale. At low input powers the output power is proportional to the input power (slope equals one). The 1 dB compression point is given by the power $P_{B,1\,dB}$ where the real output deviates 1 dB from the idealized proportional behaviour. The power of the third-order frequency component has a slope of three for low input powers. If we extrapolate the low input power behaviour for the desired and unwanted third-order component we find the third-order intercept point as the cross-point of the straight lines.

The non-linear behaviour defines the limit for the upper operating range. The lower operating range is limited by noise effects. The usable dynamic range of a mixer lies inbetween these limits. Within its dynamic range we can describe the mixer by the *conversion loss* L_C which is the input-to-output power ratio.

$$\frac{L_C}{dB} = 10\lg\left(\frac{P_B}{P_{RF}}\right) \tag{6.67}$$

The conversion loss is a positive value if the output power P_{RF} is smaller than the input power P_B. By looking at the output-to-input power ratio we get the conversion gain as $G_C/dB = -L_C/dB$. A practical mixer may show positive or negative gain.

6.12.2 Amplifiers and Oscillators

An amplifier is used to increase the amplitude of electrical signals. The output signal should show minimum non-linear distortions and low noise. The central component of a basic amplifier is a biased active element (transistor). Within a defined operating point the linear input-output behaviour of a transistor is described by a scattering matrix **S**. In order to provide sufficient amplification transistors with high transit frequencies are used, for example *Metal-Semiconductor Field-Effect Transistors* (MESFET). Furthermore,

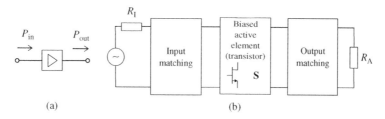

(a)   (b)

Figure 6.50 Amplifier: (a) graphical circuit symbol and (b) biased active element (transistor) with input and output matching circuits.

matching networks provide required input and output impedances (see Figure 6.50). An important aspect of amplifier design is *stability* in order to avoid unwanted oscillation.

Non-linear behaviour can be derived from analysis of detailed transistor models. The non-linear parameters are the 1 dB compression point and third-order intercept point. In order to determine the 1 dB compression point the amplifier is excited by a single-tone source and the output power is plotted over the input power. In order to determine the third-order intercept point a two-tone source with two slightly different frequencies f_1 and f_2 is applied and the third-order component is evaluated. The resulting curves to determine the 1 dB compression point and third-order intercept point are similar to the illustration in Figure 6.49.

An important parameter that describes the linear behaviour is the (power) *gain* G defined by the ratio of output-to-input power.

$$\frac{G}{dB} = 10 \lg \left(\frac{P_{out}}{P_{in}} \right) \tag{6.68}$$

Oscillators generate monofrequent signals, for example as the local oscillator (f_{LO}) for mixing purposes (see Figure 6.51). As an example oscillators can be designed by using an amplifier and a feedback network k. The frequency of oscillation is determined by a resonant element with high frequency stability. In the RF frequency range such elements are quartzes or dielectric resonators. The spectrum of a real oscillator (Figure 6.51d) differs from the spectrum of an ideal oscillator (Figure 6.51c) that produces a purely sinusoidal signal. An important parameter that describes the frequency stability in the frequency domain is *phase noise*.

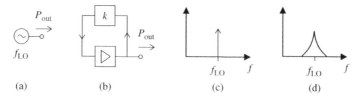

(a)   (b)   (c)   (d)

Figure 6.51 Oscillator: (a) graphical circuit symbol, (b) amplifier and feedback network, (c) spectrum of ideal oscillator and (d) spectrum of real oscillator.

6.13   RF Design Software

6.13.1   RF Circuit Simulators

RF circuit simulators are powerful tools that support engineers in the design and verification of complex circuits and systems. Usually different kinds of analyses are possible within such software tools, for example:

- Digital and analog simulation.
- Circuit simulation in the time-domain or frequency domain.
- S-parameter analysis with concentrated (lumped) elements.
- S-parameter analysis with distributed and coupled elements (transmission lines).
- Non-linear analysis.
- Noise parameter analysis.
- Parameter sweep and optimization.

Practical routine and straightforward issues such as matching, filter design, coupler design and so forth are commonly supported by design assistants (design guides) that apply proven formulas from standard literature. These standard designs can be further improved and adjusted by internal optimization algorithms.

In this book a lot of examples are calculated with a circuit simulator that is part of the RF design software ADS from Agilent [11]. Simulation examples include the time-domain analysis of voltage waves on transmission lines in Section 3.2 and the frequency-domain s-parameter simulation of filters in Section 6.4.

Circuit simulators generally use simplified mathematical models to describe component behaviour on the basis of voltages and currents at the terminals. Circuit simulators are therefore very efficient. However, the mathematical models have a limited range of usage. If applications exceed this range of usage or if no simplified model is available for a given component, a more complex approach is required.

Let us look a two examples for problems that cannot be solved within a circuit simulator. First, a circuit simulator does not consider the layout of a circuit: coupling between adjacent and distant components and the surrounding housing is not captured but can be important when – for example – EMC[1] issues are addressed. Second, antenna design requires knowledge of electric and magnetic field distributions that are not calculated within a circuit simulator. This leads us directly to EM simulation software.

6.13.2   Three-Dimensional Electromagnetic Simulators

EM simulation has become an integral part of RF design. EM simulators provide (approximate) solutions of Maxwell's equations for given three-dimensional material distributions [22–24]. From the calculated distribution of the *electromagnetic field* the software derives s-parameters, impedances, antenna parameters and so forth. Figure 6.52a shows as an example a microstrip filter in a test-fixture. The three-dimensional-simulation model is depicted in Figure 6.52b and c. Simulation results of different numerical methods are close to measurement results (Figure 6.52d).

---

[1] Electromagnetic Compatibility.

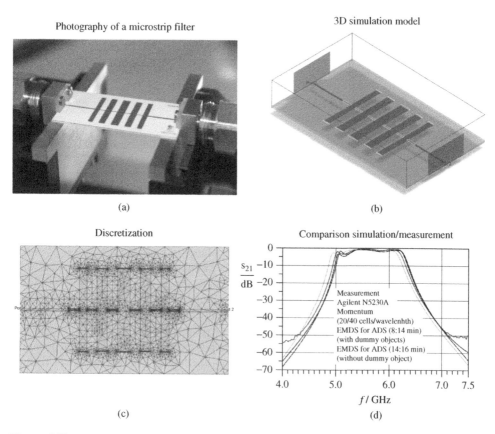

Figure 6.52 FEM and MoM simulation of a microstrip filter in ADS; comparison of s-parameter results with measurements.

By now numerous powerful and user-friendly EM simulation software packages are on the market. The software products are under continuous development. Therefore, Table 6.5 shows an incomplete list of some widely used commercial products.

EM simulation tools apply different mathematical methods to solving the electrodynamic field problem [22]:

- Finite-Element Method (FEM).
- Finite-Difference Time-Domain Method (FDTD).
- Method of Moments (MoM).

Figure 6.53 shows the typical simulation flow in a 3D EM simulator. First, the three-dimensional *geometry* is defined using a graphical user interface (GUI) similar to CAD tools for mechanical engineering. Three-dimensional objects can be drawn and manipulated by operations like addition, subtraction, intersection, rotation, extrusion, chamfering and so on. Up-to-date software furthermore supports various CAD formats in order to import CAD models from other software. Next, *material properties* (conductivity, relative permittivity, relative permeability, loss tangent, etc.) are assigned to the objects.

Table 6.5 Commercial software tools (selection)

Software package	Company
ADS (Advanced Design System)	Agilent, USA
EMPIRE	IMST, Germany
EMPro (Electromagnetic Professional)	Agilent, USA
HFSS (High-Frequency Structure Simulator)	Ansoft, USA
Microwave Studio	CST, Germany
SEMCAD	Schmidt & Partner Engineering AG, Switzerland
XFDTD	Remcom, USA

If we use the FEM or FDTD method the simulation volume must be finite. At the boundary of the simulation space *boundary conditions* are defined that determine the field behaviour. Electric and magnetic walls mimic planes of symmetry, absorbing boundaries avoid reflections and are therefore useful in simulating antennas that radiate into free space.

Furthermore, ports are defined to excite the simulation model and extract network parameters like impedance and scattering parameters. Usually we distinguish between *waveguide ports* representing a transmission line cross-section and *lumped ports* that are represented by two idealized terminals. Waveguide ports generally provide more accurate results and consider different modes on a transmission line. The excitation is further specified by a frequency range of interest (for s-parameter analysis) or a time signal (e.g. for signal integrity analysis).

The *simulation* can then be done with one of the above mentioned numerical methods. Each method requires a discretization of the geometry. FEM and FDTD use volumetric meshes (tetrahedra, cuboids). The Method of Moments uses surface meshes. Small mesh elements usually result in more accurate results. Unfortunately, the computational burden increases as the number of elements increases. Depending on the required accuracy it is essential to use meshes with adequate complexity. One method to achieve a sufficient mesh quality is to perform a sequence of simulations with advancing mesh refinement. Stable results at the end of such a process indicate an appropriate mesh. For complex structures where adaptive mesh refinement is not applicable it is important so start with an optimum mesh. The software vendors provide guidelines for setting up such meshes by the experienced user.

During the simulation process the electric and magnetic fields are determined by a recursive time-stepping algorithm (FDTD) or by the inversion of a system of equations (FEM, MoM). Systems of equations can be solved using direct or iterative methods. The simulation usually takes a few seconds (for quite simple models) up to hours or even days for very large models.

After the simulation is finished the user can visualize the results. To start with we can access electromagnetic field quantities in the near- and far-field region. Furthermore, circuit quantities (voltage, current, impedance, scattering parameters) are available. Furthermore, we can export data and perform specific post-processing tasks.

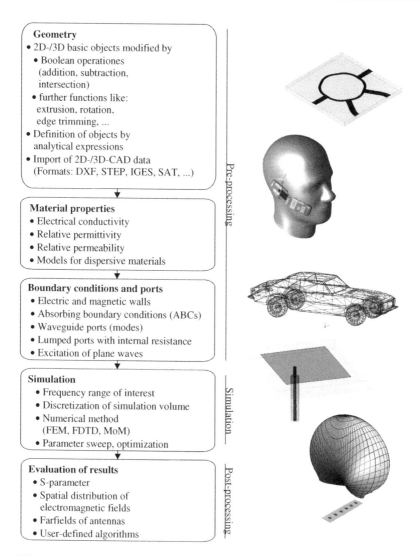

Geometry
- 2D-/3D basic objects modified by
  - Boolean operationes
    (addition, subtraction,
    intersection)
  - further functions like:
    extrusion, rotation,
    edge trimming, ...
- Definition of objects by
  analytical expressions
- Import of 2D-/3D-CAD data
  (Formats: DXF, STEP, IGES, SAT, ...)

Material properties
- Electrical conductivity
- Relative permittivity
- Relative permeability
- Models for dispersive materials

Boundary conditions and ports
- Electric and magnetic walls
- Absorbing boundary conditions (ABCs)
- Waveguide ports (modes)
- Lumped ports with internal resistance
- Excitation of plane waves

Simulation
- Frequency range of interest
- Discretization of simulation volume
- Numerical method
  (FEM, FDTD, MoM)
- Parameter sweep, optimization

Evaluation of results
- S-parameter
- Spatial distribution of
  electromagnetic fields
- Farfields of antennas
- User-defined algorithms

Pre-processing

Simulation

Post-processing

Figure 6.53 Typical simulation flow within EM simulation software (left); simulation examples: rat race coupler, mobile phone with user head, car model with external antenna, patch antenna with coaxial feed, array antenna (right; from top to bottom).

EM simulation software is widely used in the design of RF components, circuits and antennas due to a number of advantages:

- Distributions of the electromagnetic field can be visualized within the simulation volume. Field plots help understand the functional principle of the RF component or circuit and provide valuable information for further improvement.

- Numerical models can easily be parametrized. So the software may support parameter sweeps to perform a tolerance analysis. Furthermore, automatic optimization of components is commonly implemented.
- Areas of application for EM simulation software are expanding rapidly due to advances in mathematical algorithms (e.g. iterative equation solvers) as well as the computational power of modern computer hardware (e.g. parallelization, CPU speed).
- Simulations are – especially at an early stage of a design process – much faster and less cost-intensive than laboratory measurements.
- If nothing else, novices in the field of RF engineering feel motivated by the illustrative approach that relieves the burden of mathematical computations from the user. For this reason EM simulators are used intensively in engineering education.

In this book we apply EM simulation software in numerous contexts, especially if field distributions are required to discuss component behaviour. Examples are modes in waveguides (Section 4.6), waveguide filter (Section 6.5.5), nearfields and farfields of antennas (Section 7.6) and waves in free space interacting with other structures (Section 8.1). The examples are simulated with different commercial simulation tools: *Empire* from IMST [25], *ADS/Momentum/FEM* from Agilent [11] and *EMPro* from Agilent [26].

Currently, we see developments to combine RF circuit simulation and electromagnetic (EM) field simulation under a common user interface (circuit/electromagnetic co-simulation), in order to set up numerical models that consist of circuits (described by schematics and mathematical models) as well as three-dimensional structures (described by their three-dimensional geometry and Maxwell's equations). As an example we could think of an antenna with a matching network. The matching network part is solved by the circuit simulator and the antenna structure is solved by the EM simulator.

6.14   Problems

6.1 A transmission line is short-circuited at both ends. Demonstrate that the line is at resonance if it is one half-wavelength (or multiples of $\lambda/2$) in length. A transmission line is short-circuited at one end and open-circuited at the other end. Demonstrate that the line's shortest resonant length is one quarter of a wavelength.

6.2 The following real load impedances

$$Z_{A1} = 330\,\Omega \quad ; \quad Z_{A2} = 10\,\Omega \tag{6.69}$$

and complex load impedances

$$Z_{A3} = (200 + j100)\,\Omega \,; \quad Z_{A4} = (15 - j75)\,\Omega \tag{6.70}$$

shall be matched to an impedance of $Z_0 = 50\,\Omega$. Use a printed Smith chart or solve the matching problem with a computerized Smith chart tool (e.g. [4]).

6.3 A load impedance of $R_A = 1100\,\Omega$ shall be matched to a source with an internal resistance of $R_I = 50\,\Omega$ at a frequency of $f = 1\,\text{GHz}$.

1. Design a simple (single-stage) LC matching network.
2. Increase the bandwidth by designing a two-stage LC matching network.

3. Increase the bandwidth further by designing a three-stage LC matching network. Use the same impedance ratio R_A/Z_{in} for each stage. Use a Smith chart to visualize the transformation path.

6.4 Figure 6.14 shows a simple power divider with microstrip lines on an alumina substrate (height $h = 635\,\mu m$; relative permittivity $\varepsilon_r = 9.8$). The power divider shall be matched to $Z_0 = 50\,\Omega$ at a frequency of $f = 5\,GHz$.

1. Design the microstrip quarter-wave transformer (length ℓ_Q and width w_Q).
2. Use a two-stage design to increase the bandwidth. Calculate all relevant lengths and widths.

6.5 A load impedance of $Z_A = (100 + j20)\,\Omega$ shall be matched to a source resistance of $R_I = 50\,\Omega$ at a frequency of $f = 10\,GHz$. Design a matching circuit with a series line and a stub line (see Figure 6.15). Design the circuit by using a Smith chart. Realize the circuit with $50\,\Omega$-microstrip lines on a substrate with the following parameters: height $h = 254\,\mu m$; relative permittivity $\varepsilon_r = 7.8$.

6.6 Prove that the ideal Wilkinson power divider (Section 6.7) is lossy by demonstrating that the scattering matrix is no unitary matrix.

6.7 Design a branchline coupler with unequal power split at a frequency of $f = 6\,GHz$. The power ratio at the output ports 2 and 3 shall be 3:1 ($P_2/P_3 = 3/1$). The port reference impedance for all ports is $Z_0 = 50\,\Omega$. Realize a microstrip circuit using an alumina substrate (height $h = 635\,\mu m$; relative permittivity $\varepsilon_r = 9.8$).

6.8 Design a Butterworth-LC-low-pass filter with the following specifications:

1. Source impedance Z_I, load impedance Z_A and port reference impedance Z_0 equal $50\,\Omega$.
2. 3 dB cut-off frequency is $f_c = 700\,MHz$.
3. Desired minimum attenuation at twice the cut-off frequency is $A(f_s = 2f_c) \geq 17\,dB$.

6.9 Design an edge-coupled line filter using microstrip lines (substrate parameters: height $h = 400\,\mu m$; relative permittivity $\varepsilon_r = 7.8$). The filter specifications are:

1. Port reference impedance $Z_0 = 50\,\Omega$.
2. Maximum attenuation in the pass-band $A_{dB,p}^{max} \leq 3\,dB$ from $f_{p1} = 8.0\,GHz$ to $f_{p2} = 8.5\,GHz$
3. Minimum attenuation in the stop-band $A_{dB,s}^{min} \geq 20\,dB$ for $f < f_{s1} = 7.5\,GHz$ and $f > f_{s2} = 9.0\,GHz$

References

1. Detlefsen J, Siart U (2009) *Grundlagen der Hochfrequenztechnik*. Oldenbourg.
2. Heuermann H (2009) *Hochfrequenztechnik*. Vieweg.
3. Pozar DM (2005) *Microwave Engineering*. John Wiley & Sons.

4. Dellsperger F (2010) *Smith-Chart Tool* http://www.fritz.dellsperger.net/downloads.htm.
5. Bowick C (2008) *Circuit Design*. Newnes.
6. Simons RN (2001) *Coplanar Waveguide Circuits, Components and Systems*. John Wiley & Sons.
7. Ludwig R, Bogdanov G (2008) *RF Circuit Design: Theory and Applications*. Prentice Hall.
8. Fettweis A (1996) *Elemente nachrichtentechnischer Systeme*. Teubner.
9. Hong JS, Lancaster MJ (2001) *Microstrip Filters for RF/Microwave Applications*. John Wiley & Sons.
10. Matthaei GL, Young L, Jones EMT (1980) *Microwave Filters, Impedance Matching Networks, and Coupling Structures*. John Wiley & Sons.
11. Agilent Corporation (2009) *Advanced Design System (ADS)*. Agilent Corporation.
12. AWR (2003) *TX-Line Transmission Line Calculator*. AWR Corporation.
13. Jansen W (1992) *Streifenleiter und Hohlleiter*. Huethig.
14. Zinke O, Brunswig H (2000) *Hochfrequenztechnik 1*. Springer.
15. Ahn HR (2006) *Asymmetric Passive Components in Microwave Integrated Circuits*. John Wiley & Sons.
16. Meinke H, Gundlach FW (1992) *Taschenbuch der Hochfrequenztechnik*. Springer.
17. Ellinger F (2007) *Radio-Frequency Integrated Circuits and Technologies*. Springer.
18. Golio M, Golio J (2008) *The RF and Microwave Handbook*. CRC Press.
19. Hagen, JB (2009) *Radio-Frequency Electronics: Circuits and Applications*. Cambridge University Press.
20. Maas SA 1988 Nonlinear Microwave Circuits. Artech House.
21. Maas SA (1998) *The RF and Microwave Circuit Design Cookbook*. Artech House.
22. Gustrau F, Manteuffel D (2006) *EM Modeling of Antennas and RF Components for Wireless Communication Systems*. Springer.
23. Swanson DG jun, Hoefer WJR (2003) *Microwave Circuit Modeling Using Electromagnetic Field Simulation*. Artech House.
24. Weiland T, Timm M, Munteanu I (2008) Practical Guide to 3-D Simulation. *IEEE Microwave Magazine*, Vol. 9, No. (6).
25. IMST GmbH (2010) *Empire: Users Guide*. IMST GmbH.
26. Agilent Corporation (2010) *EMPro*. Agilent Corporation.

Further Reading

Baechtold W (1999) *Mikrowellentechnik*. Vieweg.

7

Antennas

Antennas are used to emit and receive electromagnetic waves that can propagate through free space. A *transmit antenna* converts power that is delivered to its circuit terminal into electromagnetic waves (see Figure 7.1a). A *receive antenna* captures power from an electromagnetic wave and provides it at its circuit terminal (see Figure 7.1b).

Figure 7.1 shows a wireless communication link with two antennas. A transmission line wave arrives at the circuit terminal (port 1) of antenna 1. Ideally the reflection coefficient associated with that port is zero and all power is accepted by the transmit antenna. The antenna produces an electromagnetic wave that propagates through free space. The mathematical treatment of the electromagnetic field in the vicinity of the antenna (*nearfield*) may be quite complex. For practical purposes it is commonly sufficient to describe the electromagnetic field in some distance of the antenna (*farfield*). In the farfield we see spherical wavefronts as discussed in Section 2.5.4. Therefore, farfield antenna parameters are commonly given in spherical coordinates (r, ϑ, φ). The receive antenna (antenna 2) is positioned far away from antenna 1 and collects power from the electromagnetic wave. The power is converted into a transmission line wave starting at port 2 and travelling down the line.

A passive antenna may be used as a transmit or receive antenna. The radiation characteristics as well as the input (port) impedance are equal in both cases (*reciprocity*). This means that an antenna that radiates (transmit case) maximum power in a certain direction is maximumly susceptible to a wave that arrives from that same direction (receive case).

In the following section we will summarize the main parameters that are used to describe radiation characteristics and circuit properties. Furthermore, we provide an overview of standard antenna types used in wireless communications.

7.1 Fundamental Parameters

7.1.1 Nearfield and Farfield

If we consider the electromagnetic fields of an antenna it makes a big difference if we look at the rather complex field distribution in the close vicinity of the antenna structure (*nearfield*) or if we look at the electromagnetic field distribution at greater distances from

RF and Microwave Engineering: Fundamentals of Wireless Communications, First Edition. Frank Gustrau.
© 2012 John Wiley & Sons, Ltd. Published 2012 by John Wiley & Sons, Ltd.

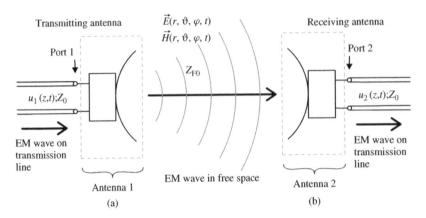

Figure 7.1 (a) Transmitting antenna converts transmission line waves to free space waves. (b) Receiving antenna captures energy from the free space waves and invokes waves on the transmission line.

the antenna (*farfield*). Usually antennas that are used to establish a wireless link are far away from each other, therefore the important radiation characteristics are derived under farfield conditions.

In Section 7.3 we mathematically derive the electromagnetic fields of a canonical antenna element (*Hertzian dipole*). We will see that the equations simplify if we progress from the nearfield region to the farfield region. Figure 7.2 shows some commonly used terms to describe field regions around an antenna. The field region closest to the antenna is known as the *reactive nearfield*. In the nearfield region fields are mainly associated with the storage of electric and magnetic energy. On the other hand, in the farfield region (*Fraunhofer zone*) the fields are associated with radial power transfer. The product of the Poynting vector $\vec{S}$ and distance squared ($\vec{S}(r, \vartheta, \varphi) \cdot r^2$) is independent of the distance r. As we will see in Section 7.1.3 this leads to a radiation pattern with an angular dependency that does not change with distance [1]. The transition area between reactive nearfield and farfield is known as the radiating nearfield (*Fresnel zone*). In the Fraunhofer zone and Fresnel zone different mathematical approximations are used. From these

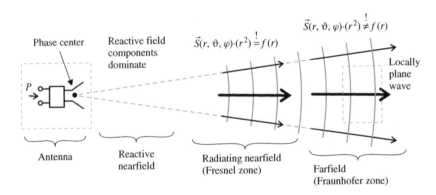

Figure 7.2 Nearfield and farfield regions of an antenna.

commonly used approximations we can derive a minimum distance for farfield conditions (*farfield distance*).

The *farfield distance* of an antenna depends on the wavelength (derived from the operational frequency) and the geometrical dimensions of the antenna structure. According to [2] the minimum farfield distance r for an *electrically small antenna* is given by

$$\boxed{r \geq 2\lambda} \quad \text{(Farfield region of an electrically small antenna)} \quad (7.1)$$

If the antenna has maximum dimensions L_{max} greater than one wavelength (*electrically large antenna*) farfield conditions apply only for greater distances

$$\boxed{r \geq \frac{2L_{max}^2}{\lambda}} \quad \text{(Farfield region of an electrically large antenna)} \quad (7.2)$$

The farfield of an antenna is characterized by the following properties.

- There is only a radial component of power density $\vec{S}$ indicating radial power transfer away from the antenna ($S_r \neq 0$). The transversal components are equal to zero ($S_\vartheta = S_\varphi = 0$).
- The magnitude of the power density $\vec{S}$ is in inverse proportion to the distance squared $|S| \sim 1/r^2$.
- The magnitude of the electric and magnetic field strengths E and H are in inverse proportion to the distance from the antenna $|E|, |H| \sim 1/r$.
- The amplitudes of the electric and magnetic field strength E and H are *in phase* and linked by the characteristic impedance Z_{F0} of free space as $E = Z_{F0}H$. The vectors $\vec{E}$ and $\vec{H}$ are perpendicular to each other and perpendicular to the radial direction of travel (TEM wave).
- The radial components of the electric and magnetic field strengths E and H are zero: $E_r = H_r = 0$. There are only components tangential to the (radial) direction of propagation.

The electric[1] farfield of an antenna can therefore be described by the following equation [2, 3].

$$\vec{E}(r, \vartheta, \varphi) = E_\vartheta(r, \vartheta, \varphi) \cdot \vec{e}_\vartheta + E_\varphi(r, \vartheta, \varphi) \cdot \vec{e}_\varphi \quad (7.3)$$

The tangential field components E_ϑ and E_φ of the electric farfield are proportional to *Green's function* of free space. Green's function describes a spherical wave that propagates in a radial direction (see Section 2.5.4).

$$E_\vartheta(r, \vartheta, \varphi) ; E_\varphi(r, \vartheta, \varphi) \sim \frac{e^{-jkr}}{r} \quad (7.4)$$

The *phase centre* of an antenna (see Figure 7.2) is defined as the origin of the spherical farfield waves. The phase centre can vary with observation angle and is therefore not necessarily a unique point in space [4].

---

[1] Likewise equations for the magnetic field strength can be given.

Spherical waves of an isotropic radiator   Directional antenna with focused beam

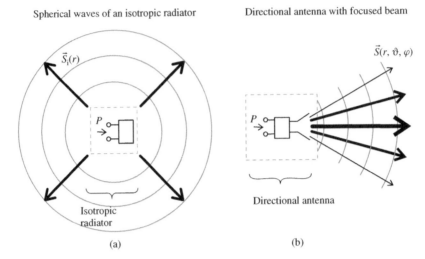

(a)   (b)

Figure 7.3 Power density distribution of (a) an isotropic radiator and (b) a directional antenna (e.g. parabolic dish antenna).

7.1.2 Isotropic Radiator

The *isotropic radiator* is a theoretical antenna that radiates power delivered to its port evenly in all angular directions (see Figure 7.3a). The power density S_i of an isotropic radiator can easily be calculated. If we look at a certain distance r, we find the radiated power P distributed uniformly on the surface $A = 4\pi r^2$ of a sphere with the radius r. Hence, the power density of an isotropic radiator is

$$\vec{S}_i = \frac{P}{4\pi r^2}\vec{e}_r = \frac{E_i^2}{Z_{F0}}\vec{e}_r = H_i^2 Z_{F0}\vec{e}_r \tag{7.5}$$

where E_i and H_i are the (effective) electric and magnetic field strength of the spherical wave.

The power density of an isotropic radiator is independent of ϑ and φ. Real antennas always show a certain directionality (see Figure 7.3b). Therefore, the power density distribution in the farfield is a function of ϑ and φ. We write $\vec{S} = S(r, \vartheta, \varphi)\vec{e}_r$.

The isotropic radiator is a theoretical concept that cannot be realized in practice. However, it represents a reference for the definition of parameters of real antennas.

7.1.3 Radiation Pattern and Related Parameters

In order to characterize the radiation behaviour of an antenna from a practical point of view antenna parameters in the farfield are defined. The directionality (angular dependency) of the farfield distribution is described by *radiation patterns* [5]. Radiation patterns may be defined either by the electric and magnetic field distributions (*field pattern* $C(\vartheta, \varphi)$) or by power distributions (power pattern $D(\vartheta, \varphi)$). We will discuss the field radiation pattern first.

The *real, scalar field pattern* C_i (*field pattern* for short) is the quotient of the electric (or magnetic) field of the antenna (in a certain angular direction ϑ and φ and at a certain distance r in the farfield of the antenna) and the electric (or magnetic) field of the isotropic radiator at the same distance r. Isotropic radiator and antenna radiate the same total power. In the farfield region C_i is independent of the distance r.

$$C_i(\vartheta, \varphi) = \left.\frac{E(r, \vartheta, \varphi)}{E_i(r)}\right|_{r \to \infty} = \left.\frac{H(r, \vartheta, \varphi)}{H_i(r)}\right|_{r \to \infty} \qquad \text{(Field pattern)} \qquad (7.6)$$

Alternatively, we can define a *normalized field pattern* C if we consider the quotient of the electric (or magnetic) field of the antenna (in a certain angular direction ϑ and φ and at a certain distance r in the farfield of the antenna) and the *maximum* electric (or magnetic) field of the antenna at the same distance r.

$$C(\vartheta, \varphi) = \left.\frac{E(r, \vartheta, \varphi)}{E_{\max}(r)}\right|_{r \to \infty} = \left.\frac{H(r, \vartheta, \varphi)}{H_{\max}(r)}\right|_{r \to \infty} \qquad \begin{array}{l}\text{(Normalized} \\ \text{field pattern)}\end{array} \qquad (7.7)$$

The field pattern C_i and the normalized field pattern C are based on the total field strength E (absolute value). It is also common to define two different field patterns for the transversal field components E_ϑ and E_φ.

In order to better resolve small values the radiation pattern is often given in logarithmic scale

$$C^\ell(\vartheta, \varphi)/\text{dB} = 20\lg(C(\vartheta, \varphi)) \quad \text{and} \quad C_i^\ell(\vartheta, \varphi)/\text{dBi} = 20\lg\left(C_i(\vartheta, \varphi)\right) \qquad (7.8)$$

where the superscript letter ℓ denotes the logarithmic scale. The letter ℓ will be dropped subsequently since there is no risk of confusing linear and logarithmic scale due to the pseudo-unit 'dB'. Logarithmic values of the field pattern C_i are given in the pseudo-unit 'dBi', where the letter 'i' indicates that the isotropic radiator is used as reference.

Figure 7.4a shows two views of a three-dimensional radiation pattern. Radiation patterns commonly show maxima and minima in different directions. A region between two minima is called a *lobe*. The lobe with the maximum value is referred to as the *main lobe*. Maximum power is transmitted in the direction of the main lobe. All other lobes are *minor lobes*. The *side-lobe level* is the ratio of maximum minor lobe amplitude to main lobe amplitude. Directional antennas show radiation patterns with a pronounced preferred direction. For a directional radio link it is desirable to have a low side-lobe level.

A three-dimensional radiation pattern provides a good illustration of the general radiation characteristics. For most antennas two-dimensional cut-planes through the three-dimensional diagram provide sufficient information for practical application. Often the cut-planes include the direction of the main lobe. Typical cut-plane orientations are *vertical* and *horizontal* (see Figure 7.4b and c). Furthermore planes where the electric field is tangential (*E*-plane) or where the magnetic field is tangential (*H*-plane) are chosen.

The *half-power beamwidth* (HPBW) represents an angle Ψ around the direction of the main lobe where the field amplitude is greater than or equal to $C_{\max}/\sqrt{2}$ (see Figure 7.4d). (In logarithmic scale a reduction of the field by a factor of $1/\sqrt{2}$ equals $-3\,\text{dB}$.) The half-power beamwidth depends on the orientation of the cut-plane through the direction of

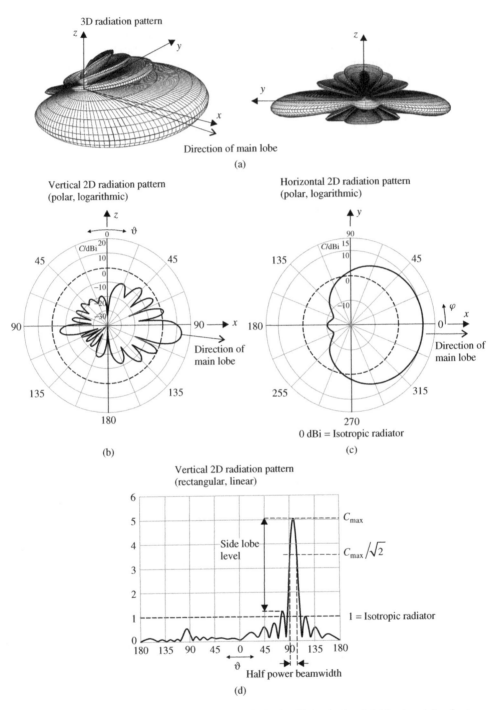

Figure 7.4 (a) Three-dimensional field pattern $C(\vartheta, \varphi)$; (b) vertical and (c) horizontal polar two-dimensional field pattern; and (d) rectangular two-dimensional field pattern.

the main lobe. As a consequence an antenna may show different half-power beamwidths in a vertical and a horizontal cut-plane.

If we consider a wireless connection with directional antennas the main lobes have to point to each other. Furthermore, in order to transmit maximum power the antennas have to be aligned with respect to their *polarization*. Therefore, the *polarization of the antenna in the direction of the main lobe* is an important parameter. The polarization of an antenna is determined by the two independent orthogonal components of the electric field E_ϑ and E_φ in the farfield. Like plane waves (see Section 2.5.2) antennas can have different polarizations (linear, circular, elliptic).

Like the field pattern C the power pattern D also describes the angular dependency of the radiation characteristic. The power pattern is the power density of the antenna normalized to the power density of the isotropic radiator.

$$D\left(\vartheta, \varphi\right) = \left.\frac{S\left(r, \vartheta, \varphi\right)}{S_i(r)}\right|_{r\to\infty} = C_i^2\left(\vartheta, \varphi\right) = DC^2\left(\vartheta, \varphi\right) \qquad \text{(Power pattern)} \qquad (7.9)$$

where the *directivity* D is the maximum of the power pattern

$$D = \max\left\{D\left(\vartheta, \varphi\right)\right\} \qquad \text{(Directivity)} \qquad (7.10)$$

The directivity of an antenna is the factor by which the power density of the antenna exceeds the power density of an isotropic radiator. Typical values of the directivity for commonly used types of antennas can be found in Table 7.1 in linear and logarithmic scale. In logarithmic scale the directivity D is given as

$$D^\ell\left(\vartheta, \varphi\right)/\text{dBi} = 10\lg\left(D\left(\vartheta, \varphi\right)\right) = C_i^\ell\left(\vartheta, \varphi\right)/\text{dBi} = 20\lg\left(C_i\left(\vartheta, \varphi\right)\right) \qquad (7.11)$$

The pseudo-unit 'dBi' distinguishes between linear and logarithmic values. Therefore, we henceforth omit the superscript letter ℓ.

Table 7.1 indicates that the directivity of a single antenna element (e.g. dipole or patch) may be small. However, combining several antennas to form a group antenna can significantly increase the directivity of the antenna arrangement (see Section 7.6).

Table 7.1 Typical values of antenna directivity

Antenna	Directivity (logarithmic scale)	Directivity (linear scale)
Isotropic radiator	0 dBi	1
Hertzian dipole	1.76 dBi	1.5
$\lambda/2$-dipole	2.15 dBi	1.64
$\lambda/4$-monopole	5.15 dBi	3.28
Patch antenna	6 dBi	4
Logarithmic periodic dipole antenna	7 dBi	5
Horn antenna	20 dBi	100
Parabolic dish antenna	> 30 dBi	> 1000

The field and power patterns illustrate the radiation behaviour in the *transmit* case. Due to the principle of reciprocity the radiation patterns also illustrate the angular dependence of the susceptibility of an antenna in the receive case. Angular directions with high power density correspond to high susceptibility in that same direction.

An antenna parameter that illustrates the receive case is the *effective capture area* A_{eff}. In order to understand this parameter we consider an incident plane wave that impinges on the antenna under the following conditions:

- The polarization of the plane wave matches the polarization of the antenna (e.g. both are vertically polarized).
- The antenna is matched (complex-conjugate match) at its terminal port, so that no power is reflected back due to mismatch and maximum power is transferred to the load impedance.
- The antenna itself is loss-less.

Under these conditions the effective capture area is the area that links the power density S of the plane wave and the power P delivered to the load impedance.

$$P = A_{eff} S \qquad (7.12)$$

The effective capture area can be determined from the directivity D and wavelength λ as

$$\boxed{A_{eff} = \frac{\lambda^2}{4\pi} D} \qquad \text{(Effective capture area)} \qquad (7.13)$$

Directivity D and effective capture area A_{eff} do not take into account losses of the antenna. However, real antennas are always lossy due to the following effects:

- Absorption in non-ideal metallic parts having finite conductivity.
- Absorption in lossy dielectric parts of the antenna.
- Reflection losses caused by non-perfect matching conditions between antenna port impedance and load impedance.

In a real situation not all power delivered to the antenna is radiated into free space. Figure 7.5 illustrates the loss mechanisms by looking at different powers. On the transmission line we find the forward travelling wave P_f from a transmitter. Due to mismatch ($|s_{11}| > 0$) a reflected wave occurs $P_r = |s_{11}|^2 P_f$. The accepted power $P_{acc} = P_f - P_r$ is partly absorbed and transformed into heat (P_{abs}). Commonly, the greater part of the accepted power is radiated into space (P_{rad}).

Based on the previous power considerations we can define an *antenna efficiency* η that summarizes metallic and dielectric losses but does not take into account reflection losses.

$$\boxed{\eta = \frac{P_{rad}}{P_{acc}} = \frac{P_{rad}}{P_f - P_r} = \frac{P_{rad}}{P_{rad} + P_{abs}} \leq 1} \qquad \text{(Antenna efficiency)} \qquad (7.14)$$

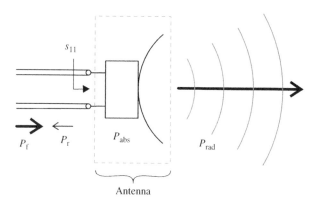

Figure 7.5 Incident, reflected, absorbed and radiated power.

If we include the reflection losses we get the *total antenna efficiency* η_{total}.

$$\eta_{\text{total}} = \frac{P_{\text{rad}}}{P_{\text{f}}} = \left(1 - |s_{11}|^2\right)\eta \le \eta \le 1 \qquad \text{(Total antenna efficiency)} \qquad (7.15)$$

Unlike the directivity D the *gain G* of an antenna includes losses and is defined as the product of directivity D and antenna efficiency η.

$$G = \eta D \qquad \text{(Gain)} \qquad (7.16)$$

The last antenna parameter to be defined is the *equivalent isotropically radiated power EIRP*. It is the product of gain G and transmit power P fed into the antenna.

$$EIRP = GP \quad \text{and} \quad EIRP/\text{dBm} = 10\lg\left(\frac{GP}{1\,\text{mW}}\right) \qquad (7.17)$$

The equivalent isotropically radiated power equals the power that must be delivered to an isotropic radiator to achieve the same power density in the direction of the main lobe as the antenna with the gain G.

Wireless communication standards commonly set limitations for maximum *EIRP* [5]. If an antenna with a gain G is used we can derive the maximum admissible transmit power P from Equation 7.17.

7.1.4 Impedance Matching and Bandwidth

At the antenna terminal we see an input impedance Z_A that in general is complex-valued and frequency-dependent. Figure 7.6a shows typical plots for real and imaginary parts of the input impedance Z_A.

$$Z_A = R_A + jX_A \qquad (7.18)$$

The resistance R_A consists of two parts. The resistance R_{rad} represents the real power emitted from the antenna and the resistance R_{abs} represents the ohmic and dielectric losses in the antenna structure.

$$R_A = R_{rad} + R_{abs} \tag{7.19}$$

If we connect the antenna to a transmission line with a characteristic impedance Z_0 we get a reflection coefficient r_A as defined in Section 3.1.10.

$$r_A = s_{11} = \frac{Z_A - Z_0}{Z_A + Z_0} \tag{7.20}$$

Figure 7.6b shows the reflection coefficient r_A for the antenna impedance Z_A given in Figure 7.6a and a port reference impedance of $Z_0 = 50\,\Omega$. In the frequency range of operation the reflection coefficient shall be small. Depending on the application reflection coefficients below $-6\,dB$, $-10\,dB$ or $-20\,dB$ are tolerable (corresponding to reflected powers of 25%, 10% or 1%, respectively). The frequency range where the impedance requirements are met is called *(impedance) bandwidth B*. Figure 7.6b shows the $-6\,dB$ and $-10\,dB$ bandwidth.

Efficient matching is often possible by taking constructional measures at the antenna feed. For example patch antennas (see Section 7.5.1) can easily be matched to different characteristic line impedances by selecting an appropriate feed point on the patch element.

Alternatively, antennas may be matched to feeding transmission lines by applying standard matching circuits as discussed in Section 6.3. However, additional matching networks may increase the complexity of the circuit.

Thus far we considered the term *bandwidth* as related to impedance requirements only *(impedance bandwidth)*. The more general concept of *(antenna) bandwidth* considers sufficient performance of all relevant antenna parameters in the frequency band of interest. A *broadband antenna* (antenna with great relative bandwidth (see Equation 5.30)) has constant antenna parameters (impedance matching, direction of main lobe, polarization, gain, etc.) in a broad frequency range.

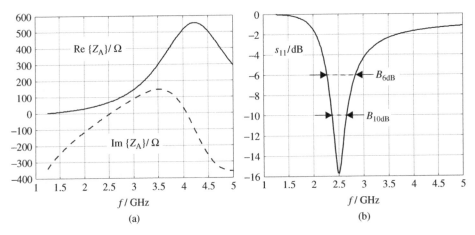

Figure 7.6 (a) Input impedance and (b) reflection coefficient of an antenna (port reference impedance is $50\,\Omega$).

7.2 Standard Types of Antennas

Antennas come in a variety of shapes and sizes since technical requirements and frequencies of operation can be very different in various applications. Electrical requirements concern antenna parameters like large bandwidth, high gain, or a certain radiation pattern with a predefined direction of the main lobe and predefined directions of nulls. Furthermore, additional requirements come into play like reduced size, low weight, low cost or predetermined manufacturing process. Figure 7.7 shows an incomplete overview of some commonly used antenna types that represent – in one way or another – good solutions in practical antenna applications.

A *dipole* antenna (see Figure 7.7a) is a basic antenna type that simply consists of two metallic rods. The antenna is fed at the centrepoint between the rods and shows an omnidirectional radiation pattern in lateral direction. (No radiation occurs in the direction of the axis of the rods.) If the overall length of the dipole is about half a wavelength the antenna is in resonance and exhibits good radiation characteristics. Furthermore, the input impedance of a *half-wave dipole* is close to technical transmission line impedances of 50 Ω and 75 Ω. We will look at dipole antennas in more detail in Section 7.4.

If we add a parallel rod in close vicinity to the dipole and connect the tops of the rod and dipole we get a *folded dipole* antenna (see Figure 7.7b). The folded dipole has an increased input impedance. This is beneficial if transmission lines with higher impedances are used or the use of a balun requires a higher input impedance due to inherent balun impedance transformation.

In order to increase the bandwidth of an antenna we use either conical arms (the cross-section of the arms increases as we move away from the feedpoint) or several thin rods forming a conical structure (see Figure 7.7c). This *biconical antenna* is commonly used in electromagnetic compatibility (EMC) measurements at lower frequencies (up to around 200 MHz).

Dipole elements can be combined to build antenna structures that have higher bandwidth or higher gain than single dipole antennas. Figure 7.7d shows a *logarithmic periodic dipole antenna* (LPDA) where dipoles of increasing length are mounted on a two-conductor line. The position and length of the individual dipole elements follow a logarithmic rule, hence the name *logarithmic periodic* dipole antenna. The antenna is fed at the foremost point on the transmission line. The wave excited at this point travels down the transmission line until the length of the dipoles is in the order of half a wavelength. The resonant elements radiate the wave into space. Hence, the radiating zone of the antenna depends on the frequency of the excitation signal. Higher frequencies are radiated near the feedpoint. Lower frequencies are radiated near the end of the transmission line, where the longer dipoles are to be found. The LPDA antenna exhibits a large bandwidth (depending on the size of the dipoles) and is used – for example – in electromagnetic compatibility (EMC) measurements at medium frequencies up to around 1 GHz.

Unlike LPDA antennas *Yagi-Uda antennas* have low bandwidth but high gain. Figure 7.7e shows a Yagi-Uda antenna with an exited folded dipole element complemented by shorter dipoles (director elements) in the direction of radiation and longer dipoles (reflector elements) in the opposite direction. The individual dipole elements interact via radiation coupling. Unlike the LPDA there is no common transmission line. Yagi-Uda antennas are used, for example, as directional outdoor television antennas for terrestrial reception and can be seen on many roof-tops.

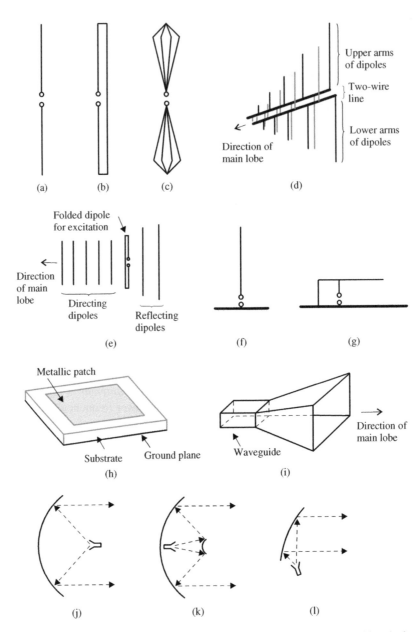

Figure 7.7 Common types of antennas: (a) dipole, (b) folded dipole, (c) biconical antenna, (d) logarithmic periodic dipole antenna (LPDA), (e) Yagi-Uda antenna, (f) monopole, (g) Inverted-F antenna, (h) patch antenna, (i) horn antenna, (j) parabolic dish antenna, (k) parabolic antenna with subreflector (Cassegrain) and (l) parabolic antenna with offset-feed.

The geometric structure of a dipole (Figure 7.7a) and its electromagnetic field show symmetry to a horizontal median cut-plane through the feedpoint. The field distribution in the upper half-space remains unchanged if we replace the plane of symmetry by an infinite electrical ground plane. We get a new antenna that is known as *monopole* (Figure 7.7f). If the original length of the dipole was half a wavelength the length of the monopole is now a quarter of a wavelength. Monopoles are often found on metallic housings or vehicles since the metallic structures represent the required ground plane.

The great height of the monopole antenna is disadvantageous if the antenna is to be integrated into a device (e.g. a notebook computer). The *inverted-F antenna* (IFA) shows a reduced height (Figure 7.7g). The inverted-F antenna is derived from a monopole that is bent. The original feedpoint of the monopole is replaced by a short to ground. Instead we feed the antenna a bit further above the rod where we find the required input impedance for matching. The inverted-F antenna can be integrated into circuits and housings and is – for instance – often used in *Bluetooth*[2] applications.

A *patch antenna* has a low profile too and is suitable for integration. The most simple patch antenna consists of a rectangular metallic strip on a dielectric substrate (see Figure 7.7h). The reverse side of the substrate is a metallic ground plane. Patch antennas may be miniaturized using thin, high permittivity substrates in a low cost manufacturing process. Compared to wire antennas patch antennas exhibit relatively low bandwidth. Patch antennas can efficiently be combined to build up group antennas (see Section 7.6), for example, for Radar and RFID applications [6–8].

Above 1 GHz or so *horn antennas*, as shown in Figure 7.7i, become an interesting option. Horn antennas are fed by a waveguide and are basically a tapered structure with a cross-section that gets larger from the feed to the aperture. The gain of a horn antenna increases with increasing length. Horn antennas are used to feed parabolic dish antennas. Furthermore, they are used in EMC measurements at higher frequencies (typically above 1 GHz).

Parabolic dish reflector antennas have a parabolic reflector area that is much larger than the operational wavelength. Therefore, in a first approach we can assume ray-like electromagnetic wave propagation (geometrical optics). In Figure 7.7j we see a horn antenna in the focus of the parabolic dish that illuminates the reflector area. Due to the laws of geometrical optics, rays that come from the focus are converted to rays parallel to the axis of the dish. Parabolic antennas exhibit high gain and are – for example – used in satellite communication links.

We can position the feed antenna near the centre of the reflector antenna by using a subreflector at the former feed antenna position. Figure 7.7k shows a convex hyperbolical subreflector (Cassegrain configuration) and the corresponding paths of rays. Alternatively, a concave elliptical subreflector can be used (Gregory configuration).

The subreflector in Figure 7.7k and the feed antenna in Figure 7.7j are located on the axis of the reflector. Therefore, they block some of the incoming or outgoing rays (shadowing effect). Figure 7.7l shows an antenna with an *offset-feed* that does not show any shadowing effect. The reflector consists of a shell-shaped parabolic segment. Offset-feed antennas are used in directional radio links.

---

[2] Bluetooth is a wireless standard that uses the 2.4 GHz ISM band for short range data exchange between e.g. mobile phones, notebooks, printers, digital cameras, etc.

7.3 Mathematical Treatment of the Hertzian Dipole

The *Hertzian dipole* is an infinitely small radiating current element. The electromagnetic field of this theoretical antenna can be calculated mathematically in closed form [2, 9, 10]. The current distribution on practical antennas – especially wire antennas – can be represented by Hertzian dipole elements. Furthermore, essential antenna parameters can be illustrated and understood from the example of the Hertzian dipole.

Figure 7.8a shows the geometry of the Hertzian dipole. A time-dependent homogeneous current I flows along the z-axis. The length ℓ of the element approaches zero, but the product of current and length $I\ell$ shall be finite. Time-dependent charges Q and $-Q$ exist at the ends of the current element. We consider harmonic time-dependency and apply complex phasor notation.

In order to calculate the electric field strength $\vec{E}$ and the magnetic field strength $\vec{H}$ we use the magnetic vector potential $\vec{A}$ as an auxiliary quantity. So, the mathematical approach is as follows: first, the magnetic vector potential $\vec{A}$ is determined from the source current density $\vec{J}$ [2, 10] by the following equation

$$\vec{A}\,(\vec{r}) = \frac{\mu_0}{4\pi} \iiint\limits_V \frac{\vec{J}\,e^{-jk|\vec{r}-\vec{r}'|}}{|\vec{r}-\vec{r}'|}\,dv' \qquad \text{(Magnetic vector potential)} \qquad (7.21)$$

Primed quantities $(\vec{r}', dv')$ indicate source points whereas the vector $\vec{r}$ indicates the observation point. Second, we use the *Lorenz*[3] gauge

$$\nabla \cdot \vec{A} = -j\omega\mu_0\varepsilon_0\varphi \qquad (7.22)$$

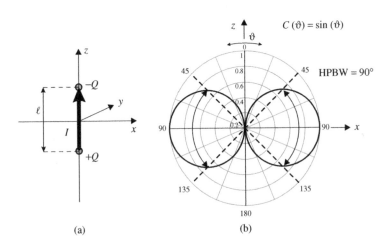

(a) (b)

Figure 7.8 Hertzian dipole: (a) linear element with uniform current and (b) vertical radiation pattern.

---

[3] Ludvig Lorenz, sometimes confused with H.A. Lorentz [11].

and the magnetic vector potential $\vec{A}$ to calculate the magnetic field strength as

$$\boxed{\vec{H} = \frac{1}{\mu_0} \nabla \times \vec{A}}$$

(7.23)

and the electric field strength as

$$\boxed{\vec{E} = -\nabla\varphi - j\omega\vec{A} = \frac{\nabla(\nabla \cdot \vec{A})}{j\omega\mu_0\varepsilon_0} - j\omega\vec{A}}$$

(7.24)

Let us get started with the calculation of the magnetic vector potential $\vec{A}$ according to Equation 7.21. The integral extends over the source volume V. For the calculation we apply Cartesian coordinates. With the point of observation $\vec{r}$ and the source point $\vec{r}'$ we write

$$\vec{r} = \begin{pmatrix} x \\ y \\ z \end{pmatrix}; \quad \vec{r}' = \begin{pmatrix} 0 \\ 0 \\ z' \end{pmatrix} \quad \rightarrow \quad |\vec{r} - \vec{r}'| = \sqrt{x^2 + y^2 + (z - z')^2}$$

(7.25)

and get

$$\vec{A}(\vec{r}) = \frac{\mu_0}{4\pi} \int\limits_{-\ell/2}^{\ell/2} \iint\limits_{A'_{xy}} \frac{\vec{J}e^{-jk\sqrt{x^2+y^2+(z-z')^2}}}{\sqrt{x^2 + y^2 + (z - z')^2}} dx'dy'dz'$$

(7.26)

where A'_{xy} is the cross-sectional area of the current element.

The Hertzian dipole has an infinitesimal length ($\ell \to 0$). Hence, the difference of observation point $\vec{r}$ and source point $\vec{r}'$ approach $\vec{r}$.

$$|\vec{r} - \vec{r}'| = \sqrt{x^2 + y^2 + (z - z')^2} \quad \rightarrow \quad |\vec{r}| = \sqrt{x^2 + y^2 + z^2} = r$$

(7.27)

Integration over the cross-sectional area of the current element yields the current of the Hertzian dipole.

$$\iint\limits_{A'_{xy}} \vec{J}dx'dy' = I\vec{e}_z$$

(7.28)

By using Equation 7.21 we calculate the magnetic vector potential $\vec{A}$ of the Hertzian dipole as

$$\boxed{\vec{A}(\vec{r}) = \frac{\mu_0 I\ell}{4\pi} \cdot \frac{e^{-jk\sqrt{x^2+y^2+z^2}}}{\sqrt{x^2 + y^2 + z^2}}\vec{e}_z = \frac{\mu_0 I\ell}{4\pi} \cdot \frac{e^{-jkr}}{r}\vec{e}_z}$$

(7.29)

Like the current density $\vec{J}$ the magnetic vector potential $\vec{A}$ only has a z-component.

In order to calculate the electric field strength $\vec{E}$ and the magnetic field strength $\vec{H}$ it is advantageous to switch from Cartesian to spherical coordinates. According to Appendix A.1.1 we can substitute the unit vector in z-direction $\vec{e}_z$ by

$$\vec{e}_z = \vec{e}_r \cos \vartheta - \vec{e}_\vartheta \sin \vartheta \tag{7.30}$$

Consequently, the magnetic vector potential has only a radial r- and a ϑ-component. We can now use Equation 7.23 and 7.24 to calculate field values. Appendix A.1.3 gives us formulas to evaluate the differential operators (curl, divergence and gradient) in these equations.

$$\nabla \times \vec{A} = \frac{1}{r \sin \vartheta} \left(\frac{\partial \left(A_\varphi \sin \vartheta \right)}{\partial \vartheta} - \frac{\partial A_\vartheta}{\partial \varphi} \right) \vec{e}_r +$$
$$\frac{1}{r} \left(\frac{1}{\sin \vartheta} \frac{\partial A_r}{\partial \varphi} - \frac{\partial \left(r A_\varphi \right)}{\partial r} \right) \vec{e}_\vartheta + \frac{1}{r} \left(\frac{\partial \left(r A_\vartheta \right)}{\partial r} - \frac{\partial A_r}{\partial \vartheta} \right) \vec{e}_\varphi \tag{7.31}$$

$$\nabla \cdot \vec{A} = \frac{1}{r^2} \frac{\partial \left(r^2 A_r \right)}{\partial r} + \frac{1}{r \sin \vartheta} \frac{\partial \left(A_\vartheta \sin \vartheta \right)}{\partial \vartheta} + \frac{1}{r \sin \vartheta} \frac{\partial A_\varphi}{\partial \varphi} \tag{7.32}$$

$$\nabla \phi = \frac{\partial \phi}{\partial r} \vec{e}_r + \frac{1}{r} \frac{\partial \phi}{\partial \vartheta} \vec{e}_\vartheta + \frac{1}{r \sin \vartheta} \frac{\partial \phi}{\partial \varphi} \vec{e}_\varphi \tag{7.33}$$

Some of the terms become zero since there is no φ-component of the magnetic vector potential. Furthermore, the magnetic vector potential has no φ-dependency. Hence, all derivatives with respect to φ are zero.

After a short computation (see Problem 7.4) we get the components of the electric and magnetic field strength as

$$H_\varphi = \frac{I \ell}{4\pi} \cdot \frac{e^{-jkr}}{r^2} (1 + jkr) \sin \vartheta$$

$$H_r = H_\vartheta = 0$$

$$E_r = \frac{I \ell}{j 2\pi \omega \varepsilon} \cdot \frac{e^{-jkr}}{r^3} (1 + jkr) \cos \vartheta \qquad \text{(Field components of} \tag{7.34}$$
$$\qquad\qquad\qquad\qquad\qquad\qquad\qquad\qquad\qquad \text{Hertzian dipole)}$$

$$E_\vartheta = \frac{I \ell}{j 4\pi \omega \varepsilon} \cdot \frac{e^{-jkr}}{r^3} \left(1 + jkr - (kr)^2 \right) \sin \vartheta$$

$$E_\varphi = 0$$

Practical antennas are commonly characterized by their field pattern in the farfield. The parenthesized expressions in Equation 7.34 contain the term $(kr)^n$ with different exponents $n \in \{0, 1, 2\}$. For large distances from the antenna $(kr \gg 1)$, the term with the highest exponent n dominates. So, the resulting magnetic field is proportional to $1/r$. The radial component of the electric field is proportional to $1/r^2$ and can be neglected compared to the ϑ-component which is proportional to $1/r$. Hence, the electric and magnetic field strength in the farfield $(kr \gg 1)$ of a Hertzian dipole are given as

$$
\begin{aligned}
H_\varphi &= j\frac{kI\ell}{4\pi} \cdot \frac{e^{-jkr}}{r} \sin \vartheta \\
H_r &= H_\vartheta = 0 \\
E_r &= 0 \\
E_\vartheta &= j\frac{I\ell k^2}{4\pi \omega \varepsilon} \cdot \frac{e^{-jkr}}{r} \sin \vartheta \\
E_\varphi &= 0
\end{aligned}
$$

(Farfield of Hertzian dipole)   (7.35)

The farfield components of the electric and magnetic field strength are linked by the characteristic impedance Z_{F0} of free space

$$
\frac{E_\vartheta}{H_\varphi} = \frac{k}{\omega \varepsilon_0} = \frac{\omega \sqrt{\mu_0 \varepsilon_0}}{\omega \varepsilon_0} = \sqrt{\frac{\mu_0}{\varepsilon_0}} = Z_{F0} = 120\pi \ \Omega \approx 377 \ \Omega
\tag{7.36}
$$

From Equation 7.35 we calculate the normalized field pattern C as

$$
\boxed{C(\vartheta) = \sin \vartheta} \qquad \text{(Field pattern)}
\tag{7.37}
$$

Figure 7.8b shows a vertical cut-plane through the field pattern. There is no radiation in the axis of the Hertzian dipole and maximum radiation in lateral direction (xy-plane). The vertical half-power beamwidth (HPBW) is $90°$.

The corresponding Poynting vector $\vec{S}$ in the farfield directs radially outwards.

$$
\vec{S} = \frac{1}{2}\vec{E} \times \vec{H}^* = \frac{1}{2}E_\vartheta H_\varphi^* \vec{e}_r
\tag{7.38}
$$

By using the farfield formulas in Equation 7.35 the radial component of the Poynting vector yields

$$
S_r(r,\vartheta) = \frac{(I\ell)^2 k^3}{2(4\pi)^2 \omega \varepsilon_0 r^2} \sin^2 \vartheta = \frac{(I\ell)^2 k^2 Z_{F0}}{2(4\pi)^2 r^2} \sin^2 \vartheta = \frac{I^2}{8r^2}\left(\frac{\ell}{\lambda}\right)^2 Z_{F0} \sin^2 \vartheta
\tag{7.39}
$$

The real part of the Poynting vector represents the real power density radiated into space. In order to determine the radiated power we integrate over the surface of a sphere around the Hertzian dipole.

$$
\begin{aligned}
P_{\text{rad}} = \oiint_{A(V)} \vec{S} \cdot d\vec{A} &= \int_0^{2\pi} \int_0^{\pi} \frac{I^2}{8r^2}\left(\frac{\ell}{\lambda}\right)^2 Z_{F0} \sin^2 \vartheta \underbrace{r^2 \sin \vartheta \, d\vartheta \, d\varphi}_{dA} \\
&= \frac{2\pi I^2}{8}\left(\frac{\ell}{\lambda}\right)^2 Z_{F0} \int_0^{\pi} \sin^2 \vartheta \, d\vartheta
\end{aligned}
$$

$$= \frac{\pi I^2}{4} \left(\frac{\ell}{\lambda}\right)^2 Z_{F0} \underbrace{\left[-\cos\vartheta + \frac{1}{3}\cos^3\vartheta\right]_0^\pi}_{4/3} = \frac{\pi I^2}{3} \left(\frac{\ell}{\lambda}\right)^2 Z_{F0} \qquad (7.40)$$

The radiated power can be represented by a *radiation resistance* R_{rad} at the antenna terminal. Next, we equate the power consumed in the resistance and the power P radiated by the Hertzian dipole.

$$P_{rad} = \frac{1}{2} R_{rad} I^2 = \frac{\pi I^2}{3} \left(\frac{\ell}{\lambda}\right)^2 Z_{F0} \qquad (7.41)$$

Hence, the radiation resistance of the Hertzian dipole becomes

$$\boxed{R_{rad} = \frac{2\pi}{3} \left(\frac{\ell}{\lambda}\right)^2 Z_{F0} \approx 790 \left(\frac{\ell}{\lambda}\right)^2 \Omega} \qquad \text{(Radiation resistance)} \qquad (7.42)$$

By using Equation 7.39 and 7.41 we determine the power pattern as

$$\boxed{D(\vartheta) = \frac{S_r(r,\vartheta)}{S_i(r)} = 4\pi r^2 \frac{S_r(r,\vartheta)}{P_{rad}} = 1.5 \sin^2 \vartheta} \qquad \text{(Power pattern)} \qquad (7.43)$$

Finally, the directivity of the Hertzian dipole is

$$\boxed{D = \max\{D(\vartheta)\} = \max\{1.5 \sin^2 \vartheta\} = 1.5 \,\hat{=}\, 1.76\,\text{dBi}} \qquad \text{(Directivity)} \qquad (7.44)$$

Unlike a Hertzian dipole a real dipole antenna has a finite length. Therefore, in the next section we illustrate what happens to the antenna parameters if we extend the length of a dipole and assume non-uniform current distribution along the dipole.

7.4 Wire Antennas

7.4.1 Half-Wave Dipole

The Hertzian dipole has an infinitesimal length with a homogeneous current distribution and cannot therefore be realized. We will look now at thin, centre-fed dipoles (see Figure 7.9a) with practical length ℓ and non-homogeneous current distribution. A dipole consists of two metallic rods and is considered to be *thin* if the diameter d of the rod wire is small compared to the dipole length ($d \ll \ell$).

To start with, we consider – as an example – a practical dipole length of ($\ell = 53\,\text{mm}$). With an EM simulator we calculate the input impedance at the centre feedpoint. Figure 7.9b and c show the real and imaginary part of the antenna input impedance Z_A as well as the Nyquist plot (locus) in the complex impedance plane. The arrow designated by the letter f indicates the course of increasing frequencies.

Electrically short dipole: If the length is small compared to wavelength ($\ell \ll \lambda$) the real part of the input impedance is small and the imaginary part is negative: the dipole shows a capacitive behaviour. The current distribution is zero at the endpoint of each arm of the dipole and increases linearly towards the centre feedpoint (see Figure 7.9d).

Half-wavelength dipole: If the length is about half a wavelength ($\ell \approx \lambda/2$) we see a first resonance (Im $\{Z_A\} = 0$). The input impedance is real and amounts to $Z_A \approx 73.2\,\Omega$ [12]. The current distribution follows a cosine curve with a maximum at the centre feedpoint. With increasing frequency ($\lambda/2 < \ell < \lambda$) the imaginary part of the input impedance becomes positive (inductive behaviour). The real part increases.

Wavelength dipole: We see a second resonance (Im $\{Z_A\} = 0$) at a length of about one wavelength ($\ell \approx \lambda$). The input impedance is much larger than in the case of the first resonance ($\ell \approx \lambda/2$). The current distribution follows a cosine curve with minima at the endpoints and at the centre point of the dipole.

One and a half wavelength dipole: At a length of about one and a half wavelengths ($\ell \approx 1.5\lambda$) a third resonance occurs with a low real input impedance. The current distribution now shows three maxima along the dipole. At the centre feedpoint the current has a maximum.

As frequency increases even further we periodically see resonances at $n\lambda/2$, where n is an integer number.

By far the most important practical dipole length is half a wavelength $\ell \approx \lambda/2$ (Figure 7.10a). Like the $\lambda/2$-transmission line resonator (Section 6.2.1) the $\lambda/2$-dipole

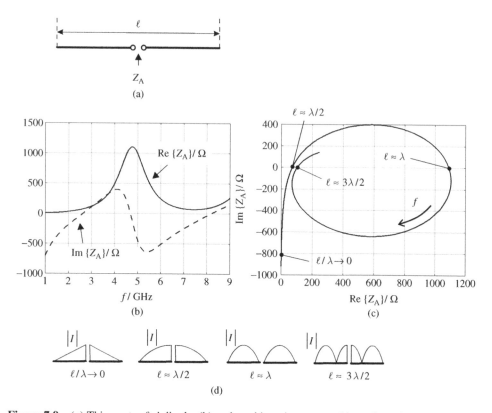

Figure 7.9 (a) Thin centre-fed dipole, (b) real- and imaginary part of input impedance, (c) Nyquist plot and (d) current distribution on dipole antennas of different lengths compared to wavelength.

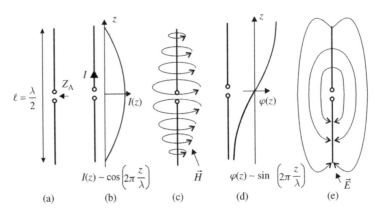

Figure 7.10 Half-wave dipole: (a) geometry, (b) current distribution on dipole, (c) magnetic field distribution, (d) electric potential along the dipole and (e) electric field distribution.

is resonant. The current distribution shows standing wave behaviour (Figure 7.10b): a cosine curve $I(z)$ with zeros at the end of the dipole and a maximum at the feedpoint. The current is encircled by a magnetic field (Figure 7.10c). The potential along the dipole follows a sinus curve with maximum potential at the ends of the dipole (Figure 7.10c). Hence, the electric field points from one dipole rod to the other rod as shown in Figure 7.10e.

Under the assumption of a cosine shaped current distribution we may calculate the electric and magnetic fields around the antenna as shown in literature (e.g. [2, 10, 13]). From the fields we derive the normalized field pattern as

$$C(\vartheta) = \frac{\cos\left(\frac{\pi}{2}\cos\vartheta\right)}{\sin\vartheta} \qquad \text{(Field pattern of } \lambda/2\text{-dipole)} \qquad (7.45)$$

and the directivity

$$D = 1.64 \cong 2.15\,\text{dBi} \qquad \text{(Directivity of } \lambda/2\text{-dipole)} \qquad (7.46)$$

Figure 7.11a shows the normalized vertical two-dimensional field pattern (E-plane) of a half-wave dipole in comparison to a Hertzian dipole. The half-wave dipole shows slightly increased directionality as can be seen by looking at the directivity (half-wave dipole $D_{\text{ha}} = 1.64$, Hertzian dipole $D_{\text{he}} = 1.5$). The three-dimensional representation of the field pattern of the half-wave dipole (Figure 7.11b) shows the omnidirectional radiation characteristic in the lateral direction.

7.4.2 Monopole

If we look at the electric field $\vec{E}$ of a half-wave dipole in Figure 7.12a we see a horizontal (median) plane of symmetry. The electric field is perpendicular to this plane (electric wall). Now we replace the plane of symmetry by an infinite ground plane. So we get a monopole

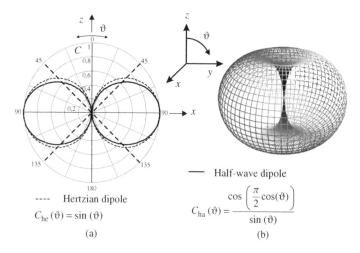

$C_{\mathrm{he}}(\vartheta) = \sin(\vartheta)$

---- Hertzian dipole

—— Half-wave dipole

$$C_{\mathrm{ha}}(\vartheta) = \frac{\cos\left(\dfrac{\pi}{2}\cos(\vartheta)\right)}{\sin(\vartheta)}$$

(a)   (b)

Figure 7.11 (a) Vertical radiation pattern of Hertzian dipole and half-wave dipole, (b) three-dimensional radiation pattern of half-wave dipole.

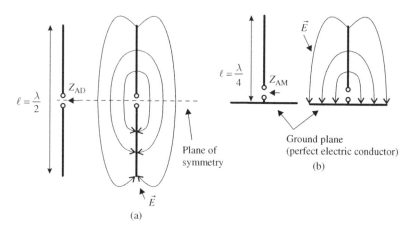

Figure 7.12 (a) Electric field distribution of half-wave dipole with plane of symmetry and (b) monopole antenna over ground plane.

above and a monopole beneath that ground plane. Since both half-spaces are independent of each other we can omit the lower monopole and consider the monopole above as a new antenna geometry. The length of the monopole is a quarter of a wavelength ($\ell = \lambda/4$). In order to function properly a sufficiently large ground plane is required in practice.

If we look at the input terminal the current I stays the same, but the voltage U is halved since the ground plane now represents the opposite terminal. Therefore, the antenna input impedance of a monopole $Z_{\mathrm{A,m}}$ is only half of the original dipole input impedance $Z_{\mathrm{A,d}}$.

$$Z_{\mathrm{A,m}} = \frac{1}{2}Z_{\mathrm{A,d}} \tag{7.47}$$

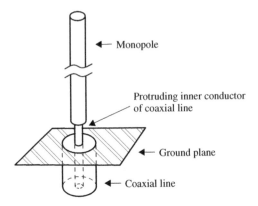

Monopole

Protruding inner conductor
of coaxial line

Ground plane

Coaxial line

Figure 7.13 Coaxial line feeding a monopole antenna through a ground plane.

By feeding the monopole with the same power as the dipole its power density is twice as high, since the monopole only radiates in the upper half-space. Consequently, the directivity D_m of the monopole is twice as large (+3 dB) as the directivity of the dipole D_d.

$$D_m = 2D_d = 2 \cdot 1.64 = 3.28 \cong 2.15\,\text{dBi} + 3\,\text{dB} = 5.15\,\text{dBi} \qquad (7.48)$$

A monopole antenna can easily be excited by a coaxial line as shown in Figure 7.13. The inner conductor of the coaxial line extends through the ground plane and contacts the monopole rod. The outer conductor is connected to the ground plane. Through the dielectric cross-sectional area between inner and outer conductor the EM wave leaves the transmission line and excites the monopole.

7.4.3 Concepts for Reducing Antenna Height

A monopole antenna has a height of a quarter wavelength and is therefore not well-suited for integration into a device like a mobile phone or vehicle. Different concepts exist to reduce the height of wire antennas in order to provide a more compact antenna structure.

Figure 7.14a shows the original *thin* (i.e. radius much smaller than a wavelength) monopole antenna with a length of $\ell = \lambda/4$. If we substantially increase the radius (Figure 7.14b) we see a slight reduction in length if we aim to obtain the same resonance frequency as the original monopole. The dominant effect, however, is an increase in bandwidth, that is the antenna is matched in a greater frequency band.

As shown in Section 7.4.1 a dipole shorter than half a wavelength is *capacitive* at its input terminal. In order to compensate for the negative imaginary part we can use an inductance at the base point (Figure 7.14c). This configuration leads to a real input impedance with a reduced antenna height.

Further wire antenna concepts with reduced height are a *T-shaped* (*top-hat-loaded*) antenna (shown in Figure 7.14d) and a *helix* antenna where the antenna wire is wound in circles (see Figure 7.14e). Especially suitable for the integration into small devices is an *inverted-F* antenna as shown in Figure 7.14f. The total length $\ell + h$ is on the order of

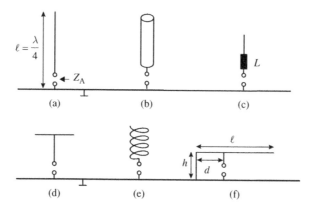

Figure 7.14 Concepts for wire antennas with reduced height: (a) thin monopole with an original length of a quarter wavelength, (b) thick monopole, (c) inductance near feedpoint, (d) capacitance on top of monopole, (e) helix structure and (f) *inverted-F* antenna.

a quarter wavelength. By varying the height h and the distance d between feedpoint and ground connection the antenna input impedance can be adjusted to the source impedance.

We will investigate some of the concepts in Problem 7.2.

7.5 Planar Antennas

Patch antennas (also referred to as *microstrip antennas*) are two-dimensional metallic strips on a dielectric substrate with a ground plane on the down side of the substrate [5, 14]. Patch antennas may differ in shape (rectangular, circular, triangular). Most commonly rectangular patch antennas are used. Figure 7.15 shows a rectangular patch with length L and width W.

Patch antennas have some useful properties: The low-cost production process already known from planar circuits can be applied to the antenna structure as well. So, antennas, matching networks, power dividers, filters and combiners can be produced in one manufacturing step. Furthermore, patch antennas are light-weight and robust and have low profile. Patch antennas can be designed to conform to curved surfaces (*conformal antennas*), for example on vehicles and aircrafts. Due to the permittivity of the substrate

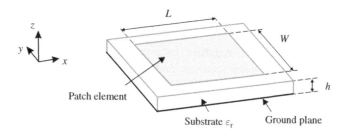

Figure 7.15 Geometry of a rectangular patch antenna.

the antenna is reduced in size. A drawback of patch antennas are their small relative bandwidth compared to dipole antennas.

7.5.1 Rectangular Patch Antenna

7.5.1.1 Radiation of a Single Patch Element

A rectangular patch antenna is a metallic strip (length L, width W) on a laterally-infinite dielectric substrate with a metallized back side. The ground plane extends to infinity as well. The substrate has a height h and a relative permittivity ε_r (Figure 7.15). Typical substrate materials are given in Section 4.3 (Table 4.1). The substrate material significantly influences the properties of the patch antenna. A great substrate height h and low relative permittivity ε_r lead to favourable antenna characteristics since the stray fields at the edges of the patch that influence the radiation become more pronounced. Thick substrates however may create losses due to the excitation of surface waves in the substrate. Furthermore, in the case of coaxial feeding (see Figure 7.17) the inductance of the inner coaxial conductor that contacts the patch element becomes unfavourably large. Also, a small relative permittivity only leads to a small reduction in size, resulting in quite a large antenna. Higher values of ε_r lead to small antennas, but also reduce the bandwidth of the antenna.

We can describe a patch antenna mathematically as an open-ended transmission line with a resonant length of about half a wavelength ($L \approx \lambda/2$). The current density distribution on the patch element follows a cosine curve lengthwise (comparable to the current distribution on a half-wavelength dipole) and is in a first approximation uniform in the transverse direction (see Figure 7.16a). This resonant mode is referred to as (1,0)-mode (i.e. *one* maximum in longitudinal direction and *no* maximum (constant distribution) in lateral direction).

The patch antenna appears electrically longer due to the stray field at the edges of the patch (see Figure 7.16b). We can compensate for this effect in our transmission line model by extending the patch element by an extra line element of length ΔL and adding magnetic walls (forcing the electric field to be vertically oriented at the ends). The radiation characteristic can be derived from the stray field (radiating slots) [2, 14]. Figure 7.16d shows the typical three-dimensional radiation pattern of a patch antenna. Due to the infinite ground plane, radiation occurs only in the upper hemisphere ($z > 0$). In the direction of the main lobe (vertical direction $\vartheta = 0$) the antenna is linearly polarized in the x-direction (i.e. direction of the current density on the patch element). Typical antenna gain for a patch antenna is on the order of 5 to 6 dBi. Typical half-power beamwidth is about $70°$ to $90°$ [14].

A patch antenna can also be regarded as a cavity resonator, where a rectangular cavity between the metallic patch element and the ground plane is in resonance (see Figure 7.16e), [2, 14]. The cavity volume is bounded by two electrical walls (patch element, ground plane) and four vertical magnetic walls (surfaces on the open lateral sides). Due to stray fields at the edges, the geometry of the patch element has to be adjusted. The four open surfaces contribute to the farfield. From the fields of the cavity mode the farfield can be calculated mathematically [2].

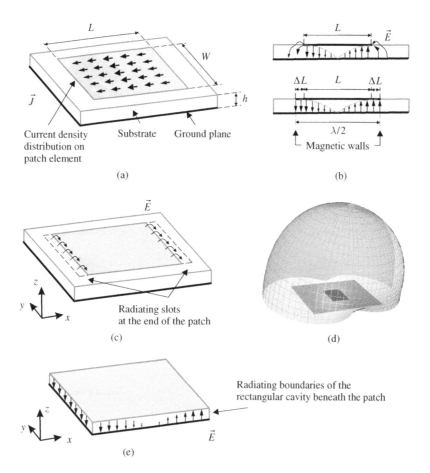

Figure 7.16 Patch antenna: (a) Geometry and current density distribution, (b) electric field distribution beneath the resonating patch, (c) radiating slots at the end of the patch, (d) three-dimensional radiation pattern and (e) patch antenna modelled as a cavity resonator with four magnetic walls.

7.5.1.2 Resonance Frequency and Patch Dimensions

A rectangular patch antenna (length L, width W) on a dielectric substrate (height h, relative permittivity ε_r) has a length of approximately half a wavelength. Due to the end-effect it is slightly shorter. The resonance frequency f_{10} of the fundamental (1,0)-mode can be estimated by the following equation [5]

$$f_{10} = \frac{c_0}{2\,(L+h)\,\sqrt{\varepsilon_{r,\text{eff}}}} \tag{7.49}$$

where $\varepsilon_{r,\text{eff}}$ is the effective relative permittivity given as

$$\varepsilon_{r,\text{eff}} = \frac{\varepsilon_r + 1}{2} + \frac{\varepsilon_r - 1}{2}\left(1 + \frac{12h}{W}\right)^{-\frac{1}{2}} \tag{7.50}$$

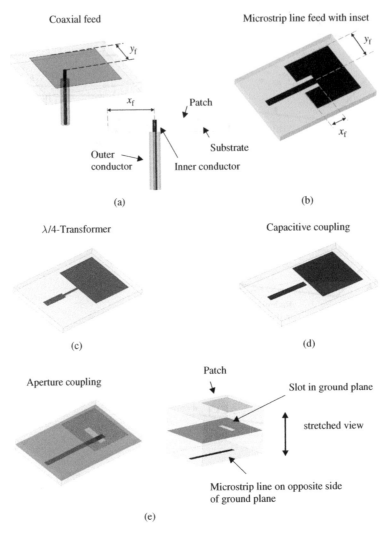

Figure 7.17 Feeding of patch antennas: (a) coaxial feed, (b) microstrip line with inset, (c) microstrip line with $\lambda/4$-transformer, (d) capacitive coupling and (e) aperture coupling.

More accurate formulas can be found in [1]. However, Equation 7.49 and 7.50 provide a good starting point for an initial design that can be further refined by EM simulations.

The width W of the patch does not significantly influence the resonance frequency of the $(1,0)$-mode. However, the width affects the input impedance and bandwidth. A greater width reduces the input resistance at resonance due to improved radiation. Furthermore, the bandwidth increases. However, patch antennas with greater width require more space and are therefore disadvantageous for group antennas. Typical values for patch width are between one and two times the patch length.

$$L \leq W \leq 2L \tag{7.51}$$

Further parameters influence the radiation and impedance characteristics of the patch antenna, for example the *finite* dimensions of substrate and ground plane, metallic and dielectric *losses* as well as the *feeding* method.

7.5.1.3 Feeding of Patch Elements

Coaxial Feed

There are different feeding methods for patch antennas. Figure 7.17a shows a coaxial feed. A coaxial line is connected to a circular opening in the ground plane. The radius of the circular opening equals the outer radius of the coaxial line. The inner conductor extends through the opening and contacts the bottom side of the patch element. The position (x_f, y_f) of the feedpoint determines the input impedance. The following equations give an approximate feedpoint position for an input impedance of $50\,\Omega$

$$y_f = \frac{W}{2} \quad \text{and} \quad x_f = \frac{L}{2\sqrt{\varepsilon_{r,\text{eff},L}}} \tag{7.52}$$

where $\varepsilon_{r,\text{eff},L}$ is the effective relative permittivity given as

$$\varepsilon_{r,\text{eff},L} = \begin{cases} \dfrac{\varepsilon_r + 1}{2} + \dfrac{\varepsilon_r - 1}{2}\left[\left(1 + \dfrac{12h}{L}\right)^{-\frac{1}{2}} + 0.04\left(1 - \dfrac{L}{h}\right)^2\right]; \text{ for } \dfrac{L}{h} \le 1 \\[4mm] \dfrac{\varepsilon_r + 1}{2} + \dfrac{\varepsilon_r - 1}{2}\left(1 + \dfrac{12h}{L}\right)^{-\frac{1}{2}}; \text{ for } \dfrac{L}{h} \ge 1 \end{cases} \tag{7.53}$$

Inset Feed

Figure 7.17b shows an alternative feeding method. The antenna is fed by a microstrip transmission line. In order to find the appropriate input impedance the microstrip feed line cuts into the patch element. The feedpoint location is determined by the inset length x_f (i.e. the length of the slots on both sides of the line) as well as the lateral position y_f. For substrate materials with $2 \le \varepsilon_r \le 10$ and an input impedance of $50\,\Omega$ the feedpoint location can be approximated by the following equation [15]

$$y_f = \frac{W}{2} \quad \text{and} \quad x_f = \frac{L}{2} \cdot \frac{c_7\varepsilon_r^7 + c_6\varepsilon_r^6 + c_5\varepsilon_r^5 + c_4\varepsilon_r^4 + c_3\varepsilon_r^3 + c_2\varepsilon_r^4 + c_1\varepsilon_r + c_0}{10^4} \tag{7.54}$$

where c_i $(i \in \{1, \ldots, 7\})$ are coefficients as given in Table 7.2. The width of the slots should correspond to the width of the feeding microstrip line.

Table 7.2 Coefficients for determination of feedpoint location (inset-feed)

c_7	c_6	c_5	c_4	c_3	c_2	c_1	c_0
0.001699	0.13761	−6.1783	93.187	−682.69	2561.9	−4043	6697

Further Feeding Methods

An alternative feeding method is depicted in Figure 7.17c. At the edge of the patch element the real part of the input impedance is quite high (usually much greater than $100\,\Omega$) since at the edge the current is low and the voltage is high. Therefore, a *quarter-wave transformer* (recognizable by the smaller line width) is used to achieve a lower input impedance of $50\,\Omega$. Quarter-wave transformers were discussed in Section 3.1.9.

Figure 7.17d shows a *capacitive feeding* by a gap between feedline and patch element. Furthermore, *aperture-coupled feeding* uses a feeding line on the back side of the substrate. The antenna is then excited via a slot in the ground plane (Figure 7.17e).

Feeding structures (inset feed, quarter-wave transformer, capacitive coupling) in the same plane as the antenna structure require extra space and may degrade the radiation pattern due to unintentional radiation from the feeding structure. Coaxial feeding and aperture-coupled feeding do not show this disadvantageous effect but involve more costly manufacturing processes.

Example 7.1 *Design of a Patch Antenna with Coaxial Feed*

We will design a patch antenna for a frequency of $f = 2.5\,\text{GHz}$. The loss-less substrate has a relative permittivity of $\varepsilon_r = 2.2$ and a height of $h = 1.524\,\text{mm}$. The width of the patch element shall be $W = 1.25L$.

With Equation 7.49 and 7.50 we calculate the length of the patch element. The equations cannot be directly solved for L. Therefore, we substitute numerical values in the equations until the desired frequency is achieved. As a result we find a length of $L = 40\,\text{mm}$ for a frequency of $f = 2.495\,\text{GHz}$. According to our specification $W = 1.25L$ we find a width of $W = 50\,\text{mm}$.

According to Equation 7.52 and 7.53 the location of the feedpoint is $y_f = 25\,\text{mm}$ and $x_f = 13.8\,\text{mm}$. Figure 7.18a shows the geometry of the designed patch antenna. The reflection coefficient in Figure 7.18b and the antenna input impedance Z_A in Figure 7.18c indicate good matching but at a slightly lower frequency of $f = 2.41\,\text{GHz}$. The frequency is about 3% below the specification. However, we can easily adjust the frequency by slightly reducing the length L of the patch element. Figure 7.18d shows the radiation pattern with a directivity of $D = 7.5\,\text{dBi}$.

Example 7.2 *Design of a Patch Antenna with Inset Feeding*

Now we will design an inset-feed patch antenna with the same parameters ($f = 2.5\,\text{GHz}$, $Z_{in} = 50\,\Omega$) and the same substrate (relative permittivity of $\varepsilon_r = 2.2$ and a height of $h = 1.524\,\text{mm}$) as in Example 7.1. Hence, the patch size is again $L = 40\,\text{mm}$ and $W = 50\,\text{mm}$.

An inset-feed patch antenna requires a $50\,\Omega$-microstrip feedline (see Figure 7.19a). Using a transmission line tool (e.g. [16]) we find a line width of $W_f = 4.65\,\text{mm}$. We use the same dimension for the width of the slots on both sides of the transmission line. Using Equation 7.54 we find a feedpoint position of $y_f = 25\,\text{mm}$ and $x_f = 9.6\,\text{mm}$. Figure 7.19a shows the geometry of the resulting patch antenna and Figure 7.19b shows a plot of the reflection coefficient s_{11}. This first design is very close to the specifications. Further improvements are possible by using EM simulation to adjust the dimensions of the patch.

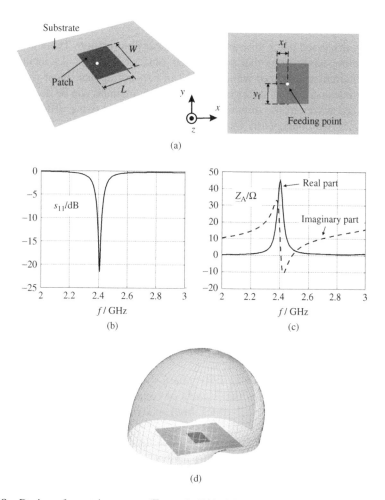

(a)

(b)

(c)

(d)

Figure 7.18 Design of a patch antenna (Example 7.1): (a) geometry, (b) reflection coefficient, (c) input impedance and (d) three-dimensional radiation pattern.

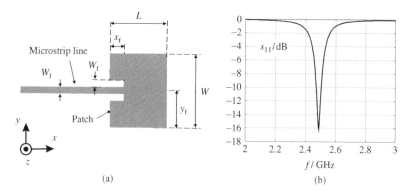

(a)

(b)

Figure 7.19 Design of a patch antenna (Example 7.2): (a) geometry and (b) reflection coefficient.

7.5.2   Circularly Polarizing Patch Antennas

The previously discussed patch antennas show a linear polarization. There are applications (e.g. GPS satellite navigation) that use circular polarized waves. As mentioned earlier (Section 2.5.2) circular polarized waves consist of two orthogonal linear-polarized waves of equal amplitude and a phase difference of $90°$. Patch antennas can be designed to radiate circular polarized waves in the direction of the main lobe by using different approaches. We will illustrate two concepts by looking at

- square patches with special feeding networks (double-fed patch).
- a modification of the patch geometry using only a single feedpoint.

In Figure 7.17c we presented an edge-feed patch antenna with a quarter-wave transformer for impedance matching to $50\,\Omega$. Now we use a square patch ($W = L$) that we feed at orthogonal edges (see Figure 7.20a) (*Double-fed patch*) [17]. At the edges we can excite two orthogonal, degenerate modes: (1,0)-mode and (0, 1)-mode. Figure 7.20a shows a white and black arrow on the patch surface that indicates the direction of the current densities associated with the different modes. A circular polarized wave is radiated if both modes resonate $90°$ out of phase. Therefore, a feeding network with the following components is used

- Two quarter-wave transformers (small lines) transform the high impedance of the patch to $50\,\Omega$.
- A $50\,\Omega$ quarter-wave line provides a $90°$ phase shift between the orthogonal modes.
- A 3 dB power splitter generates two signals of equal amplitude and phase.
- A third quarter-wave transformer (thick line) matches the power divider to the $50\,\Omega$ feedline.

Figure 7.20b shows an alternative feeding network. A 3dB branchline coupler provides the desired equal power split and $90°$ phase shift. Again quarter-wave transformers (small transmission lines) are used to match the impedance of the patch element to $50\,\Omega$. The branchline coupler is also used as duplexer. The transmit signal is fed into the upper port (TX). A received signal is available at the lower port (RX).

The two previously discussed concepts use feeding networks that excite the (1,0)-mode and (0,1)-mode by contacting the patch at two edges (*Double-fed patch*). These feeding networks require space on the substrate and may cause parasitic radiation. Therefore, concepts of interest are those that use only a single feedpoint [14, 18]. Figure 7.20c shows – as an example – a square patch element with two *truncated corners* and a single coaxial feed point. The truncated corners represent a small perturbation and cause coupling between the previously independent orthogonal modes. We can find a corner size where the modes have equal amplitude and are $90°$ out of phase. Figure 7.20d visualizes the current density distribution $\vec{J}$ at two different instances in time. At $t = 0$ the current density of mode (1,0) is maximum and points in x-direction. At $t = 0$ the current density of mode (0,1) is zero. At $t = T/4$ the current density of mode (0,1) is maximum and points in y-direction. At $t = T/4$ the current density of mode (1,0) is zero.

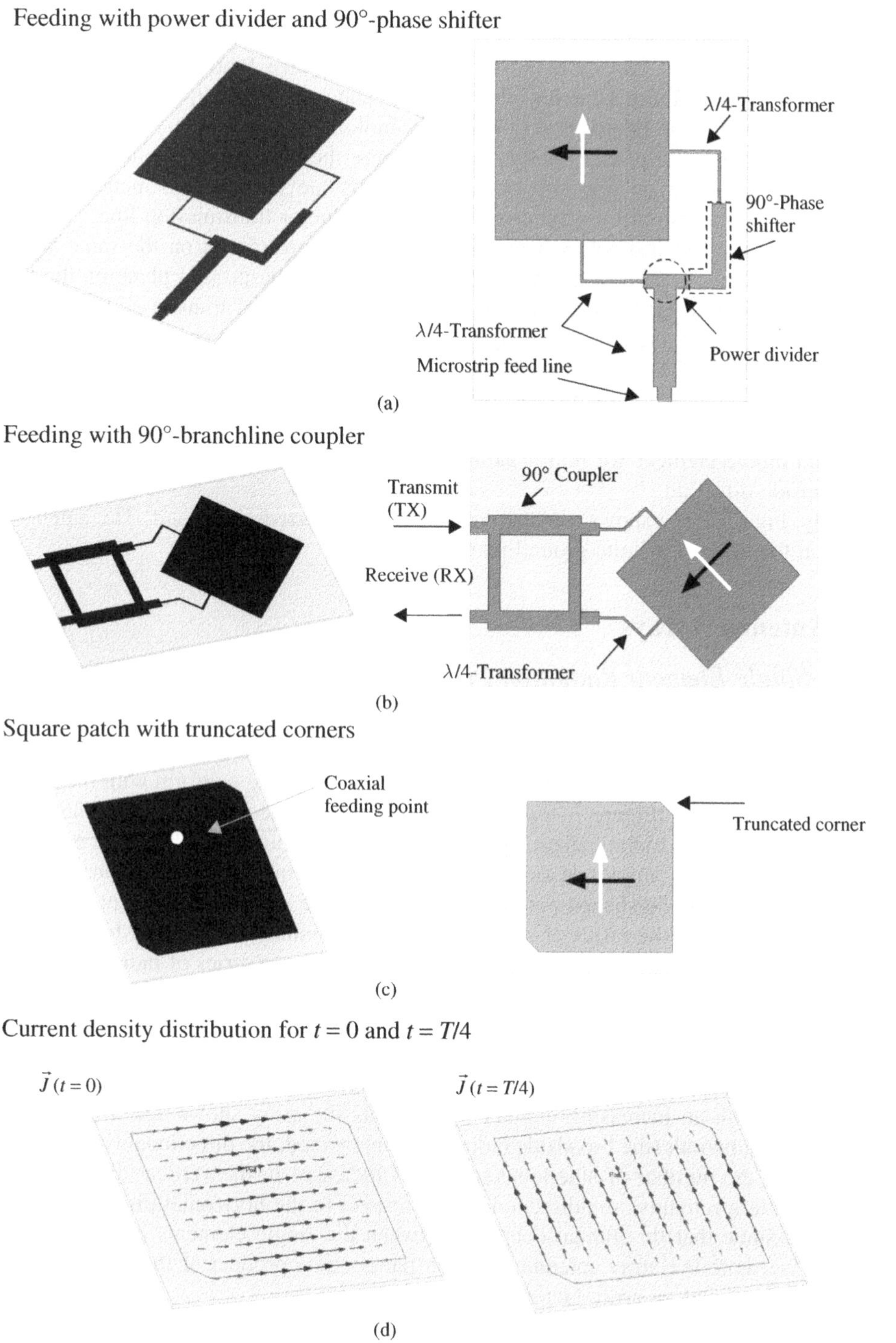
Feeding with power divider and 90°-phase shifter
λ/4-Transformer
90°-Phase shifter
λ/4-Transformer
Microstrip feed line
Power divider
(a)
Feeding with 90°-branchline coupler
Transmit (TX)
90° Coupler
Receive (RX)
λ/4-Transformer
(b)
Square patch with truncated corners
Coaxial feeding point
Truncated corner
(c)
Current density distribution for t = 0 and t = T/4
J (t = 0)
J (t = T/4)
(d)

Figure 7.20 Concepts for circularly polarized patch antennas.

7.5.3 Planar Dipole and Inverted-F Antenna

A planar patch antenna can be manufactured using low-cost fabrication technology. In order to profit from these benefits other antenna concepts (like dipole, monopole and inverted-F antenna) can be adapted to planar technology.

Figure 7.21a shows a *planar dipole antenna* where the rods of the dipole are replaced by two metallic strips on opposite sides of the substrate (there is no metallic ground plane). The antenna is fed by a symmetric ribbon conductor transmission line.

Figure 7.21b shows a dipole where both dipole strips are placed on the same side of the substrate. In the region of the antenna there is no metallic ground plane on the back side of the substrate [19]. The dipole is fed by a symmetrical transmission line (with and without ground plane). By selecting appropriate linewidth and spacing, the transition has low reflections. The symmetrical transmission line over ground is then converted to a single ended microstrip line by using a balanced-to-unbalanced transformer (see Section 6.11) and a quarter-wave transformer for impedance matching. In front of the active-fed dipole element we find a parasitic element in order to direct the radiation in the broadside direction.

Finally, Figure 7.21c shows a planar version of an *inverted-F antenna*. The antenna is located at the edge of a finite ground plane [20].

7.6 Antenna Arrays

7.6.1 Single Element Radiation Pattern and Array Factor

An antenna array consists of multiple (in most cases identical) single antenna elements, for example dipoles or patch antennas. The individual elements are arranged in a line, distributed in a plane or placed on a curved surface, for example to conform with the surface of a vehicle or aircraft (Figure 7.22). The radiating fields of the individual elements superimpose and contribute to the radiation pattern of the antenna array. With special excitation schemes (variations of amplitude and phase of different elements) the radiation pattern of the array antenna can be shaped and the main lobe can be steered in different directions.

We will investigate the effect of combining single antennas into an array by looking at the following example of stacked dipoles. Figure 7.23 shows a series of radiation patterns for antenna arrays with two to five equally spaced half-wave dipoles ($N \in \{2, \ldots, 5\}$). The distance between the vertically arranged dipoles is 82% of the free space wavelength ($d = 0.82\lambda$). All elements are excited with the same amplitude and phase. A single dipole ($N = 1$) exhibits the radiation pattern that we have already discussed in Section 7.4. With an increasing number of elements the array shows decreased vertical half-power beamwidth (the broadside radiation is intensified, the directionality increased). Furthermore, the number of side-lobes and nulls increases in the vertical direction. The radiation pattern remains omnidirectional with respect to the horizontal direction.

Let us assume that the mutual coupling between the single elements is negligible. In this case we write the array antenna radiation pattern as a product of the single element radiation pattern and an array factor [2, 5, 21].

$$\underbrace{C_G(\vartheta, \varphi)}_{\substack{\text{array antenna} \\ \text{radiation pattern}}} = \underbrace{C_E(\vartheta, \varphi)}_{\substack{\text{single element} \\ \text{radiation pattern}}} \cdot \underbrace{AF(\vartheta, \varphi)}_{\text{array factor}} \qquad (7.55)$$

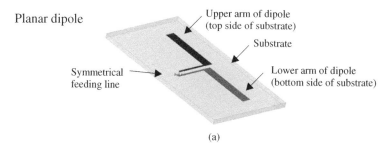

Planar dipole

Upper arm of dipole
(top side of substrate)

Substrate

Lower arm of dipole
(bottom side of substrate)

Symmetrical
feeding line

(a)

Planar quasi Yagi antenna

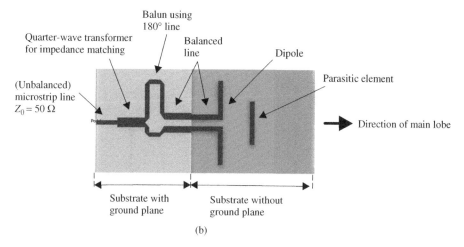

Quarter-wave transformer
for impedance matching

Balun using
180° line

Balanced
line

Dipole

Parasitic element

(Unbalanced)
microstrip line
$Z_0 = 50 \, \Omega$

Direction of main lobe

Substrate with
ground plane

Substrate without
ground plane

(b)

Planar inverted-F antenna

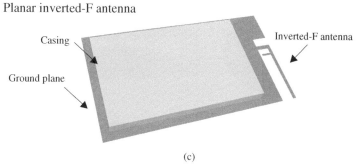

Casing

Inverted-F antenna

Ground plane

(c)

Figure 7.21 (a) Planar dipole with arms printed on different sides of the substrate, (b) feeding of a planar quasi Yagi antenna and (c) planar version of an inverted-F antenna.

In our example the array antenna consists of vertically oriented half-wave dipoles, so the single element radiation pattern is given by

$$C_E (\vartheta, \varphi) = \frac{\cos\left(\frac{\pi}{2} \cos \vartheta\right)}{\sin \vartheta} \quad (7.56)$$

For an arrangement of antenna elements in the vertical direction the array factor is a function of the distance-to-wavelength ratio d/λ as well as the vertical angle ϑ. For equally

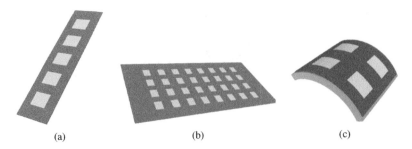

Figure 7.22 Array antenna with rectangular patch elements (a) arranged in a line, (b) distributed in a plane and (c) placed on a curved surface.

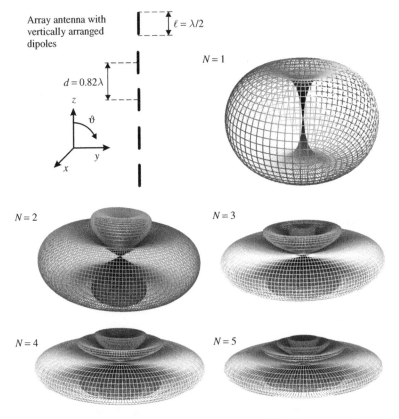

Figure 7.23 Radiation pattern series for array antennas with different numbers of vertically arranged half-wave dipoles (N = number of dipole elements).

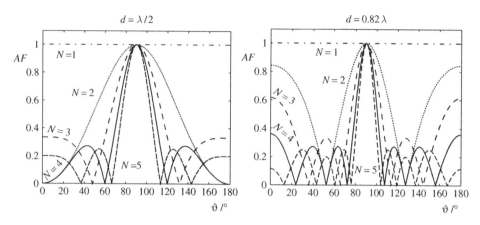

Figure 7.24 Array factors of an antenna array of vertically stacked antenna elements (N = number of antenna elements). The plots show results for two different distances d between the antenna elements ($d = \lambda/2$ and $d = 0.82\lambda$).

spaced single elements and identical complex excitation signals (all antenna elements are excited with the same amplitude and phase) we obtain an array factor of

$$AF\,(\vartheta, \varphi) = \left| \frac{\sin\left(N\pi\dfrac{d}{\lambda}\cos\vartheta\right)}{N\sin\left(\pi\dfrac{d}{\lambda}\cos\vartheta\right)} \right| \tag{7.57}$$

Figure 7.24 shows array factors AF for vertically stacked antenna elements with element distances of $d = \lambda/2$ and $d = 0.82\lambda$. The number of antenna elements varies from $N = 1$ to $N = 5$. From the graphic illustrations we draw the following conclusions:

- A higher number of elements N results in a smaller half-power beamwidth of the main lobe. The directionality of the antenna increases.
- Extending the distance between the individual elements further increases the directionality but the number of side-lobes increases too.

For a majority of applications a distance of half a wavelength is ($d = \lambda/2$) appropriate. According to [9] this distance satisfies Nyquists theorem that requires an element spacing of $\lambda/2$ or less. Interestingly, the array factor of a two-element array ($N = 2$) shows only one (main) lobe in the broadside direction ($\vartheta = 90°$) and nulls in the vertical direction ($\vartheta = 0°$ and $\vartheta = 180°$)(see left graph in Figure 7.24).

In our example of stacked half-wave dipoles a distance of $d = \lambda/2$ would result in direct contact of the dipole tips and is therefore not feasible. The distance has to be enlarged in order to reduce mutual coupling, hence we choose a distance of $d = 0.82\lambda$. Figure 7.25 illustrates the single element radiation pattern C_{E}, the array factor AF and the cumulative array radiation pattern C_{G} of a single half-wave dipole ($N = 1$) as well as vertical dipole groups with $N = 2$ to $N = 5$ radiating elements.

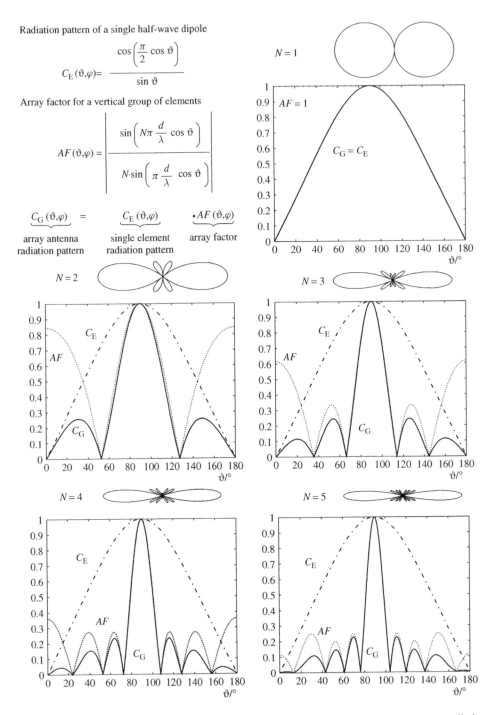

Radiation pattern of a single half-wave dipole

$$C_E(\vartheta,\varphi)= \frac{\cos\left(\dfrac{\pi}{2}\cos\vartheta\right)}{\sin\vartheta}$$

Array factor for a vertical group of elements

$$AF(\vartheta,\varphi) = \left| \frac{\sin\left(N\pi\dfrac{d}{\lambda}\cos\vartheta\right)}{N\cdot\sin\left(\pi\dfrac{d}{\lambda}\cos\vartheta\right)} \right|$$

$$\underbrace{C_G(\vartheta,\varphi)}_{\substack{\text{array antenna}\\\text{radiation pattern}}} = \underbrace{C_E(\vartheta,\varphi)}_{\substack{\text{single element}\\\text{radiation pattern}}} \cdot \underbrace{AF(\vartheta,\varphi)}_{\text{array factor}}$$

$N=1$

$AF=1$

$C_G = C_E$

$N=2$

C_E

AF

C_G

$N=3$

C_E

AF

C_G

$N=4$

C_E

AF

C_G

$N=5$

C_E

AF

C_G

Figure 7.25 Normalized single element radiation pattern, array factor and array antenna radiation pattern for vertically stacked dipoles (distance between elements $d = 0.82\lambda$).

The previous considerations require antenna elements without mutual coupling. In practice radiating elements may influence the performance of adjacent elements. 3D EM simulation software (Section 6.13.2) can be used to investigate the coupling phenomena and include the effect in the design process.

7.6.2   Phased Array Antennas

In the previous section all vertically stacked dipoles are fed with signals of equal amplitude and phase. As discussed this leads to a main lobe in the broadside direction. In a phased array antenna system the direction of the main lobe of the radiation pattern changes by applying feeding signals with non-uniform phase configurations.

Figure 7.26 shows an array of N radiating elements arranged along the z-axis. First, let us assume excitation signals with equal phase for all elements (Figure 7.26a). Each element generates a spherical head wave. According to Huygens' principle the wave fronts of the spherical waves add up constructively in the lateral direction (perpendicular to the z-axis).

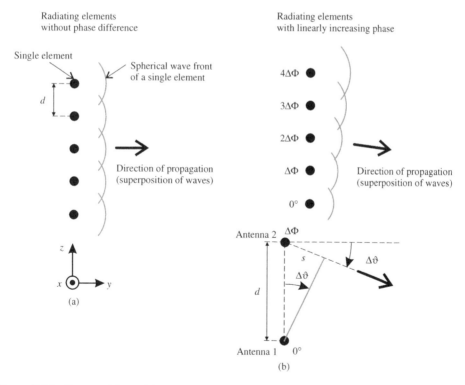

Figure 7.26   Superposition of independent head waves from equally spaced radiating elements (Huygens' principle).

Now let us consider a linear phase progression $(\Phi + n\Delta\Phi)$ along the z-axis (Figure 7.26b). Since a positive phase shift represents a temporal lead[4] the individual wave fronts have propagated different distances at a given time. Hence, the cumulative wave front is tilted by an angle of $\Delta\vartheta$.

We will now calculate the angle $\Delta\vartheta$ and the corresponding phase shift $\Delta\Phi$ from the geometric considerations given in Figure 7.26b. Let us assume a positive phase shift $\Delta\Phi$ for antenna 2, that is antenna 1 is delayed in the time domain. When the wave from antenna 1 starts, the wave from antenna 2 has already propagated a distance of

$$s = d \sin(\Delta\vartheta) \tag{7.58}$$

Electromagnetic waves travel with the speed of light c_0. From basic cinematics we calculate the time delay Δt of antenna 1 as

$$\Delta t = \frac{s}{c_0} = \frac{d \sin(\Delta\vartheta)}{c_0} \tag{7.59}$$

For a given frequency we transform this time delay Δt into a phase shift $\Delta\Phi$. The ratio of the delay time Δt and the period length T is equal to the ratio of the phase shift $\Delta\Phi$ and $360°$. With $c = \lambda f$ we obtain

$$\frac{\Delta t}{T} = \Delta t \cdot f = \Delta t \frac{c_0}{\lambda} = \frac{\Delta\Phi}{360°} \tag{7.60}$$

Hence, for a pivot angle of $\Delta\vartheta$ we calculate the phase shift $\Delta\Phi$ between the individual elements.

$$\boxed{\Delta\Phi = \frac{360°}{\lambda} d \sin(\Delta\vartheta)} \qquad \text{(Phase shift between elements)} \tag{7.61}$$

In order to tilt the main lobe of an array antenna with equally spaced elements the absolute phase varies linearly from element to element. From the distance of the n−th antenna element to the first element we determine the necessary phase shift and time delay (Figure 7.26b).

In practical applications fixed phase differences can be realized by line segments of different lengths or by passive phase shifters consisting of lumped elements. Active electronic phase shifters allow a rapid switching of the pivot angle with minimum delay.

The ability of the array antenna to steer its main lobe in a certain direction may be used to track moving objects. Furthermore, orienting the main lobe in the direction of a transmitting station may improve the signal-to-noise ratio of a communication system.

Example 7.3 *Cellular Base Station Antenna with Electrical Downtilt*
In a cellular communication network the geographical area is divided into smaller service areas called cells. Within such cells fixed frequency bands are used to provide terminal equipment (mobile phones) with access to the core network. Electromagnetic waves that

---

[4] A *negative* phase represents a *delay* in the time domain.

emanate from a base station antenna decay as they propagate through space (we take a closer look at path loss in Chapter 8), therefore the frequency bands of a cell may be re-used at an adequate distance in order to provide high capacity of the communication system. Figure 7.27a shows a schematic representation of cells. Equal cell numbers indicate equal frequency bands.

Each macrocell is illuminated by one or more base station antennas. Figure 7.27b shows a widely used configuration with three sector antennas. In order to achieve the desired coverage and to avoid reflections from obstacles such as buildings or trees the antenna is located at an elevated point.

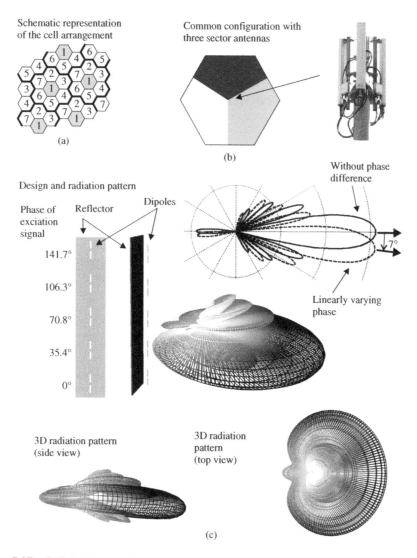

Figure 7.27   Cellular base station antenna with an electrical downtilt of $7°$ (see Example 7.3).

Figure 7.27c shows the essential parts of a GSM900 base station sector antenna. It consists of five vertically stacked dipoles placed in front of a planar metallic reflector [22]. The vertical arrangement of radiating elements leads to a small half-power beamwidth of the main lobe in the vertical direction. The metallic reflector focuses the radiation in a direction normal to the planar surface. A combination of three sector antennas would result in an omnidirectional pattern.

If we excite the dipoles with equal phase the main lobe points in a horizontal direction leading to long-range radiation. This could lead to interference with cells that use the same frequency band. There are now two approaches to limit the radio range. First, sloped mounting of the antenna tilts the main lobe downwards mechanically. The drawback of this concept is the increased antenna wind load and the more complex fastening kit. Second, straight mounting of the antenna is combined with the phased array scheme. This concept tilts the main lobe downwards electronically. In our example the distance (centre-to-centre) between the dipoles is $d = 25.5$ cm. In order to tilt the main lobe downwards by $7°$ for a frequency of $f = 950$ MHz we apply Equation 7.61. This yields a phase shift of $\Delta\Phi = 35.42° \approx 35.4°$ between adjacent elements. Figure 7.27c shows the titled main beam in a vertical cut-plane as well as the full three-dimensional radiation pattern.

The above discussed (one-dimensional) phased array concept can be generalized to more complex antenna arrangements. We will investigate the widespread configuration of a rectangular distribution of radiating elements. In order to steer the main lobe the phase relations are given by a plane equation. (In the one-dimensional case the phase relation is given by a straight line equation.)

We look at the planar group of rectangular patch antennas in Figure 7.28. The number of elements is $M \times N$. The elements are equally spaced in x- and y-direction, with the distances d_x and d_y, respectively. If we excite all patch elements with signals of uniform phase the main lobe points in the z-direction. Let us now change this direction by a vertical angle ϑ_0 and a horizontal angle of φ_0.

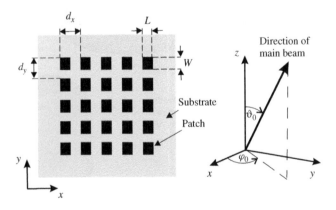

Figure 7.28 Planar group of rectangular patch antennas and definition of main beam direction.

Equations 7.62 and 7.63 provide the phase shifts $\Delta\Phi_{mn}$ and delay times Δt_{mn} depending on the locations $(m \cdot d_x, n \cdot d_y)$ of the different elements [14].

$$\Delta\Phi_{mn} = \frac{360°}{\lambda_0}\left(m \cdot d_x \cos\varphi_0 \sin\vartheta_0 + n \cdot d_y \sin\varphi_0 \sin\vartheta_0\right) \tag{7.62}$$

$$\Delta t_{mn} = \frac{\Delta\Phi_{mn}}{360°}\frac{1}{f} \tag{7.63}$$

We will investigate this formula in more detail in Problem 7.5.

Example 7.4 *Planar Array Antenna with Steered Beam*

Figure 7.29 shows a planar array antenna with 5×5 equally spaced patch antennas ($d_x = d_y = 30$ mm). If the radiating elements are fed with the same phase the main beam points in z-direction. We will change the direction by $\vartheta_0 = 30°$ and $\varphi_0 = 50°$ for a frequency of $f = 5$ GHz.

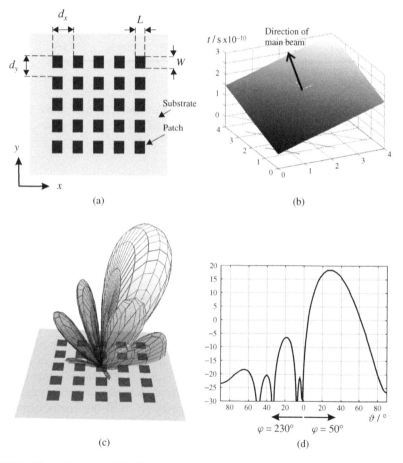

Figure 7.29 2D patch array with tilted main loop: (a) layout, (b) delay times, (c) 3D radiation pattern and (d) vertical 2D radiation pattern (Example 7.4).

Table 7.3 Delay times of antenna elements (Example 7.4)

	$n = 0$	$n = 1$	$n = 2$	$n = 3$	$n = 4$
$m = 4$	128.6 ps	166.9 ps	205.2 ps	243.5 ps	281.8 ps
$m = 3$	96.4 ps	134.7 ps	173.0 ps	211.3 ps	249.6 ps
$m = 2$	64.3 ps	102.6 ps	140.9 ps	179.2 ps	217.5 ps
$m = 1$	32.1 ps	70.4 ps	108.7 ps	147.0 ps	185.3 ps
$m = 0$	0.0 ps	38.3 ps	76.6 ps	114.9 ps	153.2 ps

Table 7.4 Phase angles of antenna elements (Example 7.4)

	$n = 0$	$n = 1$	$n = 2$	$n = 3$	$n = 4$
$m = 4$	231.5°	300.4°	369.4°	438.3°	507.2°
$m = 3$	173.6°	242.5°	311.4°	380.3°	449.3°
$m = 2$	115.7°	191.2°	253.6°	322.6°	391.5°
$m = 1$	57.9°	126.7°	195.7°	264.6°	333.5°
$m = 0$	0.0°	68.9°	137.9°	206.8°	275.8°

From Equations 7.62 and 7.63 we calculate the delay times Δt and phase shifts $\Delta \Phi$ for all elements. The results are compiled in Table 7.3 and 7.4.

7.6.3 Beam Forming

So far we have seen that grouping identical antennas with identical excitations (*amplitudes and phases* of the feeding signals are equal for all elements) increases the directionality. Furthermore, we have learnt that applying a *phase shift* between antenna elements changes the orientation of the main lobe.

Now we will investigate the effect of *different amplitudes* applied to individual elements [2, 3]. Figure 7.23 demonstrates the focusing effect of an antenna array with uniform excitation. As discussed earlier side-lobes occur that are unwanted in a variety of applications. The side-lobe level can be reduced by multiplying the uniform amplitudes with a set of weighting factors.

Let us assume a linear antenna array with a common element spacing of half a wavelength $d = \lambda/2$. The side-lobes disappear if we apply weighting factors according to the binomial expansion coefficients as shown in Figure 7.30. Pascal's triangle provides a simple way to calculate these factors. The centre elements of the antenna array are fed with the highest amplitudes. For three radiators ($N = 3$) the centre element is fed by twice the current of the edge elements. For $N = 6$ the centre feeding current is ten times higher compared to the current supplied to the edge elements. With increasing number of elements the radiation contribution of the outer elements goes down, leading to an inefficient antenna design. Furthermore, a binomial set of weighting factors increases the width of the main beam and hence reduces the directionality of the antenna.

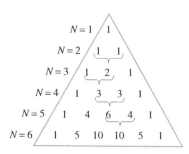

Figure 7.30 Pascal's triangle and binomial coefficients.

In practice it often is sufficient to bring down the side-lobes to a certain level, for example -20 dB or -30 dB with respect to the main lobe. This goal is achieved by the *Dolph–Chebyshev method*. Since the calculation of the weighting coefficients is complex we refer to standard antenna literature, for example [5]. Compared to the binomial weights the Dolph–Chebyshev method provides weighting factors that are less divergent, leading to improved efficiency due to higher contribution of the outer radiating elements. Furthermore, the influence on the directionality of the antenna is less significant.

In the following example we apply the beam forming concepts to a linear patch antenna array with element spacing of $d = \lambda/2$.

Example 7.5 *Linear Array Antenna with Six Patch Elements*

Figure 7.31a shows a group of six patch antennas with a spacing of $d = \lambda/2 = 6.25$ mm for a frequency of $f = 24$ GHz. The patches ($L = 2.15$ mm and $W = 2.8$ mm) are positioned on a substrate layer ($h = 200 \, \mu m$, $\varepsilon_r = 7.8$). Each patch has a separate coaxial feed.

We apply excitation signals to the individual patch elements according to the weighting factors of the above mentioned schemes:

- *uniform weights (equal amplitudes, no beam forming),*
- *binomial weights,*
- *Dolph–Chebyshev weights (side-lobe level of -30 dB).*

The weighting factors are listed in Figure 7.31b. Figure 7.31c shows radiation pattern cuts (xz-cut plane) in linear and logarithmic scales. Uniform amplitudes lead to minimum main lobe width (maximum directionality) but significant side-lobes. Binomial weighting factors cancel out side-lobes completely but the main lobe is enlarged considerably. Dolph–Chebyshev weights provide a compromise for both quantities (acceptable side-lobes and moderate reduction of directionality). Small asymmetries in the plots result from the asymmetric feeding of the patches.

Figure 7.31d depicts three-dimensional radiation patterns in linear scale. If we look at the Dolph–Chebyshev weight-related radiation pattern and compare it to the binomial weight-related pattern it is virtually impossible to distinguish between the diagrams visually in linear scale.

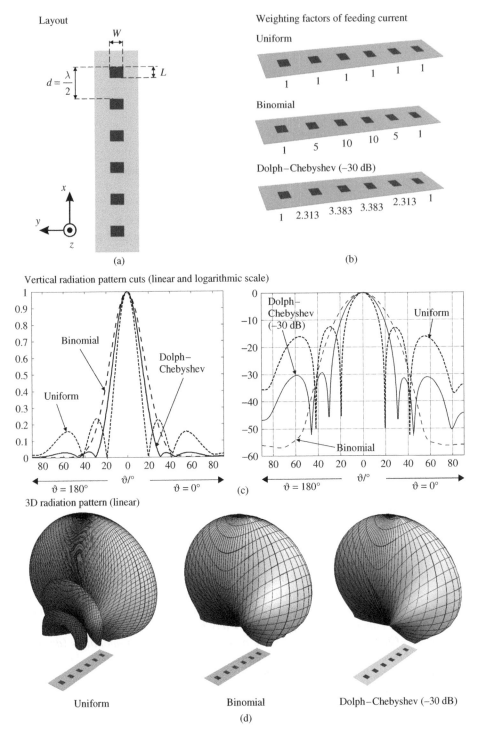

Figure 7.31 Array antenna with six equally spaced patch elements (uniform, binomial and Dolph–Chebyshev weighting of feeding signals).

7.7   Modern Antenna Concepts

As seen before a group of antennas can be used to shape the radiation pattern or steer the direction of the main lobe. Further developments are possible when multi-antenna systems are combined with signal processing algorithms. These algorithms are beyond the scope of this introductory book. However, we will briefly mention some interesting concepts to encourage the reader to study this interesting topic elsewhere in the literature.

- A *smart antenna* uses a group of antennas to *dynamically* adapt the radiation pattern in order to optimize a communication link. The radiation pattern can be formed by adjusting the amplitude and phase of each individual antenna element. A smart antenna can steer the direction of the main lobe towards a certain wireless transmitter to increase the signal-to-noise ratio. Furthermore, by producing nulls in the radiation pattern, interfering sources can be suppressed. The *smartness* of a smart antenna comes from its adaptive algorithms rather than from the antenna hardware [21].
- *Diversity* is another method to use more than one antenna at the receiver or transmitter site to improve a wireless communication link. There are different types of diversity: using antennas with different polarizations (e.g. vertical and horizontal polarization) is called *polarization diversity*. Using different antennas at different points in space is called *space diversity*. As we will discuss in Section 8.3.1, radio wave propagation in multipath environments leads to *fading* effects due to constructive and destructive superposition of waves. Points in space that are less than a wavelength apart may show significant deviation in receiving field strength. The use of more than one antenna can mitigate this effect. If the receiving field strength of an antenna (antenna 1) is not sufficient (due to destruction of incoming waves at the point in space where antenna 1 is placed) a second antenna that is located at another point in space may receive a sufficient signal. Therefore, switching between antennas can improve the received field strength and increase the reliability of the wireless communication link. In addition to switching between antennas, signals may be combined to increase signal-to-noise ratio [23].
- *Multiple-Input Multiple-Output* systems (MIMO) use multiple antennas at the transmitter and receiver site [23]. This methods aims for maximum transmission rate by opening a number of independent communication channels between transmitter and receiver, that is there is a transmission of several data streams in parallel (spatial multiplexing) [24]. Signal processing algorithms are used to separate the different data streams at the receiver site. MIMO is an integral part of the fourth-generation cellular system 3GPP Long Term Evolution (LTE).

7.8   Problems

7.1 We consider a communication standard with a maximum transmit power of 100 mW (EIRP). Transmitter and receiver amplifiers are connected via low-loss transmission lines (attenuation $a_1 = 2.5$ dB) to antennas with a gain of $G_1 = 5$ dBi.

In order to extend the range of the wireless system the transmission lines are substituted by shorter transmission lines with an attenuation of $a_2 = 1$ dB. Furthermore, antennas with larger gain are used $G_2 = 15$ dBi. What is the maximum permissible transmit power of the transmitter amplifier before and after the system modification? Why is there a range extension at all?

7.2 Investigate some of the ideas to reduce the height of monopole antennas shown in Section 7.4.3 by using an EM simulator of your choice. Select an operating frequency of $f = 2.45$ GHz.

7.3 Design a coaxial-feed patch antenna for a frequency of $f = 4$ GHz. The substrate has a relative permittivity of $\varepsilon_r = 3.38$ and a height of $h = 1.6$ mm. The width shall be $W = 1.5L$, where L is the resonant length of the patch. Check the performance of your design with an EM simulator of your choice.

7.4 Demonstrate that the electric and magnetic field strength of a Hertzian dipole given in Equation 7.34 can be derived from Equation 7.29 by using Equations 7.23 and 7.24.

7.5 Derive the relation in Equation 7.62 that determines the phase of antenna elements of a group antenna in order to steer the direction of the main lobe.

References

1. Meinke H, Gundlach, FW (1992) *Taschenbuch der Hochfrequenztechnik*. Springer.
2. Kark K (2010) *Antennen und Strahlungsfelder*. Vieweg.
3. Geng N, Wiesbeck W (1998) *Planungsmethoden für die Mobilkommunikation*. Springer.
4. von Milligan TA (2005) *Modern Antenna Design*. John Wiley & Sons.
5. Balanis CA (2008) *Modern Antennas Handbook*. John Wiley & Sons.
6. Dobkin DM (2008) *The RF in RFID*. Elsevier.
7. Finkenzeller K (2008) *RFID Handbuch*. Hanser.
8. Skolnick M (2008) *Radar Handbook*. McGraw Hill.
9. Balanis CA (2005) *Antenna Theory*. John Wiley & Sons.
10. Zinke O, Brunswig H (2000) *Hochfrequenztechnik 1*. Springer.
11. van Bladel JG (2007) *Electromagnetic fields*. John Wiley & Sons.
12. Voges E (2004) *Hochfrequenztechnik*. Huethig.
13. Detlefsen J, Siart U (2009) *Grundlagen der Hochfrequenztechnik*. Oldenbourg.
14. Garg R, Bhartia P, Bahl I, Ittipiboon A (2001) *Microstrip Antenna Design Handbook*. Artech House.
15. Ramesh M, Yip K (2003) Design Formula for Inset Fed Microstrip Patch Antenna. *Journal of Microwave and Optoelectronics*. Vol. 3.
16. AWR (2003) *TX-Line Transmission Line Calculator*. AWR Corporation.
17. Sainati RA (1996) *CAD of Microstrip Antennas for Wireless Applications*. Artech House.
18. Crawford D (2004) Numerische Analyse planarer Antennen. *D&V Kompendium 2004/2005*, pp. 109–111, Publish-industry Verlag.
19. Clenet M (2005) *Design and Analysis of Yagi-like Antenna Element Buried in LTCC Material for AEHF Communication Systems*. Defence R& D Canada.
20. IMST GmbH (2010) *Empire: Users Guide*. IMST.
21. Gross F (2005) *Smart Antennas for Wireless Communications*. McGraw-Hill.
22. Gustrau F, Manteuffel D (2006) *EM Modeling of Antennas and RF Components for Wireless Communication Systems*. Springer.
23. Saunders SR, Aragón-Zavala A (2007) *Antennas and Propagation for Wireless Communication Systems*. John Wiley & Sons.
24. Molisch AF (2011) *Wireless Communications*. John Wiley & Sons.

Further Reading

von Kriezis EE, Chrissoulidis DP, Papagiannakis AG (1992) *Electromagnetics and Optics*. World Scientific Pub.

8

Radio Wave Propagation

Spherical electromagnetic waves that emanate from a transmitting antenna experience a reduction in field strength as the distance to the antenna increases. Furthermore, in terrestrial wireless communication links, EM waves interact with obstacles (buildings, walls, trees, hills, etc.) leading to absorption, diffraction and scattering of waves. Consequently, the field strength decreases even more quickly than in free space. In order to design and operate wireless communication systems it is essential to estimate the *path loss* from transmitter to receiver. For a given transmit power the path loss determines the received power and hence the coverage of the wireless system.

In this chapter we start by illustrating physical aspects of electromagnetic wave propagation. Furthermore, fundamental mathematical models that predict path loss in simple environments are derived from basic theory. Finally, we present more complex models to predict path loss in real environments.

8.1 Propagation Mechanisms

8.1.1 Reflection and Refraction

In free space plane waves and spherical waves may propagate. In Section 2.5.3 we looked at *reflection* and *refraction* of plane waves interacting with material interfaces. In the following we will summarize the basics and include further physical effects of wave propagation that are important when investigating path loss.

Figure 8.1a shows a plane wave (starting in medium 1) that impinges on a material interface under an angle of ϑ_f (oblique incidence). There are two physical effects to be observed:

Reflection: A plane wave that hits a material surface is partly *reflected back* into medium 1. The angle of reflection ϑ_r equals the angle of incidence ϑ_f.

Refraction: A plane wave that hits a material surface is partly *transmitted through* the material interface into medium 2. Due to different phase velocities in the different media, the plane wave changes its direction of travel. The angle of transmission ϑ_t may be smaller or greater than the angle of incidence ϑ_f (see Section 2.5.3).

RF and Microwave Engineering: Fundamentals of Wireless Communications, First Edition. Frank Gustrau.
© 2012 John Wiley & Sons, Ltd. Published 2012 by John Wiley & Sons, Ltd.

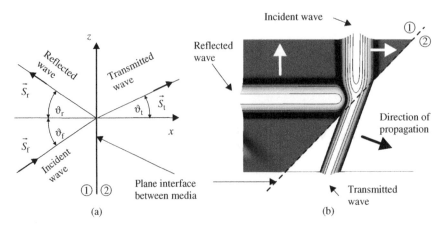

Figure 8.1   Reflection and refraction of a plane wave that impinges on a plane interface of two media.

Figure 8.1b illustrates the situation by looking at a single, positive wavefront at a certain instant in time. The wavefront in medium 1 hits the dielectric material interface at an angle of $\vartheta_f = 45°$. The relative permittivity of medium 2 is greater than the relative permittivity of medium 1 ($\varepsilon_{r2} > \varepsilon_{r1}$). We see the wave fronts and the direction of travel of the incident wave, the reflected wave and the transmitted wave.

8.1.2   Absorption

In the case of *lossy* materials, we observe a decreasing amplitude along the direction of travel. Energy is absorbed and converted into heat. The absorbed energy is no longer available for the transportation of EM signals.

> **Absorption/attenuation:** A plane wave that travels through *lossy material* is attenuated. We see an *exponential decay* of the magnitude as the wave propagates. Electromagnetic energy is absorbed and converted into thermal energy (heat).

8.1.3   Diffraction

At the edges of objects electromagnetic waves are *diffracted*, that is the waves are bent around the edges and propagate into the geometric shadow region. As an example we look at Figure 8.2a that shows an electromagnetic wavefront impinging upon an impenetrable object (metallic wall). The geometric shadow region behind the metallic wall is marked by dashed lines. If electromagnetic waves would propagate in a ray-like manner (following a straight line as assumed in geometric optics) we would expect no wave in this shadow region. Due to the *wave behaviour* of electromagnetic radiation we see diffracted waves at the edges. In Figure 8.2b the wave front reaches the object. In the centre region – where the object is located – the wave is totally reflected back. On the upper and lower zone – where no object is located – the waves run past the object. In

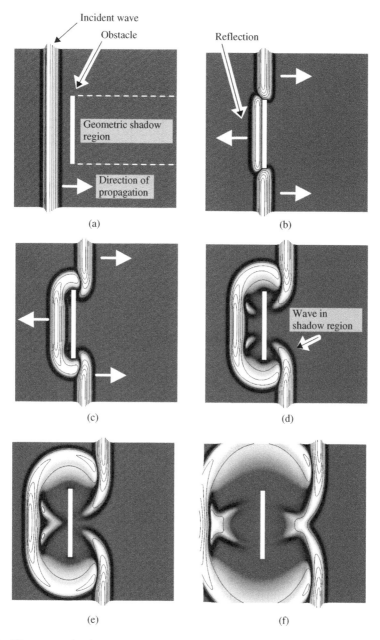

Figure 8.2 Plane wave impinges upon an impenetrable obstacle: diffraction at the edges leads to EM fields inside the geometric shadow region.

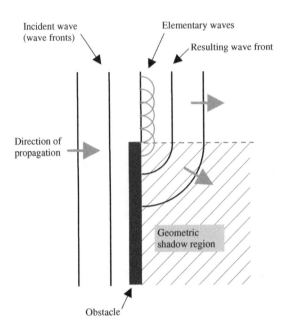

Figure 8.3 Diffraction at an edge of an obstacle illustrated by Huygens' principle.

Figure 8.2c and d we see the reflected wave running backwards. Furthermore, a diffracted wave penetrates the geometric shadow region. The effects continue in Figure 8.2e and f.

The diffraction of waves can be explained by *Huygens' principle*. Figure 8.3 shows a plane wave that hits an impenetrable obstacle. Huygens' principle tells us that each point of a wave front creates an *elementary spherical wave*. The superposition of all elementary waves leads to a new wave front. From Huygens' principle it is immediately clear that waves propagate into the geometric shadow region.

> **Diffraction:** Electromagnetic waves are bent around the edges of objects and enter the geometric shadow region. The effect can be understood from Huygens' principle.

Diffraction of electromagnetic waves is essential for wireless communication. It enables radio reception behind obstacles like buildings or hills. A canonical problem that considers an infinite half-space to estimate radio reception in the shadow region of objects is the *knife-edge* problem [1]. We will look at knife-edge diffraction in more depth when we talk about directional radio links (Section 8.2.4).

8.1.4 Scattering

An electromagnetic wave that interacts with a number of small objects (e.g. rain drops in the atmosphere or branches and leaves of vegetation) produces a diffuse response. At each individual object we see reflection, diffraction, refraction and absorption. The superimposition of all these effects is known as *scattering*.

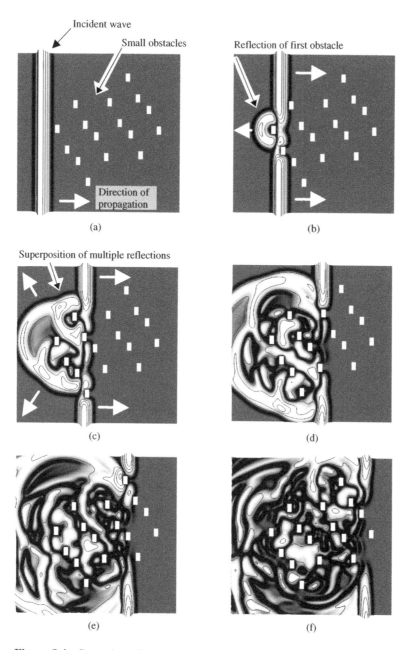

Figure 8.4 Scattering of a plane wave by a great number of small objects.

Scattering: An electromagnetic wave interacts with *a great number of small objects*. The interaction mechanisms include reflection, diffraction, refraction and absorption. The total response is quite complex due to the number of objects involved.

Figure 8.4a shows a wave front that impinges on a number of small metallic objects. In Figure 8.4b we see that the wave front interacts with the first object (reflection, diffraction). As the wave front moves further other objects produce reflected and diffracted waves too. The reflected and diffracted waves are then reflected and diffracted by other objects and so forth. In the end the total response is quite irregular and diffuse.

8.1.5 Doppler Effect

The interaction of electromagnetic waves with objects can be used in a variety of applications. As an example we look at a typical radar (*Radio Detection and Ranging*) scenario. Figure 8.6 shows a monostatic radar where transmitter and receiver are positioned at the same location. Radar systems determine positions and velocities of distant objects (e.g. aircrafts or vehicles) by evaluating *propagation times* and *frequency shifts* due to the Doppler effect.

Electromagnetic waves in vacuum (and air) travel with the (finite) speed of light c_0. Hence, an electromagnetic wave that travels from transmitter to the object and back again to the receiver needs the time Δt. The distance s can be derived from the propagation time as

$$s = \frac{1}{2}c_0\Delta t \qquad \text{(Distance between transmitter and scattering object)} \qquad (8.1)$$

Objects show different backscattered signals depending on their size, speed, material and geometric features. Therefore, in radar applications it is possible – within certain limits – to classify objects by their response. In automotive radar applications – for example – we can distinguish between pedestrians and vehicles. The amplitude of the backscattered signal is determined by the *radar cross-section* σ (RCS) or *echo area* [2]. This quantity in general shows an angular dependency, since the reflection depends on the angle of incidence. If we put an object with a radar cross-section in the direction of the main lobe, the received power P_{RX} is given as [2, 3]

$$P_{RX} = P_{TX}\frac{G_{TX}G_{RX}\lambda_0^2}{(4\pi)^3 s^4}\sigma \qquad \text{(Received power)} \qquad (8.2)$$

where G_{TX} and G_{RX} are the gains of the transmitting and receiving antennas, respectively.

Moving radar targets produce a frequency shift in the receiving signals due to the *Doppler effect*. The effect is a consequence of the constant propagation velocity of waves in media. A receiver that moves towards a stationary source 'sees' a higher frequency since wave fronts arrive at shorter time intervals than they are transmitted due to decreasing distance between transmitter and receiver with time (see Figure 8.5a). On the other hand, a stationary receiver that listens to a transmitter that moves towards the receiver, receives wave fronts at shorter time intervals than they are transmitted, due to decreasing distance between transmitter and receiver with time (see Figure 8.5b).

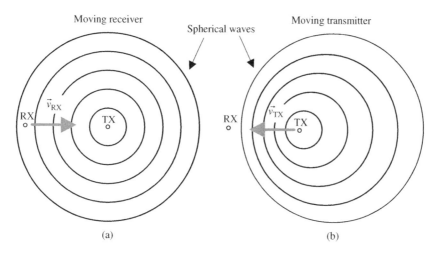

Figure 8.5   Doppler effect: (a) moving receiver and (b) moving transmitter.

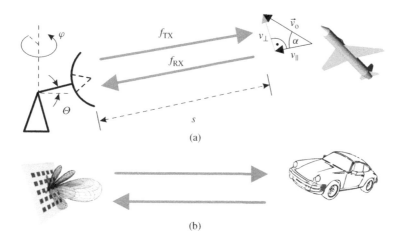

Figure 8.6   Aerospace and automotive radar applications.

If the object moves towards the radar antenna, the receiving signal frequency f_{RX} is higher than the frequency f_{TX} originally transmitted (see Figure 8.6a). The difference in frequency is known as the Doppler frequency f_D and is given as

$$f_D = f_{RX} - f_{TX} = f_{TX}\frac{2v_\parallel}{c_0} = \frac{2v_\parallel}{\lambda_0} \qquad \text{(Doppler frequency shift in radar)} \qquad (8.3)$$

where $v_\parallel$ is the component of the velocity towards the radar antenna [3]. If the object moves away from the radar antenna the frequency of the receiving signal is lower then the transmitted signal.

Doppler effect: Moving a transmitter changes the wavelength of waves in media. Hence, a stationary receiver observes a signal that is shifted in frequency. Furthermore, a receiver that moves through the field distribution of a stationary transmitter observes a signal shifted in frequency. The Doppler effect is a consequence of constant propagation velocity of waves in media.

With an antenna that can be steered mechanically or electronically (see Section 7.6) the position of an object can be determined from

- the elevation (or vertical) angle Θ;
- the azimuth (or horizontal) angle φ;
- the distance s.

Figure 8.6a shows a mechanical radar antenna and Figure 8.6b shows an array antenna for electronic beam steering. If we are only interested in the horizontal angle (*look direction*) a rotating antenna will suffice. In order to provide a good angular resolution in azimuthal direction the horizontal half-power beamwidth should be small. On the other hand, in order to collect signals of objects at different heights the vertical half-power beamwidth should be large.

Example 8.1 *Doppler Frequency*
  A vehicle moves at an angle of $\alpha = 45°$ and with a velocity of $v_0 = 60$ km/h towards a radar system that operates at a frequency of $f = 24$ GHz (see Figure 8.6). By using Equation 8.3 we get a Doppler frequency shift of

$$f_D = f_{RX} - f_{TX} = f_{TX}\frac{2}{c_0}v_\| = 24\,\text{GHz}\frac{2}{3\cdot 10^8\,\text{m/s}}\cos(\alpha)\frac{60}{3.6}\frac{\text{m}}{\text{s}} \approx +1.886\,\text{kHz} \quad (8.4)$$

The unit of the velocity is changed from km/h to m/s by considering a factor of $1/3.6$. The angle α reduces the effective velocity by a factor of $\cos(\alpha) = 1/\sqrt{2}$.

8.2 Basic Propagation Models

8.2.1 Free Space Loss

Figure 8.7a shows the power transmission from a transmitter (transmitted power P_{TX}, antenna gain G_{TX}) to a receiver (received power P_{RX}, antenna gain G_{RX}). The antennas are assumed to be perfectly matched and the distance r is large enough to consider the antennas to be in the farfield region, which is virtually always the case in practical communication links. Furthermore, the main lobes of the antennas point exactly at each other and the antennas exhibit the same polarization.

In free space there is a direct path for the electromagnetic wave from transmitter to receiver and the receiver power is given by the *Friis*-equation as

$$\boxed{P_{RX} = P_{TX}G_{RX}G_{TX}\left(\frac{\lambda_0}{4\pi r}\right)^2} \qquad \text{(Free space progapation)} \qquad (8.5)$$

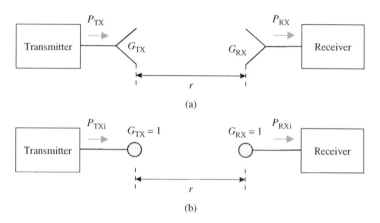

Figure 8.7 Free space loss: wireless link with (a) real (non-isotropic) antennas and (b) isotropic radiators.

We reformulate the Friis equation by using logarithmic representation of the quantities.

$$\frac{P_{RX}}{dBm} = \frac{P_{TX}}{dBm} + \frac{G_{TX}}{dBi} + \frac{G_{RX}}{dBi} - \underbrace{20\lg\left(\frac{4\pi r}{\lambda_0}\right)}_{L_{F0}/dB}$$ (8.6)

where L_{F0} is the *free space loss*.

The free space loss in decibels (dB) gives us the ratio of transmitted power P_{TXi} and transmitted power P_{RXi} in the case of isotropic radiators (see Figure 8.7b).

$$\boxed{\frac{L_{F0}}{dB} = 10\lg\left(\frac{P_{TXi}}{P_{RXi}}\right) = 20\lg\left(\frac{4\pi r f}{c_0}\right)}\qquad\text{(Free space loss)}$$ (8.7)

In the case of real antennas that have a certain gain we write for the isotropic transmit power $P_{TXi} = P_{TX}G_{TX} = EIRP$ and for the isotropic receive power $P_{RXi} = P_{RX}/G_{RX}$ (see Figure 8.7a).

$$\frac{L_{F0}}{dB} = 10\lg\left(\frac{P_{TXi}}{P_{RXi}}\right) = 10\lg\left(\frac{P_{TX}G_{TX}}{P_{RX}/G_{RX}}\right) = 10\lg\left(\frac{P_{TX}}{P_{RX}}G_{TX}G_{RX}\right)$$ (8.8)

Hence, the free space loss L_{F0} is independent of the antenna gains and describes the power loss due to spherical wave propagation.

We can use the following logarithmic *equation between quantities* to determine the free space loss

$$\frac{L_{F0}}{dB} = -147.56 + 20\lg\left(\frac{r}{m}\right) + 20\lg\left(\frac{f}{Hz}\right)$$ (8.9)

$$= 32.4 + 20\lg\left(\frac{r}{km}\right) + 20\lg\left(\frac{f}{MHz}\right)$$

where we substitute the distance in either metres (m) or kilometres (km) and the frequency in Hertz (Hz) or Megahertz (MHz).

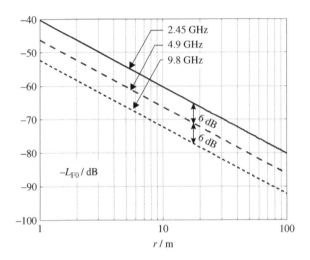

Figure 8.8 Free space loss as a function of distance for three different frequencies (see Example 8.2).

The free space loss L_{F0} increases with distance r and frequency f. The increase amounts to 20 dB per decade (tenfold frequency) that is 6 dB per octave (twofold frequency).

Figure 8.8 shows a logarithmic plot of the free space loss L_{F0} over distance r for three different frequencies $f_1 = 2.45\,\text{GHz}$, $f_2 = 2f_1 = 4.9\,\text{GHz}$ and $f_3 = 4f_1 = 9.8\,\text{GHz}$. Commonly we plot the negative value of the free space loss $-L_{F0}$ to illustrate that receive power decreases as the distance increases. As the frequency is doubled the receiver power is a quarter of the original value. Hence, the curve for frequency $f_2 = 2f_1$ is 6 dB lower than the curve for f_1.

Example 8.2 *Received Power in the Free Space Scenario*
Two antennas with gains of $G_{TX} = G_{RX} = 1.5 \,\widehat{=}\, 1.76\,\text{dBi}$ *are separated by a distance of* $r = 20\,\text{m}$. *For a transmit power of* $P_{TX} = 50\,\text{mW} \,\widehat{=}\, 17\,\text{dBm}$ *and a frequency of* $f = 2.45\,\text{GHz}$ *we calculate a receive power* P_{RX} *of*

$$P_{RX} = \left(\frac{\lambda_0}{4\pi r}\right)^2 G_{RX}G_{TX}P_{TX} = \left(\frac{c_0}{4\pi r f}\right)^2 G_{RX}G_{TX}P_{TX} \tag{8.10}$$

$$= 26.7\,\text{nW} \,\widehat{=}\, -45.73\,\text{dBm}$$

Alternatively we can express all values in logarithmic scale and write

$$P_{RX} = P_{TX} + G_{TX} + G_{RX} - \underbrace{20\lg\left(\frac{4\pi r}{\lambda_0}\right)}_{L_{F0}} \tag{8.11}$$

$$= 17\,\text{dBm} + 1.76\,\text{dBi} + 1.76\,\text{dBi} - 66.25\,\text{dB} = -45.73\,\text{dBm}$$

where $L_{F0} = 66.25$ dB *is the free space loss. The value can be graphically read from Figure 8.8 at* $r = 20$ m *we find* $-L_{F0} \approx -66$ dB.

8.2.2 Attenuation of Air

So far we have considered free space loss, which describes the attenuation associated with decreasing amplitudes of spherical waves due to distance. Electromagnetic waves that travel through air undergo an additional attenuation due to absorption from atmospheric gases and weather phenomena like rain, fog or snow. As an example we will look at the attenuation (absorption) from atmospheric gases [4].

Dry air mainly shows absorption from *oxygen* molecules. Oxygen has several resonances around 60 GHz resulting in strong absorption of electromagnetic waves. The dashed line in Figure 8.9a shows the attenuation constant α_{da} of dry air. The curves are for normal conditions at sea level, that is pressure 1013 hPa, temperature 15° C. High absorption (i.e. limited range) can be advantageous for short range communication or short range automotive radar since it minimizes interference with other users. For wide range communication however, frequencies are selected that show small attenuation.

Water vapour is another important gas that contributes to the absorption of electromagnetic waves. The solid line in Figure 8.9a shows the attenuation constant α_{wv} due to water vapour in air (water vapour density 7.5 g/m^3). The attenuation constant generally shows an upward trend with maxima at resonance frequencies. A first resonance occurs around 20 GHz.

If we consider both absorption mechanisms we get the attenuation constant α_{air} in Figure 8.9b with two significant resonances below 100 GHz. The additional path loss of a terrestrial link due to absorption can be calculated as

$$L_{air} = (\alpha_{da} + \alpha_{wv})r = \alpha_{air}r \qquad \text{(additional path loss)} \qquad (8.12)$$

where r is the distance between the antennas. For frequencies up to around $f = 10$ GHz the attenuation of air is quite small compared to free space path loss. With increasing frequency the path loss contribution of air becomes more pronounced.

8.2.3 Plane Earth Loss

Free space propagation – discussed in Section 8.2.1 – is an ideal case that is seldom encountered in practical applications since it ignores the interaction of waves with surrounding objects and structures. In wireless communication the ground effect is often particularly important.

Figure 8.10a shows that there are two paths for electromagnetic waves between transmit and receive antenna:

- The *direct path* with a path length r that equals the distance d_0 between the antennas $(r = d_0)$.
- The *indirect path* due to ground reflection with a path length of $r = d_1 + d_2 > d_0$.

The length of the indirect path $d_1 + d_2$ depends on the antenna heights over ground h_{TX} and h_{RX}. Since the indirect path is longer than the direct path $(d_1 + d_2 > d_0)$ the waves that

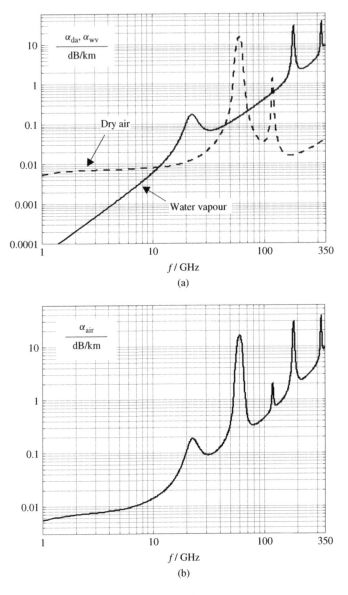

Figure 8.9 Specific attenuation of (a) dry air, water vapour and (b) normal air.

arrive at the receiver antenna show greater attenuation and phase shift. Furthermore, the reflection at ground under a certain angle has to be considered by an appropriate *reflection coefficient* r_{ground} (see Section 2.5.3). Consequently, at the location of the receive antenna the waves of the direct and indirect path may superimpose *constructively* or *destructively* depending on the phase difference of the signals. The resulting signal may therefore be larger or smaller than the direct (free space) signal alone.

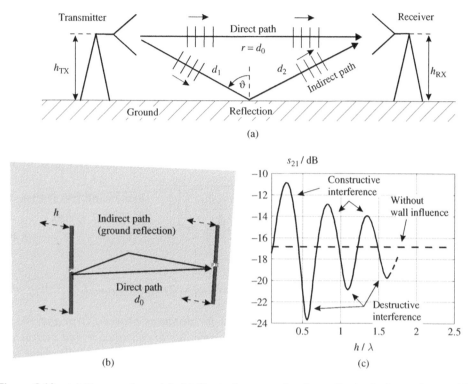

Figure 8.10 (a) Plane earth model, (b) illustrative example of two dipoles in front of a metallic wall and (c) resulting transmission coefficient as a function of antenna height compared to wavelength h/λ.

As an illustrative example we look at two dipole antennas that are separated by a distance d_0 and positioned in a distance h in front of a metallic wall. Figure 8.10b shows the antennas as well as the direct and indirect path. In order to study the wall influence we vary the distance h between wall and dipoles. The dashed horizontal line in Figure 8.10c represents as a reference the transmission coefficient *without* the wall, that is only the direct path is considered. When the wall is present, on the other hand, we see – depending on the distance between dipoles and wall – alternating ranges with constructive and destructive interference. At small distances h/λ the effect is very obvious. As the distance between the wall and the dipoles increases the direct path becomes more dominant and the influence of the indirect path decreases showing less pronounced transmission maxima and minima.

In order to provide a workable and simple formula for path loss estimations we focus on a special case often encountered in practical applications considering wireless links over flat terrain [1]. Our model is based on the following assumptions:

- The distance between the antennas d_0 shall be much larger than the antenna heights $(d \gg h_{\mathrm{TX}}, h_{\mathrm{RX}})$.

- Due to the large distance d_0 compared to the antenna height over the ground the angle of incidence ϑ at ground approaches $90°$. For flat incidence ($\vartheta \to 90°$) the absolute value of the reflection coefficient approaches one ($|r_{ground}| \to 1$; see Section 2.5.3) [5].
- *Amplitudes* of waves that travel on the direct and indirect path can be assumed to be equal since the path lengths are comparable $d_0 \approx d_1 + d_2$.
- *Phase differences* of waves that propagate on different paths have to be considered in more detail. We will look at it in Problem 8.1.
- The antennas shall be isotropic radiators ($G_{TX} = G_{RX} = 1$).

As a result we get the following equation that describes the path loss over ground [5].

$$\boxed{\frac{1}{L_{PEL}} = \frac{P_{RX}}{P_{TX}} = 2\left(\frac{\lambda_0}{4\pi r}\right)^2 \left[1 - \cos\left(\frac{4\pi h_{TX} h_{RX}}{\lambda_0 r}\right)\right]}$$
(Plane earth loss (PEL), for $r \gg h_{TX}, h_{RX}$)

(8.13)

Example 8.3 *Plane Earth Loss*

We consider antennas with heights of $h_{TX} = h_{RX} = 2\,m$ over ground operating at a frequency of $f = 2\,GHz$.

Figure 8.11 shows the plane earth path loss for distances between $10\,m$ and $10\,km$. For ranges up to $100\,m$ we see minima and maxima due to constructive and destructive interference of direct and indirect signals. The maxima reach values of $+6\,dB$ above free space loss. Free space loss is plotted as a dashed straight line. The minima are zero under the above mentioned modelling assumptions ($-\infty\,dB$). For larger distances (in our

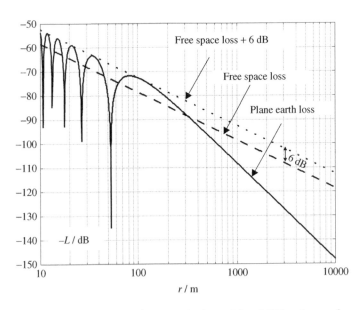

Figure 8.11 Free space loss vs. plane earth loss ($f = 2\,GHz$, $h_{TX} = h_{RX} = 2\,m$) (see Example 8.3).

example above 100 m) *the plane earth loss approaches*

$$\frac{1}{L_{\text{PEL}}} = \frac{P_{\text{RX}}}{P_{\text{TX}}} \approx \frac{h_{\text{TX}}^2 h_{\text{RX}}^2}{r^4} \qquad (\text{for } r > d_{\text{break}}) \tag{8.14}$$

*where d_{break} is the minimum distance (*break point distance*) for using Equation 8.14. According to [6] the break point distance d_{break} is given by*

$$d_{\text{break}} = \frac{4h_{\text{TX}}h_{\text{RX}}}{\lambda} \tag{8.15}$$

Hence, for distances larger than d_{break} the receiver power under *plane earth loss* conditions decays as 40 dB/decade ($\sim 1/r^4$) whereas under *free space loss* conditions it decays as 20 dB/decade ($\sim 1/r^2$).

In macrocells[1] of cellular telephony networks we commonly find a path loss increasing with r^n (receiver power decays with $1/r^n$) where n is close to 4. The exponent n is referred to as *path loss exponent*. We look at more detailed models in Section 8.3.

Example 8.4 *Plane Earth Loss with Different Antenna Heights*

In order to illustrate the influence of the antenna height over ground we look at two different antenna heights (2 m and 10 m). The frequency of operation is again $f = 2\,\text{GHz}$. Figure 8.12 shows the plane earth loss for antenna heights of $h_{\text{TX}} = h_{\text{RX}} = 2\,\text{m}$ already seen in Example 8.3. Now we increase the position of the transmit antenna

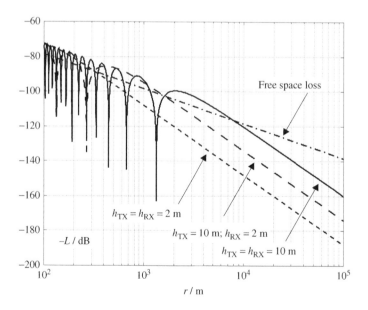

Figure 8.12 Free space loss vs. plane earth loss for different antenna heights (see Example 8.4).

---

[1] Macrocells of cellular networks have typical radii of 2 km to 30 km.

($h_{TX} = 10$ m, $h_{RX} = 2$ m). Obviously this increases the receive power by around 14 dB *if the receiver is far away from the transmitter. A further increase in receive power is reached if we set the receive antenna at a higher position too ($h_{TX} = h_{RX} = 10$ m).*

8.2.4 Point-to-Point Radio Links

Directional radio links between fixed stations use high gain antennas and high antenna locations to minimize ground effects. In order to consider free space propagation the space between the antennas has to be free of obstacles. A visual connection (i.e. an idealized direct line-of-sight) however is not sufficient, since electromagnetic waves do not propagate in a ray-like manner but show diffraction at objects. Therefore, a certain clearance around the direct visual line is required.

In literature (e.g. [1]) we find analytical models that describe the influence of a diffracting half-plane between transmitter and receiver (*knife-edge model*). We consider two distant antennas (TX and RX) that are targeted at each other as shown in Figure 8.13a.

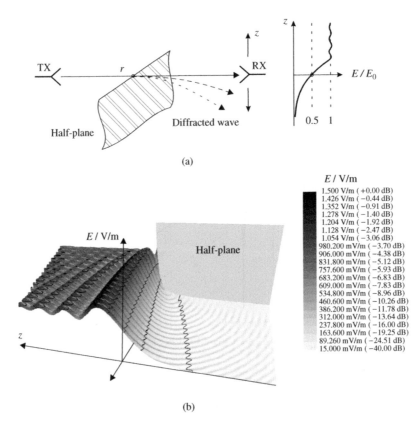

(a)

(b)

Figure 8.13 Diffraction by a half-plane (knife-edge model): (a) geometry and electric field strength along the z-axis and (b) EM simulation of diffracted wave behind the half-space.

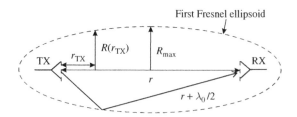

Figure 8.14   First Fresnel zone.

The antennas have visual contact. However, the lower region is covered by an impenetrable half-space. The received field strength E is halved compared to the free space field strength E_0 (without half-space), that is $E = 0.5 E_0$, due to the diffraction of the edge. If the electric field strength is divided by a factor of two only a quarter of the free space power is available at the receiver. Figure 8.13b shows the distribution of the electric field strength behind the half-space.

As we move the receiver antenna into the geometric shadow region (negative z-direction) we see a monotonic decaying signal. Moving the receiver antenna upwards (positive z-direction) the incident electric field approaches free space conditions. Alternatively, as we push away the half-space from the line of sight the incident field E approaches E_0.

In order to describe the influence of interfering objects on communication links we introduce the concept of *Fresnel zones*. Fresnel zones are ellipsoids with the antenna positions being the focal points. The boundary of the n-th Fresnel zone is defined by a path length that is a distance of $n \cdot \lambda_0/2$ (n-fold of half a wavelength) greater than the direct path r from one antenna to the other. For the first Fresnel zone the path length difference is $\lambda_0/2$ (see Figure 8.14).

From the knife-edge model we see a considerable effect on the wireless link if objects reach into the *first Fresnel-zone*. Moving the half-plane outside the first Fresnel ellipsoid reduces the influence sufficiently for practical designs. Therefore, we design directional radio links with an object-free first Fresnel zone.

In practical applications we find that the distance r between the antenna is much larger than the radius R_{max} of the ellipsoid. In this case we approximate the radius at any point along the direct path as

$$R(r_{\text{TX}}) = \sqrt{\lambda_0 \frac{r_{\text{TX}}(r - r_{\text{TX}})}{r}} \qquad (8.16)$$

where r_{TX} is the distance from the transmit antenna. The maximum radius is given as

$$R_{\text{max}} = \sqrt{\lambda_0 \frac{r}{4}} \qquad (8.17)$$

Example 8.5 *First Fresnel Zone*

A directional radio link ($f = 6\,\text{GHz}$) over flat terrain has two antennas at a distance of $r = 20\,\text{km}$. The antennas shall be mounted on towers with a height h. At the midway *point between the antennas is a complex of buildings with a height of $h_b = 10$ m. What is the minimum height of the antennas on the towers to keep the buildings away from the first Fresnel ellipsoid?*

Since the buildings are located midway between the antennas we can use Equation 8.17 to calculate the radius of the first Fresnel ellipsoid as

$$R_{max} = \sqrt{\lambda_0 \frac{r}{4}} = \sqrt{\frac{c_0}{f} \cdot \frac{r}{4}} = 15.8 \text{ m} \qquad (8.18)$$

We find the minimum antenna height h by adding up the radius of the ellipsoid R_{max} and the building height h_b, so the minimum antenna height for a free first Fresnel ellipsoid is $h = R_{max} + h_b = 25.8$ m.

8.2.5 Layered Media

In buildings electromagnetic waves propagate through walls and floors. In a first approximation these obstacles can be considered planar. Therefore, studying wave propagation in *layered media* can reveal interesting aspects of indoor radio propagation.

Things are most simple if we consider the normal incidence of a plane wave on a structure of n-layered dielectric lossy slabs (see Figure 8.15a). In this case we encounter a one-dimensional problem that can be solved with the transmission line theory already known from Section 3.1. Plane waves in air behave like TEM waves on air-filled transmission lines. Furthermore, plane waves in *layered media* behave like TEM waves on *cascading* transmission line segments (see Figure 8.15b). In order to calculate the transmission and reflection by transmission line segments the following steps are necessary:

- The length of each transmission line segment ℓ_i corresponds to the thickness of the slab d_i.
- The dielectric material (relative permittivity ε_{ri}, loss tangent $\tan \delta_{\varepsilon i}$) inside each line segment corresponds to the material of the slab.

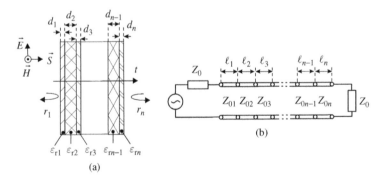

(a)

(b)

Figure 8.15 (a) Normal incidence of a plane wave on layered media and (b) line segments for efficient calculation of transmission and reflection coefficients.

- The source and load impedances as well as the characteristic impedances of *all* transmission line segments (*when filled with air*) equal Z_0. Of course the characteristic impedances differ when the transmission lines are assumed to be material-filled. The characteristic impedances of the material-filled transmission line segments are $Z_{0i} = Z_0/\sqrt{\varepsilon_{ri}}$.

The calculation can easily be performed by a circuit simulator (s-parameter simulation) using coaxial line segments (inner radius R_i, outer radius R_o). If we use load and source impedances of $Z_0 = 50\,\Omega$ the coaxial line segments should have a ratio of $R_o/R_i = 2.3$ to represent a $50\,\Omega$-line, when air-filled. The transmission coefficient t describes the wall path loss L_{wall}

$$t = \frac{E_t}{E_f} \qquad \rightarrow \qquad L_{wall} = 20\,\lg\left(\frac{1}{t}\right) \tag{8.19}$$

where E_t and E_f are the electric field strength amplitudes of the *transmitted* and *incident* (forward propagating) waves, respectively.

Example 8.6 *Loss-less Single Layer Wall*
Let us consider a plane wave that impinges normally on a single dielectric slab (relative permittivity $\varepsilon_r = 4$, loss tangent $\tan\delta_\varepsilon = 0$, thickness $d = 25\,mm$). With a circuit simulator we calculate the reflection coefficient r and transmission coefficient t in the frequency range from $f = 0.1\,GHz$ to $10\,GHz$ (see Figure 8.16).

The reflection and transmission coefficients show a periodic frequency-dependence. At integer multiples of $f = 3\,GHz$ the transmission coefficient equals one, the reflection coefficient is zero. Hence, there is no path loss, all power is transmitted through the dielectric wall. At a frequency of $f = 3\,GHz$ the wavelength inside the dielectric slab is

$$\lambda(3\,GHz) = \frac{c_0}{f\sqrt{\varepsilon_r}} = 50\,mm \tag{8.20}$$

So, at $3\,GHz$ the wall thickness d equals half a wavelength $d = \lambda/2$. In Section 3.1.9 we discussed a half-wave transformer where the input impedance equals the output impedance independent of the characteristic line impedance. From transmission line

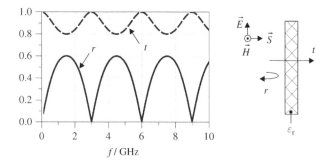

Figure 8.16 Reflection and transmission coefficient of a loss-less single layer wall (see Example 8.6).

theory we conclude that, if a homogeneous loss-less dielectric wall has a thickness of half a wavelength – or multiples thereof – the wall appears transparent, *that is all power is transmitted* ($t = 1$, $L_{\text{wall}} = 0\,\text{dB}$).

Maximum attenuation occurs if the wall has a thickness of a quarter wavelength $d = \lambda/4$ *(in our example at a frequency of* $f = 1.5\,\text{GHz}$). *From Figure 8.16 we read a transmission coefficient of* $t = 0.8$, *that is* $L_{\text{wall}} = 1.94\,\text{dB}$.

Example 8.7 *Lossy Three-Layer Wall*

Next we consider a more realistic multilayer wall model with lossy materials. The three layers are defined by the following parameter:

- *First layer:* $d_1 = 30\,\text{mm}$, $\varepsilon_{r1} = 3$, $\tan \delta_{\varepsilon 1} = 0.01$.
- *Second layer:* $d_2 = 115\,\text{mm}$, $\varepsilon_{r2} = 4.5$, $\tan \delta_{\varepsilon 2} = 0.05$.
- *Third layer:* $d_3 = 30\,\text{mm}$, $\varepsilon_{r3} = 3$, $\tan \delta_{\varepsilon 3} = 0.01$.

Figure 8.17 shows the transmission coefficient t and the reflection coefficients r_1 *and* r_2. *The transmission behaviour is quite complex with resonances at lower frequencies and an increasing attenuation with increasing frequency.*

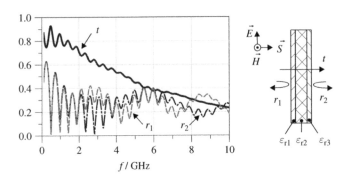

Figure 8.17 Reflection and transmission coefficient of a lossy three-layer wall (see Example 8.7).

8.3 Path Loss Models

The previously discussed basic propagation models consider isolated effects like attenuation of air, plane earth reflection and one-dimensional propagation through a layered wall. In most practical propagation scenarios (e.g. macrocells of cellular networks in urban environments or indoor coverage of wireless networks) more refined models are needed.

8.3.1 Multipath Environment

Figure 8.18 shows a mobile receiver and a fixed transmitter in an urban environment. Electromagnetic waves may propagate on different paths from transmitter to receiver

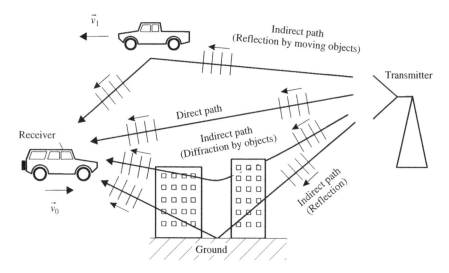

Figure 8.18 Multipath propagation in complex environment.

(*multipath environment*). First, a direct path (line of sight, LOS) may be possible. Next, reflections from fixed and mobile objects and ground may contribute to the receive signal. Furthermore, waves are diffracted or attenuated by objects (e.g. buildings or hills). At the receiver location all these signals from different paths superimpose constructively or destructively due to different path lengths and phase shifts. Finally, objects, transmitter and receiver can generally be mobile, resulting in a time-dependent and frequency-selective communication channel.

If a receiver in a rural or urban environment moves through the coverage area of a base station antenna the receiving signals vary with distance due to general path loss and interfering objects. In general we can distinguish between three different effects [5]:

- First, as shown in Figure 8.19a we see general path loss L due to increasing distance. This effect is given by our plane earth path loss or in some cases by free space loss.
- Large objects (buildings, hills, woods) block signals and lead to *slow* variations as the receiver moves (Figure 8.19b). This effect is known as *slow fading* or *shadowing*.
- Due to *multipath propagation* the receive signal is the superposition of signals that arrive under different angles and travel over different distances. As the receiver moves the changing path lengths of different ways lead to changing phases of the incoming signals. Depending on the phase relations we see a fluctuating receive power. This effect is known as *fast fading* or *multipath fading*. At distances of around half a wavelength we find deep signal dips (Figure 8.19c). (In radio network planning we are generally interested in *local mean* values of path loss. Therefore, path loss models do not consider fast fading effects. In measurements *local mean values* may be estimated by averaging measurement results over sufficiently large areas with a radius of a couple of wavelengths.)

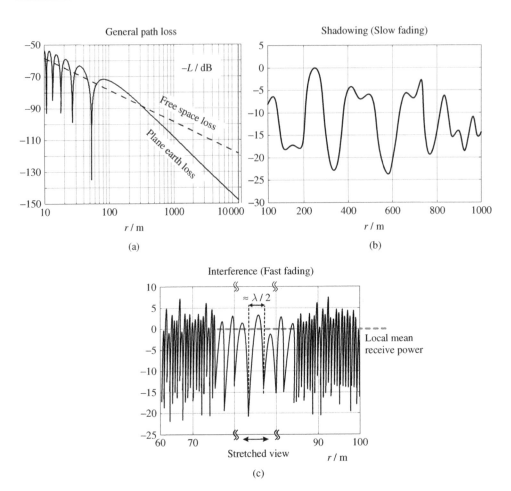

Figure 8.19 (a) General path loss due to distance; (b) shadowing by large objects between transmitter and receiver and (c) interference due to multipath propagation [5].

There are different approaches to deal with these complex scenarios.

Empirical models are based on measurements that are performed in different environments and at different frequencies. From these measurements best-fit curves are derived that are used for the estimation of radio coverage in environments that are similar to the one where the measurements were originally taken. In the following we look at a simple *clutter factor model* and the classical *Okumura–Hata model*. To apply these models pixel-oriented raster databases help to identify land cover (clutter), such as urban area, forest and so on.

Semi-empirical models are based on measurement data too, but include physical aspects of wave propagation. For example topographic data can be used to consider diffraction and shadowing from landscape elements like hills and mountains. Considering side specific data increases the reliability of the estimation. However, the improved accuracy comes at the expense of increased computational burden. Semi-empirical models that represent a combination of empirical and physical models are widely used for planing macrocells of cellular networks. A widely accepted model for macrocells is the COST231/Walfish–Ikegami model that is based on diffraction models including empirical correction factors [5, 7].

Physical models are based on physical aspects of electromagnetic wave propagation only. The methods require detailed knowledge of the side under investigation, for example terrain elevation data and vector models of buildings. Using these models may involve a significant (or even tremendous) computational effort. The models follow different numerical approaches and include simple *over the rooftop diffraction models*, *ray tracing* (and ray launching) tools that consider reflection, transmission and diffraction in outdoor and indoor environments. Finally, *EM simulation* tools may be used to calculate electromagnetic field distributions in complex environments.

8.3.2 Clutter Factor Model

The plane earth model in Section 8.2.3 is a rather simple model that does not account for any morphological details (buildings, vegetation, etc.). It is quite obvious that EM waves undergo *extra attenuation* as they propagate through vegetation. Furthermore, in urban environments buildings block the direct (line of sight) path and diffracted waves with *reduced* power arrive at observation points in that specific environment.

Therefore, *land use* (or *clutter*) databases may be used to incorporate *additional path loss* associated with a specific clutter category at the transmitter and receiver site. Typical clutter categories are urban, park land, industrial zone, forest and so on [8]. Databases on land coverage are commonly available in pixel (raster) format with resolutions of – for example – 30 m × 30 m.

Although the clutter (land cover) classification may seem very coarse the use of clutter databases can significantly improve the accuracy of the predicted path loss if reliable path loss factors (in dB) are available for different clutter types. The total path loss L_{total} may be expressed by summing up general path loss – for example given by plane earth loss (PEL) L_{PEL} – and a clutter loss factor (in dB) that accounts for losses due to land cover at the location of the mobile user.

$$L_{total} = L_{PEL} + L_{Clutter} \tag{8.21}$$

These clutter loss factors $L_{Clutter}$ may be found by appropriate measurements in reference scenarios [9]. Alternatively, mathematical models are available from literature [8] for commonly used clutter types.

8.3.3 Okumura–Hata Model

The Okumura–Hata model is an *empirical model* that is based on *measurements* conducted in an area around Tokyo in the 1960s. The measurement results have been converted into a set of formulas to approximate the path loss in different environments [5, 10, 12]. The Okumura–Hata model distinguishes between *open area*, *suburban area* and *urban area*.

The approximation formulas are valid under the following conditions:

- Frequency range: $150\,\text{MHz} \le f \le 1500\,\text{MHz}$;
- Distance r between transmitter and receiver: $r \ge 1\,\text{km}$;
- Base station antenna height: $30\,\text{m} \le h_b \le 200\,\text{m}$; and
- Mobile station antenna height: $1\,\text{m} \le h_m \le 10\,\text{m}$.

The path loss L (in decibels) depends on the area type and is given as

$$L/\text{dB} = A + B\lg(r) - E \qquad \text{(Urban areas)} \tag{8.22}$$

$$L/\text{dB} = A + B\lg(r) - C \qquad \text{(Suburban areas)} \tag{8.23}$$

$$L/\text{dB} = A + B\lg(r) - D \qquad \text{(Open areas)} \tag{8.24}$$

where r is the distance between transmitter and receiver in km.
The coefficients A, B, C, D and E are given by

$$A = 69.55 + 26.16\lg(f) - 13.82\lg(h_b) \tag{8.25}$$

$$B = 44.9 - 6.55\lg(h_b) \tag{8.26}$$

$$C = 2\,(\lg(f/28))^2 + 5.4 \tag{8.27}$$

$$D = 4.78\,(\lg(f))^2 - 18.33\lg(f) + 40.94 \tag{8.28}$$

$$E = (1.1\lg(f) - 0.7)\,h_m - (1.56\lg(f) - 0.8)$$

$$\text{(for medium \& small cities)} \tag{8.29}$$

$$E = 8.29\,(\lg(1.54h_m))^2 - 1.1 \qquad \text{(For large cities and } f < 300\,\text{MHz)} \tag{8.30}$$

$$E = 3.2\,(\lg(11.75h_m))^2 - 4.97 \qquad \text{(For large cities and } f \ge 300\,\text{MHz)} \tag{8.31}$$

where f is the operating frequency in MHz, h_b is the height of the fixed base station antenna in metres (m) and h_m is the height of the mobile station antenna in metres (m).

The Okumura–Hata model is restricted to frequencies up to $f = 1.5\,\text{GHz}$. An extension for frequencies up to $2\,\text{GHz}$ has been developed by a COST (European Cooperation in Science and Technology) project [7].

Example 8.8 *Okumura–Hata Path Loss Prediction*

At this point we will look at an example using the Okumura–Hata model. We assume a base station antenna height of $h_b = 40\,\text{m}$ and a mobile station antenna height of $h_b = 1.8\,\text{m}$. The frequency of operation shall be $f = 950\,\text{MHz}$. Let us first consider an open area and estimate the path loss in distances between $r = 1\,\text{km}$ and $r = 30\,\text{km}$. For comparison we look at plane earth path loss. Additionally, we use the simplified path loss formula for greater distances (Equation 8.14, $r > d_{\text{break}}$) because the distances are beyond the break point distance. With Equation 8.15 we calculate a break point distance of $d_{\text{break}} = 1.01\,\text{km}$.

The Okumura–Hata model for open areas predicts slightly higher path loss than the plane earth loss model (Figure 8.20). The path loss exponent of the plane earth model is

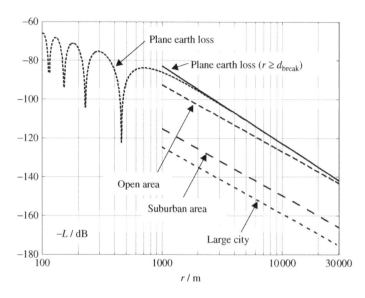

Figure 8.20 Okumura–Hata path loss predictions in comparison to plane earth loss for a frequency of $f = 950$ MHz and antenna heights of $h_b = 40$ m and $h_m = 1.8$ m (Example 8.8).

$n = 4$ *as discussed in Section 8.2.3. The Okumura–Hata model gives a pass loss exponent of* $n = B/10 = 3.44$ *resulting in flatter slope of the curve.*

In suburban areas – and all the more in large cities – we find increased path loss due to shadowing from densely concentrated buildings.

8.3.4 Physical Models and Numerical Methods

Canonical propagation scenarios may by solved *analytically* as we have already discussed in Section 8.2 on basic propagation models:

- Free space path loss (Section 8.2.1).
- Plane earth path loss (Section 8.2.3).
- Path loss due to diffraction from edges (Section 8.2.4).
- Path loss due to wave propagation through layered walls (Section 8.2.5).

In macrocell scenarios wave diffraction over a greater number of buildings and hills is of particular importance. Consequently, refined physical models have been developed that consider multiple knife-edge diffraction (rooftop diffraction). These models are beyond the scope of this introductory book. Therefore, the reader is referred to the standard literature for further physical models, for example [5].

A rigorous full-wave *EM simulation* (see Section 6.13.2) of real *macrocell* propagation scenarios is practically impossible. Firstly, EM simulation requires segmentation of the computational volume into sub-wavelength-sized cells. The resulting mathematical problems include such large numbers of unknowns that they cannot be solved in practice.

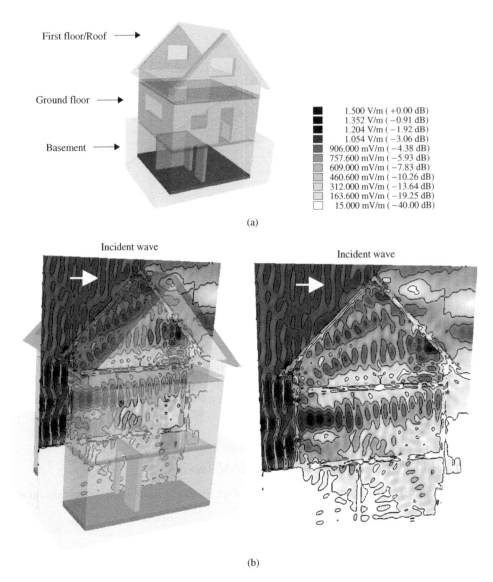

(a)

(b)

Figure 8.21 Outdoor-to-indoor propagation at $f = 450\,\text{MHz}$: (a) generic building model and (b) electric field strength (magnitude) in a vertical cut-plane through the building (Example 8.9).

Secondly, such a numerical model would require much more data than is commonly available including detailed building features, material parameters, locations of trees, vehicles, people and so on.

The previously discussed empirical and semi-empirical models provide reasonable accuracy for macrocells at reasonable cost (considering terrain and land cover (clutter) databases available, computational effort and calibration/tuning [13] of models by own measurements).

However, things change as we look at *picocells* of cellular networks. Picocells only cover an area of – for example – 100 m × 100 m. If vector data of buildings and indoor objects is available than ray-tracing software tools become an interesting alternative.

EM simulation may be applicable for path loss predictions if we look at short range communication. Wave propagation between devices in a room inside a building or in adjacent rooms or floors can be directly investigated for commonly used frequencies. Example 8.9 shows full-wave EM simulation at 450 MHz.

Example 8.9 *Outdoor-to-Indoor Propagation for Trunked Radio Applications at 450 MHz*
As an example for using full-wave EM simulation software in path loss prediction we look at outdoor-to-indoor propagation at $f = 450$ MHz. Figure 8.21a shows a generic building model. A vertically polarized plane wave with a normalized electric field amplitude of $E_z = 1$ V/m represents the incoming wave from a remote base station antenna. Figure 8.21b illustrates wave propagation by showing the magnitude of the electric field strength in a vertical cut-plane through the building. The representation of the magnitude of the electric field strength allows us to make realistic estimations for outdoor-to-indoor path loss. In order to provide local mean values the field strength should be averaged over a sufficiently large volume.

From a didactic point of view EM simulations are extremely valuable. EM simulation tools visualize wave propagation and make physical effects easier to understand. Engineers are often visually oriented people, so the coloured plots suit them better than lengthy mathematical treatments. The field plots of our example show all the physical effects like shadowing of a building (lower field values behind the building), standing waves in front of the building due to reflection and attenuation of waves due to transmission through concrete walls and windows.

8.4 Problems

8.1 Derive the plane earth path loss equation (Equation 8.13) for electromagnetic wave propagation over ground.

8.2 Two antennas ($f = 400$ MHz) are located at the same height $h_{TX} = h_{RX} = 5$ m at a distance of $r_1 = 2$ km. We assume plane earth loss conditions.
   Now, the distance between the antennas is increased to $r_2 = 3$ km. To what height must the antennas be set to get the same receive power as in the previous case?

8.3 Let us consider a communication link ($f = 10$ GHz) between a ground station and a satellite in geostationary earth orbit (GEO). The distance between a GEO satellite and the ground station is $r = 36\,000$ km. The antenna gains for ground station and satellite are $G_{gs} = 30$ dBi and $G_{sat} = 20$ dBi, respectively.

   a) Determine the free space loss L_{F0}
   b) Calculate the receive power P_{RX} for a transmit power of $P_{TX} = 1$ W.

References

1. Geng N, Wiesbeck W (1998) *Planungsmethoden fuer die Mobilkommunikation*. Springer.
2. Balanis CA (2005) *Antenna Theory*. John Wiley & Sons.

3. Pehl E (1992) *Mikrowellen in der Anwendung*. Huethig.
4. International Telecommunications Union (2006) *Recommendation ITU-R P.676-7 Attenuation by Atmospheric Gases*. ITU.
5. Saunders SR, Aragón-Zavala A (2007) *Antennas and Propagation for Wireless Communication Systems*. John Wiley & Sons.
6. Molisch AF (2010) *Wireless Communications*. John Wiley & Sons.
7. Commission of the European Communities (1999) *Digital Mobile Radio: COST 231 View on the Evolution towards 3rd Generation Systems*. Commission of the European Communities.
8. International Telecommunications Union (2009) *Recommendation ITU-R P.452-14 Prediction Procedure for the Evaluation of Interference Between Stations on the Surface of the Earth at Frequencies above 0.1 GHz*. ITU.
9. Anderson HR (2003) *Fixed Broadband Wireless System Design*. John Wiley & Sons.
10. Hata M (1980) Empirical Formula for Propagation Loss in Land Mobile Radio Services. *IEEE Trans. on Vehicular Technology*, Vol. 29.
11. International Telecommunications Union (1997) *Recommendation ITU-R P.529-2 Prediction Methods for the Terrestrial Land Mobile Service in the VHF and UHF Bands*. ITU.
12. Okumura Y, Ohmori E, Kawano T, Fukuda K (1968) Field Strength and Its Variability in VHF and UHF Land-Mobile Radio Service. *Review of the Electrical Communication Laboratory*, 16.
13. Graham AW, Kirkman NC, Paul PM (2007) *Mobile Radio Network Design in the VHF and UHF Bands - A Practical Approach*. John Wiley & Sons.

Further Reading

Blaunstein N, Christodoulou C (2006) *Radio Propagation and Adaptive Antennas for Wireless Communication Links: Terrestrial, Atmospheric and Ionospheric*. John Wiley & Sons.
IMST GmbH (2010) *Empire: Users Guide*. IMST GmbH.
Kark K (2010) *Antennen und Strahlungsfelder*. Vieweg.
Meinke H, Gundlach FW (1992) *Taschenbuch der Hochfrequenztechnik*. Springer.
Zinke O, Brunswig H (2000) *Hochfrequenztechnik 1*. Springer.

Appendix A

Coordinate systems are used to define the location of objects in space (see Figure A.1). The most commonly used coordinate system is the *Cartesian coordinate system*. For problems that have cylindrical or spherical geometry it may be advantageous to use *cylindrical* or *spherical* coordinates.

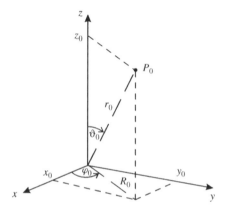

Figure A.1   Definition of Cartesian, cylindrical and spherical coordinates.

A.1   Coordinate Systems

A.1.1   Cartesian Coordinate System

Cartesian coordinates: x, y, z
Incremental line, surface and volume element:

$$\text{Line element:}\quad \mathrm{d}\vec{s} = \mathrm{d}x\,\vec{e}_x + \mathrm{d}y\,\vec{e}_y + \mathrm{d}z\,\vec{e}_z \tag{A.1}$$

$$\text{Surface element:}\quad \mathrm{d}\vec{A} = \mathrm{d}y\,\mathrm{d}z\,\vec{e}_x + \mathrm{d}x\,\mathrm{d}z\,\vec{e}_y + \mathrm{d}x\,\mathrm{d}y\,\vec{e}_z \tag{A.2}$$

$$\text{Volume element:}\quad \mathrm{d}v = \mathrm{d}x\,\mathrm{d}y\,\mathrm{d}z \tag{A.3}$$

RF and Microwave Engineering: Fundamentals of Wireless Communications, First Edition. Frank Gustrau.
© 2012 John Wiley & Sons, Ltd. Published 2012 by John Wiley & Sons, Ltd.

Vector differential operators:

Gradient operator: $\mathrm{grad}\,\phi = \nabla\phi = \dfrac{\partial\phi}{\partial x}\vec{e}_x + \dfrac{\partial\phi}{\partial y}\vec{e}_y + \dfrac{\partial\phi}{\partial z}\vec{e}_z$ (A.4)

Divergence operator: $\mathrm{div}\,\vec{V} = \nabla\cdot\vec{V} = \dfrac{\partial V_x}{\partial x} + \dfrac{\partial V_y}{\partial y} + \dfrac{\partial V_z}{\partial z}$ (A.5)

Curl operator: $\mathrm{curl}\,\vec{V} = \nabla\times\vec{V} = \left(\dfrac{\partial V_z}{\partial y} - \dfrac{\partial V_y}{\partial z}\right)\vec{e}_x$

$$+ \left(\dfrac{\partial V_x}{\partial z} - \dfrac{\partial V_z}{\partial x}\right)\vec{e}_y + \left(\dfrac{\partial V_y}{\partial x} - \dfrac{\partial V_x}{\partial y}\right)\vec{e}_z \quad \text{(A.6)}$$

Scalar Laplace operator: $\Delta\phi = \nabla^2\phi = \dfrac{\partial^2\phi}{\partial x^2} + \dfrac{\partial^2\phi}{\partial y^2} + \dfrac{\partial^2\phi}{\partial z^2}$ (A.7)

Vector Laplace operator: $\Delta\vec{V} = \left(\nabla^2 V_x\right)\vec{e}_x + \left(\nabla^2 V_y\right)\vec{e}_y + \left(\nabla^2 V_z\right)\vec{e}_z$ (A.8)

where $\nabla^2 V_x = \dfrac{\partial^2 V_x}{\partial x^2} + \dfrac{\partial^2 V_x}{\partial y^2} + \dfrac{\partial^2 V_x}{\partial z^2}$ (A.9)

and $\nabla^2 V_y = \dfrac{\partial^2 V_y}{\partial x^2} + \dfrac{\partial^2 V_y}{\partial y^2} + \dfrac{\partial^2 V_y}{\partial z^2}$ (A.10)

and $\nabla^2 V_z = \dfrac{\partial^2 V_z}{\partial x^2} + \dfrac{\partial^2 V_z}{\partial y^2} + \dfrac{\partial^2 V_z}{\partial z^2}$ (A.11)

Relations between unit vectors:

$\vec{e}_x = \vec{e}_R\cos\varphi - \vec{e}_\varphi\sin\varphi = \vec{e}_r\sin\vartheta\cos\varphi + \vec{e}_\vartheta\cos\vartheta\cos\varphi - \vec{e}_\varphi\sin\varphi$ (A.12)

$\vec{e}_y = \vec{e}_R\sin\varphi + \vec{e}_\varphi\cos\varphi = \vec{e}_r\sin\vartheta\sin\varphi + \vec{e}_\vartheta\cos\vartheta\sin\varphi + \vec{e}_\varphi\cos\varphi$ (A.13)

$\vec{e}_z = \vec{e}_z = \vec{e}_r\cos\vartheta - \vec{e}_\vartheta\sin\vartheta$ (A.14)

A.1.2 Cylindrical Coordinate System

Cylindrical coordinates: R, φ, z
Incremental line, surface and volume element:

Line element: $\mathrm{d}\vec{s} = \mathrm{d}R\,\vec{e}_R + R\mathrm{d}\varphi\,\vec{e}_\varphi + \mathrm{d}z\,\vec{e}_z$ (A.15)

Surface element: $\mathrm{d}\vec{A} = R\mathrm{d}\varphi\,\mathrm{d}z\,\vec{e}_R + \mathrm{d}R\,\mathrm{d}z\,\vec{e}_\varphi + R\mathrm{d}R\,\mathrm{d}\varphi\,\vec{e}_z$ (A.16)

Volume element: $\mathrm{d}v = R\mathrm{d}R\,\mathrm{d}\varphi\,\mathrm{d}z$ (A.17)

Vector differential operators:

Gradient operator: $\mathrm{grad}\,\phi = \nabla\phi = \dfrac{\partial\phi}{\partial R}\vec{e}_R + \dfrac{1}{R}\dfrac{\partial\phi}{\partial\varphi}\vec{e}_\varphi + \dfrac{\partial\phi}{\partial z}\vec{e}_z$ (A.18)

Divergence operator: $\text{div}\,\vec{V} = \nabla \cdot \vec{V} = \dfrac{1}{R}\dfrac{\partial\left(RV_R\right)}{\partial R} + \dfrac{1}{R}\dfrac{\partial V_\varphi}{\partial\varphi} + \dfrac{\partial V_z}{\partial z}$ (A.19)

Curl operator: $\text{curl}\,\vec{V} = \nabla \times \vec{V} = \left(\dfrac{1}{R}\dfrac{\partial V_z}{\partial\varphi} - \dfrac{\partial V_\varphi}{\partial z}\right)\vec{e}_R$

$$+ \left(\dfrac{\partial V_R}{\partial z} - \dfrac{\partial V_z}{\partial R}\right)\vec{e}_\varphi \qquad\qquad\text{(A.20)}$$

$$+ \dfrac{1}{R}\left(\dfrac{\partial\left(RV_\varphi\right)}{\partial R} - \dfrac{\partial V_R}{\partial\varphi}\right)\vec{e}_z \qquad\text{(A.21)}$$

Scalar Laplace operator: $\Delta\phi = \nabla^2\phi = \dfrac{1}{R}\dfrac{\partial}{\partial R}\left(R\dfrac{\partial\phi}{\partial R}\right)$

$$+ \dfrac{1}{R^2}\dfrac{\partial^2\phi}{\partial\varphi^2} + \dfrac{\partial^2\phi}{\partial z^2} \qquad\qquad\text{(A.22)}$$

Vector Laplace operator: $\Delta\vec{V} = \nabla^2\vec{V} = \nabla\left(\nabla\cdot\vec{V}\right) - \nabla\times\left(\nabla\times\vec{V}\right)$ (A.23)

Transformation from cylindrical to Cartesian coordinates:

$$x = R\cos\varphi \qquad\qquad\text{(A.24)}$$

$$y = R\sin\varphi \qquad\qquad\text{(A.25)}$$

$$z = z \qquad\qquad\text{(A.26)}$$

Relations between unit vectors:

$$\vec{e}_R = \vec{e}_x\cos\varphi + \vec{e}_y\sin\varphi = \vec{e}_r\sin\vartheta + \vec{e}_\vartheta\cos\vartheta \qquad\text{(A.27)}$$

$$\vec{e}_\varphi = -\vec{e}_x\sin\varphi + \vec{e}_y\cos\varphi = \vec{e}_\varphi \qquad\qquad\text{(A.28)}$$

$$\vec{e}_z = \vec{e}_z = \vec{e}_r\cos\vartheta - \vec{e}_\vartheta\sin\vartheta \qquad\qquad\text{(A.29)}$$

A.1.3 Spherical Coordinate System

Spherical coordinates: r, ϑ, φ
Incremental line, surface and volume element:

Line element: $d\vec{s} = dr\,\vec{e}_r + r\,d\vartheta\,\vec{e}_\vartheta + r\sin\vartheta\,d\varphi\,\vec{e}_\varphi$ (A.30)

Surface element: $d\vec{A} = r^2\sin\vartheta\,d\vartheta\,d\varphi\,\vec{e}_r + r\sin\vartheta\,dr\,d\varphi\,\vec{e}_\vartheta + r\,dr\,d\vartheta\,\vec{e}_\varphi$ (A.31)

Volume element: $dv = r^2\sin\vartheta\,dr\,d\vartheta\,d\varphi$ (A.32)

Vector differential operators:

Gradient operator: $\text{grad}\,\phi = \nabla\phi = \dfrac{\partial\phi}{\partial r}\vec{e}_r + \dfrac{1}{r}\dfrac{\partial\phi}{\partial\vartheta}\vec{e}_\vartheta + \dfrac{1}{r\sin\vartheta}\dfrac{\partial\phi}{\partial\varphi}\vec{e}_\varphi$ (A.33)

Divergence operator: $\text{div } \vec{V} = \nabla \cdot \vec{V} = \dfrac{1}{r^2} \dfrac{\partial \left(r^2 V_r \right)}{\partial r}$

$$+ \frac{1}{r \sin \vartheta} \frac{\partial \left(V_\vartheta \sin \vartheta \right)}{\partial \vartheta} + \frac{1}{r \sin \vartheta} \frac{\partial V_\varphi}{\partial \varphi} \tag{A.34}$$

Curl operator: $\text{curl } \vec{V} = \nabla \times \vec{V} = \dfrac{1}{r \sin \vartheta} \left(\dfrac{\partial \left(V_\varphi \sin \vartheta \right)}{\partial \vartheta} - \dfrac{\partial V_\vartheta}{\partial \varphi} \right) \vec{e}_r$

$$+ \frac{1}{r} \left(\frac{1}{\sin \vartheta} \frac{\partial V_r}{\partial \varphi} - \frac{\partial \left(r V_\varphi \right)}{\partial r} \right) \vec{e}_\vartheta \tag{A.35}$$

$$+ \frac{1}{r} \left(\frac{\partial \left(r V_\vartheta \right)}{\partial r} - \frac{\partial V_r}{\partial \vartheta} \right) \vec{e}_\varphi$$

Scalar Laplace operator: $\Delta \phi = \dfrac{1}{r^2} \dfrac{\partial}{\partial r} \left(r^2 \dfrac{\partial \phi}{\partial r} \right) + \dfrac{1}{r^2 \sin \vartheta} \dfrac{\partial}{\partial \vartheta} \left(\sin \vartheta \dfrac{\partial \phi}{\partial \vartheta} \right)$

$$+ \frac{1}{r^2 \sin^2 \vartheta} \frac{\partial^2 \phi}{\partial \varphi^2} \tag{A.36}$$

Vector Laplace operator: $\Delta \vec{V} = \nabla^2 \vec{V} = \nabla \left(\nabla \cdot \vec{V} \right) - \nabla \times \left(\nabla \times \vec{V} \right)$ (A.37)

Transformation from spherical to Cartesian coordinates:

$$x = r \cos \varphi \sin \vartheta \tag{A.38}$$

$$y = r \sin \varphi \sin \vartheta \tag{A.39}$$

$$z = r \cos \vartheta \tag{A.40}$$

Relations between unit vectors:

$$\vec{e}_r = \vec{e}_x \sin \vartheta \cos \varphi + \vec{e}_y \sin \vartheta \sin \varphi + \vec{e}_z \cos \vartheta = \vec{e}_R \sin \vartheta + \vec{e}_z \cos \vartheta \tag{A.41}$$

$$\vec{e}_\vartheta = \vec{e}_x \cos \vartheta \cos \varphi + \vec{e}_y \cos \vartheta \sin \varphi - \vec{e}_z \sin \vartheta = \vec{e}_R \cos \vartheta - \vec{e}_z \sin \vartheta \tag{A.42}$$

$$\vec{e}_\varphi = -\vec{e}_x \sin \varphi + \vec{e}_y \cos \varphi = \vec{e}_\varphi \tag{A.43}$$

A.2   Logarithmic Representation

A.2.1   Dimensionless Quantities

Dimensionless real and positive quantities (like antenna gain G) and absolute values of complex-valued quantities (like absolute values of scattering parameter, i.e. $|s_{ij}|$) are often given in logarithmic scale. Logarithmic representations are advantageous if quantities vary over several orders of magnitude. Logarithmic values maintain a good resolution for both small and large values.

Table A.1 Conversion between linear and logarithmic scale

Logarithmic scale dB	Linear scale (Voltage ratio)	Linear scale (Power ratio)
+40	100	$10\,000 = 10^4$
+30	≈ 31.6	$1\,000 = 10^3$
+20	10	$100 = 10^2$
+10	≈ 3.16	$10 = 10^1$
+6	≈ 2	≈ 4
+3	≈ 1.41	≈ 2
0	1	1
-3	≈ 0.707	≈ 0.5
-6	≈ 0.5	≈ 0.25
-10	≈ 0.316	$0.1 = 10^{-1}$
-20	0.1	$0.01 = 10^{-2}$
-30	≈ 0.0316	$0.001 = 10^{-3}$
-40	0.01	$0.0001 = 10^{-4}$

For power-based quantities (like antenna gain) a factor of 10 is used, whereas for voltage, current or field strength-based values (like scattering parameters) a factor of 20 is used. So, the logarithmic values are given as

$$G/\mathrm{dB} = 10\lg G \qquad \text{and} \qquad s_{ij}/\mathrm{dB} = 20\lg |s_{ij}| \qquad (A.44)$$

where $\lg = \log_{10}$ is the *common logarithm*. We do not use different symbols for linear and logarithmic representation. The pseudo-unit 'dB' (decibel) indicates the logarithmic scale and avoids confusing linear and logarithmic values: for example a gain of $G = 1$ in linear scale equals a gain of $G = 0\,\mathrm{dB}$ in logarithmic scale. Table A.1 correlates commonly used linear and logarithmic values.

In some special cases the *natural logarithm* $\ln = \log_e$ may be used to define logarithmic values. If we use the natural logarithm it is indicated by the pseudo-unit 'Np' (Neper). For power-based quantities a factor of $1/2$ is used, whereas for voltage, current or field strength-based values a factor of 1 is used. So, the logarithmic values are given as

$$G/\mathrm{Np} = \frac{1}{2}\ln G \qquad \text{and} \qquad s_{ij}/\mathrm{Np} = \ln |s_{ij}| \qquad (A.45)$$

We may convert between Neper and decibel by using the following relation

$$1\,\mathrm{Np} = 20/\ln(10)\,\mathrm{dB} \approx 8.686\,\mathrm{dB} \qquad (A.46)$$

A.2.2   Relative and Absolute Ratios

In order to give dimensionful (non-dimensionless) quantities like voltage U or power P in logarithmic scale the values have to be normalized. Normalization can be done with respect to a maximum or reference value with the same physical unit, for example a reference power of P_0. For power and power density levels a factor of 10 is used, whereas for voltage, current or field strength levels a factor of 20 is used. Hence, *relative levels* are given as

$$\frac{U}{\text{dB}} = 20 \lg\left(\frac{U}{U_0}\right) \quad \text{and} \quad \frac{P}{\text{dB}} = 10 \lg\left(\frac{P}{P_0}\right) \quad \text{(Relative level)} \quad \text{(A.47)}$$

Furthermore, we can normalize to a *fixed* physical value, for example a power of $P_0 = 1\,\text{mW}$. In order to indicate the *fixed reference value* (e.g. $1\,\mu\text{V}$) the pseudo unit 'dB' is complemented by a fixed reference value itself (e.g. dBμV). So, *absolute levels* are given as

$$\frac{U}{\text{dB}\mu\text{V}} = 20 \lg\left(\frac{U}{1\,\mu\text{V}}\right) \quad \text{and} \quad \frac{P}{\text{dBmW}} = 10 \lg\left(\frac{P}{1\text{mW}}\right) \quad \text{(Absolute level)} \quad \text{(A.48)}$$

The pseudo-unit 'dBmW' indicating a fixed reference power of $1\,\text{mW}$ is commonly written in its abbreviated form which is 'dBm'. A power of $P = 3\,\text{dBm}$ equals a power of $P = 2\,\text{mW}$.

Relative levels are used to describe transfer functions of two-port (or more general *n*-port) networks, for example by looking at the output to input power ratio we can define a logarithmic power amplification *a* as

$$\frac{a}{\text{dB}} = 10 \lg\left(\frac{P_{\text{out}}}{P_{\text{in}}}\right) \quad \text{(A.49)}$$

A.2.3　Link Budget

Logarithmic values are suitable for plotting *link budgets*. Figure A.2 shows – as an example – a communication link with generator, mixers, amplifiers, transmission lines, detector and a wireless link with two antennas. The generator produces a signal with a

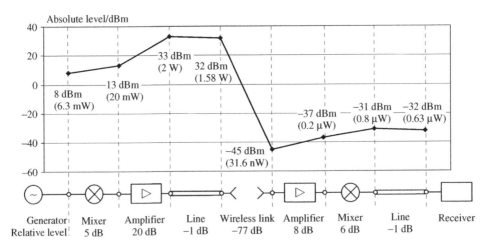

Figure A.2　Link budget for a communication path.

power of 8 dBm. The transfer functions of the active components in the circuit (mixers, amplifiers) are described by a positive relative level, thus lifting up the absolute power level along the path. The transmission lines and the wireless link decrease the absolute power level. The wireless link itself may consist of two antennas with gains of $G_1 = 10$ dBi and $G_2 = 3$ dBi and an isotropic path loss of $L = -90$ dB resulting in an overall relative level of $G_1 + G_2 + L = -77$ dB.

If all components are matched to their port reference impedance ($Z_0 = 50\,\Omega$ is commonly used) we can easily add up logarithmic values to determine the receive power. So, in our example, we get a receive power of $P = -32$ dBm. By using logarithmic values and a graphic representation power levels at different locations on the transmission path can easily be estimated.

Finally, we should note, that in *multipath environments* the path loss models represent *mean* values. Therefore, the receive power we derive from our link budget represents *median* power. The actual receive power may be lower due to fading effects. In order to minimize the risk for loss of connection usually an extra *fade margin* is added.

Index

RF and Microwave Engineering: Fundamentals of Wireless Communications, First Edition. Frank Gustrau.
© 2012 John Wiley & Sons, Ltd. Published 2012 by John Wiley & Sons, Ltd.

Printed and bound by CPI Group (UK) Ltd, Croydon, CR0 4YY

27/10/2024

14580152-0003